Advanced Materials and Nano Systems: Theory and Experiment

(*Part 3*)

Edited by

Dibya Prakash Rai
Department of Physics
Mizoram University
Aizawl, India

Kingsley O. Obodo
ICTP-East African Institute for Fundamental Research, Kigali, Rwanda
Center for Space Research, North-West University Potchefstroom, South Africa
National Institute of Theoretical and Computational Sciences, Johannesburg, South Africa

&

Jitendra Pal Singh
Department of Sciences (Physics)
Manav Rachna University
Faridabad, India

Advanced Materials and Nano Systems: Theory and Experiment

(Part 3)

Editors: Dibya Prakash Rai, Kingsley O. Obodo & Jitendra Pal Singh

ISBN (Online): 978-981-5223-10-1

ISBN (Print): 978-981-5223-11-8

ISBN (Paperback): 978-981-5223-12-5

First published in 2024.

Bentham Science Publishers Pte. Ltd.
80 Robinson Road #02-00
Singapore 068898
Singapore
Email: subscriptions@benthamscience.net

BENTHAM SCIENCE

CONTENTS

FOREWORD

First of all, I would like to congratulate Dr. Dibya Prakash Rai for successfully publishing a series of two books, Part 1 & 2, entitled “Advanced Materials and Nano Systems: Theory and Experiment”. Following the success of the previous two books, the editor is motivated for the third edition. I believe this third edition of the book will receive grand success with the support of two dynamic researchers Dr. Kingsley O. Obodo and Dr. Jitendra Pal Singh. The subjects covered in this book range from bulk to nanoscale materials at all applied levels, including some of the most significant and current issues in materials science and nanosystems. It includes introductory chapters, discussions of fundamentals, and problems with in-depth solutions. It covers practically all of the subject matter in the concerned discipline. This book walks readers through using the most basic features or utilities of a program or method. I have had the pleasure of reading several extremely fascinating chapters on the creation of nanoparticles. These chapters concentrate on the potential for reducing environmental pollution and the application of nanosystems for energy harvesting. This book also addresses the existing constraints and difficulties in the development of nanotechnology. It offers a perspective on its potential future development and application in overcoming the current energy crisis, in medicine, and the evolution of digital technology, *etc*. Over the past 20 years, there has been a tremendous increase in nanotechnology research, which focuses on the science, technology and engineering of nanoscale materials. Nanotechnology has seen a rise in global investment from all sectors, including the public, businesses, governments, institutes, *etc*. Developed nations, including the United States, Germany, the United Kingdom, France, and others, are working to advance research and development in this crucial sector and creating the scientific understanding required to address queries and worries about the potential effects of this nanotechnology on the environment, health, and safety. In both clinical care and biological research, nanotechnology is gaining importance and prominence. Nanotechnology is anticipated to have a bright future in many facets of medicine, much like other sciences. Nanoparticles can deliver medications into cells to reach illness areas because of their tiny size, which raises local drug concentrations and lowers systemic toxicities. The diversity and promise of nanotechnology products thrill me. Still, I am also conscious of the challenging research required to provide solid answers to important queries regarding their potential effects on human health and the environment. To consider the possible influence on human health that this new and promising kind of material science represents, many of the established assumptions of safety and health must be re-examined, if not modified. Technological advancements have made it possible for humans to view atoms, manipulate them individually, alter their positions one at a time, and utilize them to generate new codes. The challenge was and continues to be, how to handle the system at an atomic scale. This is a challenging undertaking, but it is still feasible. Electronics experts who work with integrated circuits participated in the race for shrinking as well as moving circuitry from micro to submicron levels and reaching ever-closer to the nanoscale. The concept of electronics will need to be completely rethought if researchers ultimately get to the point where a single electron serves as the foundation for electronics (nano-transistor). Since other scientific disciplines like mechanics, optics, chemistry, and biology have also begun to create their nanoworlds, which we now refer to as nanosystems, this evolution is not simply affecting electronics. The silicon airbag's accelerometer in automobiles refers to the example of a mass-produced nanosystem. As far as I am concerned, the application of nanosystems as nanomotors, nanorobots or nano-transistor has not been realized yet. This is because development from lab-scale to commercial-scale will take some time. In actuality, the energy situation is what is most concerning. Rapid energy use eventually threatens the viability of life and may even bring about the end of the contemporary planet one day. As a result, one should seek renewable energy sources or locations where energy may be produced without harming the environment. We may find various promising energy harvesting technologies in the field of materials science, such as thermoelectricity, piezoelectricity, solid-state batteries, *etc*. Since revolutions and technological advancement do not come naturally to people or communities, it is to be expected that certain worries will accompany these developments. We must be cognizant of the current difficulties and the issues at hand. I want to invite the readers to continue reading this book. Let's explore this new universe in all the forms of materials science while maintaining the realism of an engineer and the creative progress.

Victor Sunday Aigbodion
Department of Metallurgy and Materials Engineering
Faculty of Engineering, University of Nigeria
Nsukka, Nigeria

PREFACE

The first two volumes of the book "Advanced Materials and Nano Systems: Theory and Experiment Parts 1 and 2" were a great success. I received numerous overwhelming responses and best wishes from both the readers and well-wishers. This has motivated me a lot and encouraged me to work on the third volume of "Advanced Materials and Nano Systems: Theory and Experiment". Furthermore, I am supported by two dynamic and experienced scientists who are working in the field of nanoscience and nanotechnology, Dr. Kingsley O. Obodo from South Africa and Dr. Jitendra Pal Singh from India. They helped me in the selection and scrutiny of the best 11 chapters. The best 11 selected chapters contain very popular topics related to novel nanomaterials for drug delivery, environmental safety, industrial biocatalysts, energy harvest, electronics, *etc*. As usual with previous volumes of the book, this upcoming edition will be very exciting and will include interesting chapters in the said field from very expert authors and researchers from all over the world. The title of the book, "Advanced Materials and Nano Systems: Theory and Experiment", is fascinating and truly subject-specific on an international level, which easily captures the attention of readers who are working in the field of nanoscience and nanotechnology. This book also covers a wide range of scientific topics, ranging from bulk materials to nanomaterials and their applications. Advances in materials science *via* nanotechnology have resulted in significant advancements in a variety of disciplines, such as biomedicine, biomaterials, biosensors, nanoelectronics, energy production, construction, packaging, food, health care, automotive, and defense, among others. Knowing the rapid popularity of nanotechnology and its benefits, we have decided to compile this third edited book; hence, we have invited numerous chapters from active researchers who are working in the thrust areas of nanomaterials.

The opening of Chapter 1 by Shomaila Khanam and Sanjeeb Kumar Rout contributes to the role of plasmonic metal-semiconductor heterostructure in photocatalytic hydrolysis and degradation of toxic dyes. In this experimental work, they have reported the efficiency of plasmonic metal-semiconductor heterostructure as a superior material for photocatalytic water splitting activity for clean and cheap energy generation *via* the localized surface plasmon resonance (LSPR) method. The LSPR generation by these kinds of materials has proved to be very efficient in the photocatalytic hydrolysis of the hydrogen-rich compound, photocatalytic water splitting, and photocatalytic degradation of organic dyes.

Chapter 2 gives details on $BaZrO_3$-based ceramics and composites as smart materials for advanced applications. This chapter's introductory section provides a brief history of ceramic materials and their schematic development. Ceramic materials have been used in everything from pottery to pellets, ammunition to antennas, and electrolytes to electronics. This chapter focuses on the barium zirconate ($BaZrO_3$)-based ceramic composite as smart materials for advanced applications targeting the energy sector, photocatalyst for hydrogen production, a smart bullet in defense, and microwave dielectric resonators for wireless communication systems, respectively.

In Chapter 3, Houmad *et al.* used density functional theory (DFT) to investigate the 2D graphene-like nanomaterial Sili-graphene as a high-capacity anode material for lithium-ion batteries. Herein, they predict silicon-doped graphene (Siligraphene) as an efficient anode material for potential Li-ion batteries (LIBs). They reported the most stable lithium adsorption at different sites in the hexagonal structure of the silica sheet. Also, they have calculated a very important parameter, the Li-ion migration efficiency in siligraphene, for the LIBs. The reported diffusivity was found to be about 0.095 eV for Li on top of silicon atoms and about

0.223 eV for oxygen, indicating rapid charging and discharging processes. They also report a minute variation in the voltage essential for its potential applications in LIBs.

In Chapter 4, Singh and Devi present an introduction to the fabrication of white light-emitting diodes. They have highlighted the fascinating and eco-friendly sources of white light-emitting diodes due to their high energy savings in various fields, including lighting, architecture, medicine, *etc*. In this study, they focus on the development and fabrication of cheap and eco-friendly light sources (white light emitting diodes), as the lighting sector is one of the most important and attractive fields because it consumes a large amount of electricity, about 15–20% of total electricity production in the world. They emphasize the two fabrication methods for white light-emitting diodes. The luminous efficiency and rendering index of a lamp are influenced by the type of fabrication. In this chapter, the different fabrication methods for white light-emitting diodes are discussed. The chapter is divided into three parts. The first part is devoted to the general introduction of the light-emitting diode (LED), its working principle, and its applications. In the second part, the characteristics of light, including CIE, colour temperature, and rendering index, are briefly discussed. In the last part, the different methods of fabrication and their advantages and disadvantages are discussed.

Chhana *et al*. investigated the electronic and piezoelectric properties of non-metal doped II-VI monolayer compounds in Chapter 5. In this study, they have used spin-polarized density functional theory (DFT) to calculate the electronic and piezoelectric properties of the dynamically stable II–VI monolayers: ZnO, ZnS, CdO, and CdS. They also doped the monolayer system with boron and carbon to analyze the variation of these properties for their integration into potential technological applications. They report the half-metallicity of the above-stated monolayers when doped with B and C atoms. The presence of half-metallicity in doped monolayers is promising for spintronic applications. The B-doped ZnO and ZnS monolayer results in negative structural stiffness and negative piezoelectric tensors, while C-doping provides stable and enhanced elastic as well as piezoelectric properties of the monolayer.

Habib Rached and Ismail Ouadha contributed to Chapter 6. They reported a theoretical calculation on the new quaternary MAX-phase compounds $(Zr1\text{-}xTix)_3AlC_2$ (where x = 0-1). They mainly studied the structural, electronic, mechanical, and thermodynamic properties of the MAX phases $(Zr_{1-x}Ti_x)_3AlC_2$ compounds by using the full-potential plane-wave FP-LAPW method as implemented in the Wien2k code. They have reported the importance of exchange-correlation functionally, in which they have adopted the most commonly adopted exchange-correlation (XC) energy of all electrons within the Perdew-Burke-Ernzerhof parametrization. They reported an agreement between their theoretical findings with the experimental results. They studied the formation energy to check the ground-state stability of the system. The negative value of formation energy confirms the stability of their compounds. The calculation of the electronic structure in the MAX phase found it to be metallic with prominent *p-d* hybridization. They have also studied the elastic properties by calculating the elastic constants using the Hex-elastic package. They report the mechanical stability of the MAX-phase compound. This chapter also includes a study of the temperature and pressure effects on bulk modulus, Debye temperature, and heat capacity at constant volume (CV) and constant pressure (CP), respectively.

Chapter 7 highlights the work of Obodo *et al*. Here, they studied the surface segregation in Pt_3Nb and Pt_3Ti using density-functional-based methods. In their DFT-based studies, they reported the surface segregation of the stoichiometrically ordered Pt_3X system because, due to the endothermic process of direct exchange in Pt_3Ti, the exchange of surface atoms is forbidden in the ground state. Direct exchange in Pt3Nb was also reported as exothermic surface segregation. They describe an intriguing phenomenon in which an overlayer forms on the Pt_3X system as a result of PtX migration into an antisite defect in an off-stoichiometric configuration. Doped surfaces, rather than pristine structures, have the optimal oxygen adsorption energy for Pt(111). The instability was found in the Pt_3X system with the X-skin surface configuration. While the Pt-skin configuration improved the adsorption energy, the inclusion of the antisite configuration improved ORR activity for both Pt_3X systems. Finally, they found that Pt3Nb had better catalytic performance than Pt_3Ti.

Chapter 8: This chapter is contributed by Roy and his group, and their work is a review of the nanoparticles and their manipulation for capturing the toxic elements from the air to keep the environment healthy. The high surface area to volume ratio of extremely small particles (nanoparticle size between 1-100 nm) accounts for their distinct behavior. They report the importance of artificially synthesized and manipulated nanoparticles as per the specific needs. Alarming impact on the environment as a result of the massive use of nanoparticle-based industrial products degrades environmental health. Environmental health is determined by the quality of the air, water, and soil, which is reflected in a healthy ecosystem and its biodiversity. The effects of nanoparticles released into the environment vary depending on their type, surface coating, and degree of environmental transformation. Some nanoparticles are harmful to the environment, and some are beneficial. Some of the nanoparticles in the environment get bioaccumulated in plants and animals, disturbing their growth and productivity. Recognizing the potential of nanotechnology in manipulating nanoparticles, remediation has proven to be an effective technique for removing some toxic elements and compounds from the environment, thereby providing a way to reduce pollution efficiently. In this review, they have presented an overview of the sources, fate, and effects of nanoparticles available in air, water, and soil. They also promise the long-term assessment of nanoparticles and the formulation of strict guidelines for their usage by the concerned industries for better environmental health and, in turn, a healthy ecosystem.

Chapter 9 includes the research work of Lalhumhima *et al.*, who reported the optimal site for the adsorption and population effect of lithium on a silicene monolayer. This is a purely theoretical work based on density functional theory's first principles. They report the most stable site for adsorbing the lithium ions in the silicene monolayer. They discovered the active site by calculating the minimum adsorption energy required to place the lithium atom in various possible locations. They reported that the center of the hexagonal structure in silicene was the most favorable site. Using transition state search (TSS), the optimized stable state and a low diffusion energy barrier (DEB) of 0.348 eV were indicated. The metallic property of pristine silicene is maintained throughout Li atom adsorption.

Chapter 10 details the methods for synthesizing the metal oxide nanoparticles as well as the difficulties. They have emphasized the significance of nanoscience and nanotechnology in raising the standard of living. The advancement of nanotechnology is facilitated by the new class of materials known as nanoparticles (NPs). The NPS should have at least one particle dimension between 1 and 100 nm. The synthesis methods can change the size and structure of NPs, which are important in molecular biology, physics, organic and inorganic chemistry, biology, medicine, and materials research. Synthesis techniques can produce NPs with high surface areas, increasing their value and supplying necessary features like surface reactivity. They discussed the various methods for creating nanoparticles. They concentrate on the two parts, top-down and bottom-up, which are categorized according to the raw materials used to create the NPs in this study. An overview of synthesis techniques and their uses in nanotechnology and nanoscience was given in this review. The electrochemical discharge procedure, a revolutionary method for creating NPs, is covered in detail. Tables are used to discuss the synthesized materials, including ZnO, carbon, graphene, and other metal oxides, as well as their composites. Finally, the difficulties, benefits, drawbacks, conclusions, and synthesis of NPs are examined.

The final chapter, 11 by Singh and Devi, discusses the Heterogeneous semiconductor photocatalysis for water purification: Basic mechanism and advanced strategies. In this chapter, they have discussed the different techniques based on the semiconductor photocatalysis for water purification. They specially emphasize the contribution of water pollution *via* anthropogenic activities, human waste, utilization of toxic chemicals in agricultural activities *etc*., and the solution of purification of water to make it drinkable with the help of nanotechnology. Heterogeneous semiconductor photocatalysis is the most effective green method in this regard because it enables it to degrade into nonhazardous products like CO_2 and H_2O without releasing any harmful residue. Therefore, understanding the knowledge of photocatalysis mechanism is very significant to enable further improvement. Hence, this chapter presents the basic mechanism of photocatalysis, its drawbacks and the advanced strategies to improve the catalytic efficiency. Finally, some of the important factors that influence catalytic activity have also been discussed.

In the current book, various computational and experimental techniques of recent advances in materials science are presented, with a primary emphasis on the synthesis, characterization, and analysis of functional properties of novel nanomaterials to address a variety of problems encountered on a daily basis. This book's timely release should be of interest to scholars, professors, and students in the field of nanotechnology. Given the increasing demand for nanomaterials in various human activities, it is anticipated that the topic of advanced materials and nanosystems from theory and experimental perspectives will continue to be at the forefront of cutting-edge research in science and technology for the foreseeable future. With a focus on the synthesis, characterization, and analysis of functional properties of innovative nanomaterials to solve a variety of problems encountered on a daily basis, the current book presents numerous computational and experimental approaches to recent breakthroughs in materials science.

The timely publication of this book (Part 3) should be of great interest to professionals, educators, researchers and students in the field of materials science and nanotechnology. It is projected that the subject of nanomaterials and nanotechnology will remain at the forefront of cutting-edge research in science and technology for the foreseeable future, given the growing integration of nanomaterials in numerous human activities. I also want to express my gratitude to the editorial team of the Bentham Books for their prompt release of Parts 1 and 2 of "Advanced Materials and Nano Systems: Theory and Experiment", which gave me the confidence I needed to quickly collect the chapters compile them and finish Part 3.

Dibya Prakash Rai
Department of Physics
Mizoram University
Aizawl, India

Kingsley O. Obodo
ICTP-East African Institute for Fundamental
Research, Kigali, Rwanda
Center for Space Research, North-West University
Potchefstroom, South Africa
National Institute of Theoretical and Computational
Sciences, Johannesburg, South Africa

&

Jitendra Pal Singh
Department of Sciences (Physics)
Manav Rachna University
Faridabad, India

List of Contributors

A. El Kenz	Laboratory of Condensed Matter and Interdisciplinary Sciences (LaMScI), Faculty of Science, Mohammed V University of Rabat, Rabat, Morocco
A. Benyoussef	Laboratory of Condensed Matter and Interdisciplinary Sciences (LaMScI), Faculty of Science, Mohammed V University of Rabat, Rabat, Morocco
Alok Kumar Das	Indian Institute of Technology, Indian School of Mines, Dhanbad, India
Adedapo S. Adeyinka	Research Centre for Synthesis and Catalysis, Department of Chemical Sciences, University of Johannesburg, Johannesburg, South Africa
Aparna Roy	Department of Zoology , University of Calcutta, Kolkata, West Bengal, India
Biswajit Roy	CSIR-Centre for Cellular and Molecular Biology, Hyderabad, Telangana, India
Dibya Prakash Rai	Department of Physics, Mizoram University, Aizawl, India
Divya Rai	Department of Zoology , University of Calcutta, Kolkata, West Bengal, India
D. Vignesh	Birla Institute of Technology, Mesra, Ranchi, Jharkhand, India
Ela Rout	Birla Institute of Technology, Mesra, Ranchi, Jharkhand, India
Harish Bishwakarma	Indian Institute of Technology, Indian School of Mines, Dhanbad, India
Habib Rached	Magnetic Materials Laboratory, Faculty of Exact-Sciences, Djillali-Liabès University of Sidi-Bel-Abbès, Sidi-Bel-Abbès, Algeria
Ismail Ouadha	Magnetic Materials Laboratory, Faculty of Exact-Sciences, Djillali-Liabès University of Sidi-Bel-Abbès, Sidi-Bel-Abbès, Algeria
Kumaresh Mandal	Department of Zoology , University of Calcutta, Kolkata, West Bengal, India
Kingsley O. Obodo	ICTP-East African Institute for Fundamental Research, Kigali, Rwanda Center for Space Research, North-West University Potchefstroom, South Africa National Institute of Theoretical and Computational Sciences, Johannesburg, South Africa
Lalrinthara Pachuau	Department of Physics, Physical Sciences Research Centre (PSRC), Pachhunga University College, Aizawl, Mizoram, India
Lalmuanpuia Vanchhawng	Department of Physics, Physical Sciences Research Centre (PSRC), Pachhunga University College, Aizawl, Mizoram, India
Lalmuan Chhana	Department of Physics, School of Physical Sciences, Mizoram University, Aizawl, Mizoram, India Physical Sciences Research Centre (PSRC), Department of Physics, Pachhunga University College, Mizoram University, Aizawl, Mizoram, India
Lalhriat Zuala	Physical Sciences Research Centre (PSRC), Department of Physics, Pachhunga University College, Mizoram University, Aizawl, Mizoram, India

Lalrin Kima	Department of Physics, Mizoram University, Aizawl, India Department of Physics, Pachhunga University College, Mizoram University, Aizawl, India
Lalhumhima	Department of Physics, Physical Sciences Research Centre (PSRC), Pachhunga University College, Aizawl, Mizoram, India Department of Physics, Mizoram University, Aizawl, Mizoram, India
Lalmuanchhana	Department of Physics, Physical Sciences Research Centre (PSRC), Pachhunga University College, Aizawl, Mizoram, India Department of Physics, Mizoram University, Aizawl, Mizoram, India
Mukul Anand	Indian Institute of Technology, Indian School of Mines, Dhanbad, India
M. Houmad	Laboratory of Condensed Matter and Interdisciplinary Sciences (LaMScI), Faculty of Science, Mohammed V University of Rabat, Rabat, Morocco
M. Khuili	Superior School of Technology (EST-Khenifra), University of Sultan Moulay Slimane, Khenifra, 54000, Morocco
Naorem Premjit Singh	A. R. S. D. College, University of Delhi, Dhaula Kuan, New Delhi, India
Rakesh Tamang	Department of Zoology , University of Calcutta, Kolkata, West Bengal, India
Ramesh Chandra Tiwari	Department of Physics, School of Physical Sciences, Mizoram University, Aizawl, Mizoram, India
Shivraj Gurung	Physical Sciences Research Centre (PSRC), Department of Physics, Pachhunga University College, Mizoram University, Aizawl, Mizoram, India
Shishir Tamang	Department of Zoology , University of Calcutta, Kolkata, West Bengal, India Department of Zoology, Darjeeling Government College, Darjeeling, West Bengal, India
Soni Subba	Department of Zoology , University of Calcutta, Kolkata, West Bengal, India
Saptaparni De	Department of Zoology , University of Calcutta, Kolkata, West Bengal, India
Shomaila Khanam	Department of Physics, Birla Institute of Technology, Mesra, Ranchi, Jharkhand, India
Sanjeeb Kumar Rout	Department of Physics, Birla Institute of Technology, Mesra, Ranchi, Jharkhand, India
Yengkhom Rangeela Devi	Department of Physics, Pachhunga University College (PUC), Aizawl, Mizoram, India
Zodinmawia	Department of Physics, Mizoram University, Aizawl, Mizoram, India
Zodinmawia	Department of Physics, School of Physical Sciences, Mizoram University, Aizawl, Mizoram, India

CHAPTER 1

Role of Plasmonic Metal-semiconductor Heterostructure in Photo Catalytic Hydrolysis and Degradation of Toxic Dyes

Shomaila Khanam[1] and **Sanjeeb Kumar Rout**[1,*]

[1] *Department of Physics, Birla Institute of Technology, Mesra, Ranchi, Jharkhand, India*

Abstract: Plasmonic metal-semiconductor heterostructure has become the most prominent content for water splitting by photocatalytic means. It is thought to be an effective, clean, and affordable energy source. Hydrolysis, water splitting, and destruction of organic dyes have all demonstrated the high efficiency of LSPR formation by these materials. A noble metal combined with a low bandgap semiconductor makes for the perfect photocatalyst. In this case, both semiconductors and noble metals can absorb visible light. They are prone to producing positive and negative pairs and inhibit their recombination, causing the resulting electron-hole pairs to interact with the chemicals in the immediate environment, thereby increasing photocatalytic activity. The strong SPR's combined effect with the efficient separation of photogenerated electrons and holes supported by noble metal particles can be credited with the increased photocatalytic activity. It has become a useful method for overcoming the limitations of conventional photocatalysts and promoting photocatalytic mechanisms.

This book chapter has three main goals: briefly describing plasmonic dynamics, explaining the preparation techniques, analyzing the key characteristics of the plasmonic metal nanostructure that influence photocatalysis, summarizing the reported literature, and offering an in-depth explanation of the four fundamental plasmonic energy transfer process.

Keywords: Ammonia borane, Localized surface, Plasmon resonance effect, Photocatalytic degradation, Plasmonic, Photocatalytic hydrolysis.

INTRODUCTION

Hydrogen is recognized as a green fuel on a worldwide scale because of its high energy density, ability to emit no harmful waste products, and status as an alternative to petroleum distillates. In recent years, hydrogen has emerged as a

* **Corresponding author Sanjeeb Kumar Rout:** Department of Physics, Birla Institute of Technology, Mesra, Ranchi, Jharkhand, India; E-mail: skrout@bitmesra.ac.in

Dibya Prakash Rai, Kingsley O. Obodo, & Jitendra Pal Singh (Eds.)

crucial fuel, satisfying society's need for fuel with minimal environmental impact. Sadly, there is still a roadblock in the way hydrogen becomes a practical source of energy due to its safe and affordable storage. In addition, applications for hydrogen outside of transportation are as important to take into account; these include its use in fertilizer production, metal treatment, food processing, petroleum refining, and food processing. Hydrogen has high gravimetric and volumetric density, and can be stored in a variety of solid-state hydrogen storage materials, including ammonia borane (NH_3BH_3, AB), sodium borohydride, ammonium borohydride, decaborane diammoniate of diborane, and others. Ammonia borane is regarded as a standout candidate for hydrogen applications [1 - 3] because of its many benefits, like its sufficient hydrogen content (19.6% wt), lower molecular weight (30.7 g mol-1), stability in solution, solubility in water, and tendency to undergo hydrolysis reaction. In order to generate three equivalents of hydrogen from one mole of AB by photocatalysis, several homogeneous catalysts have been produced. For instance, Goldberg *et al.* observed that an iridium catalyst from AB released hydrogen quickly within 20 minutes of being at ambient temperature [2, 4]. By using photocatalytic hydrolysis of AB, Baker and colleagues experimented with the acid-induced hydrolysis of AB [5] and the effectiveness of a nickel-based catalyst in the release of 94% hydrogen after 3 hours. Numerous academic papers by Xu and Chandra [6], Manners [7], and Jagirdar [8] proved the heterogeneous photocatalysis of ammonia borane for hydrogen release through combined noble metals and nanoceramics composites is very effective. Plasmonic photocatalyst is the name given to the heterogeneous catalyst consisting of noble metals and semiconductors. When the light wave of plasmonic resonance strikes them, they begin to oscillate, which eventually results in LSPR (a bound electromagnetic field). The best examples of these materials with negative real permittivity are silver and gold. Plasmonic nanoceramics have demonstrated their use for everything from organ imaging to sensors. Plasmonic materials have a special ability to create an increased electric field at the metal-semiconductor interface, which makes them ideal photocatalysts. Its ability to generate photoacoustic waves and heat is beneficial in the medical sector, in which it is employed to treat cancer and image organs. The usage of metal-semiconductor composite as a photocatalyst is very common. When exposed to light, photocatalysts can accelerate photochemical processes and start the reduction and oxidation process. It may remove a variety of toxic colours from water while also serving as a disinfectant. By starting the dehydrogenation reaction and releasing hydrogen from the hydrogen-stored compounds, it aids in the creation of hydrogen by splitting water. It can provide us with sustainable water and air purification as well as green and clean energy.

Three processes—hydrolysis, pyrolysis, and solvolysis—can liberate hydrogen from ammonium borane [2, 7, 9, 10]. Hydrolysis is the most effective technique for removing hydrogen from AB out of the three. It has low energy consumption and needs a sufficient catalyst at room temperature [7]. In contrast, pyrolysis exhausts energy and results in byproducts that are difficult to get rid of [9]. Moreover, solvolysis calls for organic solvent, which is incompatible with the need for green hydrogen [10]. In the presence of a catalyst, 1 mol of AB can produce 3 mol of hydrogen through hydrolysis. The hydrolysis equation is:

$$NH_3BH_3 + 2H_2O \rightarrow NH^{4+} + BO^{2-} + 3H_2$$

We will learn more about the function of noble metals and nanoceramics hybrid in the effective release of hydrogen from AB in this book chapter. The combination of noble metals and semiconductor catalysts has the capacity to control the rate of the reaction, release hydrogen from ammonia borane and degrade toxic dyes. By hydrolyzing ammonia borane, noble metals like Pt, Ag, and Au and their nanocomposites produce a large amount of hydrogen. The Pt-based photocatalysts can also complete the process in under a minute. However, taking into account cost-effectiveness, the less expensive metals, such as Ag and Pd, along with semiconductors, may be useful in releasing hydrogen from AB.

PLASMONIC DYNAMICS

Hot electrons can pass from metal to semiconductor and vice versa, thanks to the plasmonic structures' LSPR effect at their Schottky junction [2, 11, 12]. Due to its dependence on their structure and size, it has a confined effect [13]. In an electromagnetic wave moving in the direction of plasmonic nanocrystals, there are two types of light waves that can be observed: TE (transverse electric) modes and TM (transverse magnetic) modes. The electric field component of the transverse electric mode is parallel to the wave propagation vector. The electric field in the transverse magnetic mode runs parallel to the direction in which light travels. The surface plasmonic wave travels down the x-axis before dissipating in the z-axis.

The electric field and magnetic field equation for TE wave is written as:

$$Z>0, \qquad \tilde{E} = E_1\vec{y}e^{i(k_x x-\omega t)}e^{-a_1 z} \tag{1}$$

$$Z<0, \qquad \tilde{E} = E_2\vec{y}e^{i(k_x x-\omega t)}e^{a_2 z} \tag{2}$$

We have from Maxwell's equation,

$$\nabla \times E = -\mu_o \frac{\partial H}{\partial t} \qquad \xrightarrow{\frac{\partial}{\partial t} = i\omega} \qquad \nabla \times E = \mu_o i\omega \vec{H} \tag{3}$$

The magnetic equations are:

$$Z>0, \qquad H_1 = \frac{1}{\mu_o i\omega}[a_1\hat{x} + ik_x\hat{z}]E_1 e^{i(k_x\hat{x}-\omega t)}e^{-a_1 z} \tag{4}$$

$$Z<0, \qquad H_2 = \frac{1}{\mu_o i\omega}[-a_2\hat{x} + ik_x\hat{z}]E_2 e^{i(k_x\hat{x}-\omega t)}e^{a_2 z} \tag{5}$$

These equations do not meet the boundary condition. Thus, surface Plasmon resonance and TE mode are incompatible. The magnetic field equation for the TM mode is given by:

$$Z>0, \qquad \vec{H}_1 = H_1\hat{y}e^{i(k_x\vec{x}-\omega t)}e^{-a_1\vec{z}} \tag{6}$$

$$Z<0, \qquad \vec{H}_2 = H_2\hat{y}e^{i(k_x\vec{x}-\omega t)}e^{-a_2\vec{z}} \tag{7}$$

From the Maxwell equation,

$$\nabla \times H = \varepsilon \frac{\partial E}{\partial t} \tag{8}$$

The solution of the above equations is:

$$Z>0, \qquad E_1 = \frac{1}{\varepsilon i\omega}[- a_1\vec{x} - ik_x\vec{z}]H_1\, e^{i(k_x\vec{x}-\omega t)}e^{-a_1\vec{z}} \tag{9}$$

$$Z<0, \qquad E_2 = \frac{1}{\varepsilon i\omega}[a_2\vec{x} - ik_x\vec{z}]H_2\, e^{i(k_x\vec{x}-\omega t)}e^{-a_2\vec{z}} \tag{10}$$

At Z=O, $H_1 = H_2$, TM polarised light can produce surface plasmon oscillation as the boundary condition is satisfied. TM wave also require $E_1=E_2$. Therefore,

$$\frac{a_1}{\varepsilon_1} = -\frac{a_2}{\varepsilon_2} \tag{11}$$

Thus, E_1 *or* E_2 <0. This equation demonstrates that the surface wave remains only at the interface of materials with dielectric properties that have opposite polarity in the real part. One more condition must be met based on the Maxwell equation.

$$a^2 - k_x^2 = \frac{\omega^2}{c^2}\varepsilon_r\mu_r \qquad \textbf{(12)}$$

$$z > 0, \qquad a_1^2 - k_x^2 = \frac{\omega^2}{c^2}\varepsilon_1\mu_1 \qquad \textbf{(13)}$$

$$z < 0, \qquad a_2^2 - k_x^2 = \frac{\omega^2}{c^2}\varepsilon_2\mu_2 \qquad \textbf{(14)}$$

Combining and solving the above equation, we get the dispersion relation of the surface plasmons to be:

$$K_x = \frac{c}{\omega}\sqrt{\frac{\varepsilon_1 + \varepsilon_2}{\varepsilon_1\varepsilon_2}}. \qquad \textbf{(15)}$$

While propagating, plasma waves have a greater wave factor. For there to be a coupling between the evanescent wave and the surface plasmon, the propagation constants of the wave and the surface plasmon must be equal. Drude's model provides the greatest justification for the LSPR mechanism in plasmonic nanocrystals. Drude's model explains the electrodynamic properties of metals. In metals, most electrons are free because they are not bound to electrons. For this reason, the restoring force is negligible, and there is no natural frequency $\omega_o = 0$. This model gives a better understanding of metals. The Drude model for metals is derived from the Lorentz model by setting $\omega_o = 0$. Metal's Drude's permittivity is expressed as a complicated number.

$$\varepsilon(\omega) = \acute{\varepsilon} + i\ddot{\varepsilon} \text{ and } \varepsilon(\omega) = \varepsilon_\infty - \frac{\omega_p^2}{\omega^2+\gamma^2} + i\frac{\omega_p^2}{\omega(\omega^2+\gamma^2)}\gamma \qquad \textbf{(16)}$$

Where ω_p is the plasma frequency:

$$\omega_p^2 = Ne^2/m\varepsilon_0 \qquad \textbf{(17)}$$

and γ is the damping factor, which is proportional to fermi velocity $\gamma = v/l$. Here, v is fermi velocity, and l is the mean free path of an electron between successive oscillation events. The material's overall reaction to the external field is measured by its permittivity. The imaginary portion of permittivity, which increases with decreasing frequency, depicts material losses, while the real portion describes how light is refracted [2, 3]. To support the plasmonic effect, the imaginary part of the permittivity must be close to zero, and the real part must be negative. This circumstance produces strong resonance and significant light absorption. The overall field produced around the plasmonic nanocrystals when the TM polarized light shines on them is the combination of the crystal response field and external field. As a result, the electrons will be drawn out of the nucleus and into equilibrium. The same coulomb field that was created when the electrons and nucleus were removed but in the opposite direction will cancel the two other fields and prevent any field from escaping, creating the shield-like electric field at the metal and dielectric interface as illustrated in Fig. (**1a**) and (**1b**) [2]. Zhao *et al.* groundbreaking research from 2009 demonstrated that plasmonic nanoparticles have the singular ability to produce an evanescent electric field when electromagnetic light with a wavelength greater than the particle's size and plasma frequency $\omega=\omega_p$ is made to an incident on the surface of a noble metal. Metal electrons begin to oscillate at a resonant frequency. A plasmon is the name given to this pulsating electron. Surface Plasmon Resonance refers to the coupling of polaritons (photons and electrons) at the metal-dielectric contact. Strong absorption and mild dispersion of incoming light are made possible by the SPR [2]. This can be determined by looking at the UV spectrometer's absorbance intensity. The characteristics of the particle, such as its form, structure, metal type, size, and dielectric material, affect the SPR band's intensity and wavelength. The two fundamental excitations are Surface Propagation Plasmons (SPP) and LSPR (Localized Surface Plasmons). SPP stands for scattering plasmons. A traveling surface wave is produced along the surface when it is in a resonant condition. A localized Surface Plasmon is restricted geometrically to a small object, such as a point or a nanoparticle. The hybrid of noble metals and semiconductors also exhibits a promising LSPR.

SYNTHESIS PROCESS OF METAL-SEMICONDUCTOR PLASMONIC PHOTOCATALYST

The process of preparation is crucial in describing the qualities of a superior plasmonic photocatalyst. More charge pairs and the least amount of electron recombination should be produced by a potential plasmonic photocatalyst with a high LSPR. Yet, the geometry- shape, size, and level of metal-semiconductor content completely affect the LSPR impact of a plasmonic photocatalyst. As the radius increases, so does the wavelength at which Plasmon resonance occurs, revealing the redshift in LSPR spectra. The resonance wavelength depends on the object's radius and material composition. According to Fig. (**2a**) and (**2b**), the particle in gold nanorods exhibits two SPR wavelengths: transverse and longitudinal [2]. The plasmonic gold nanorod's longitudinal SPR occurs at a longer wavelength, demonstrating its ease in polarization. The longitudinal SPR is more responsive to aspect ratio as compared to transverse SPR. By adapting to longer aspect ratios, the longitudinal LSPR of gold nanorods may be controlled between 500 nm and 2000 nm, while the transverse LSPR stays the same between

510 nm and 520nm. Fig. (**2**) depict the longitudinal and transverse SPR of a gold nanorod, respectively.

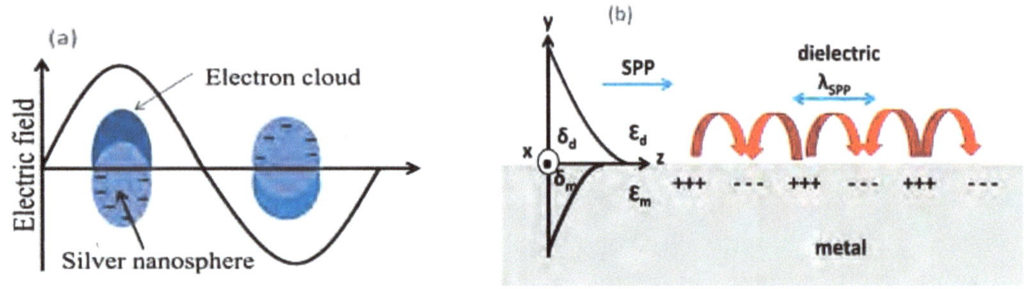

Fig. (1). Show the generation of electric field in a nanosphere and in a metal rod, respectively.

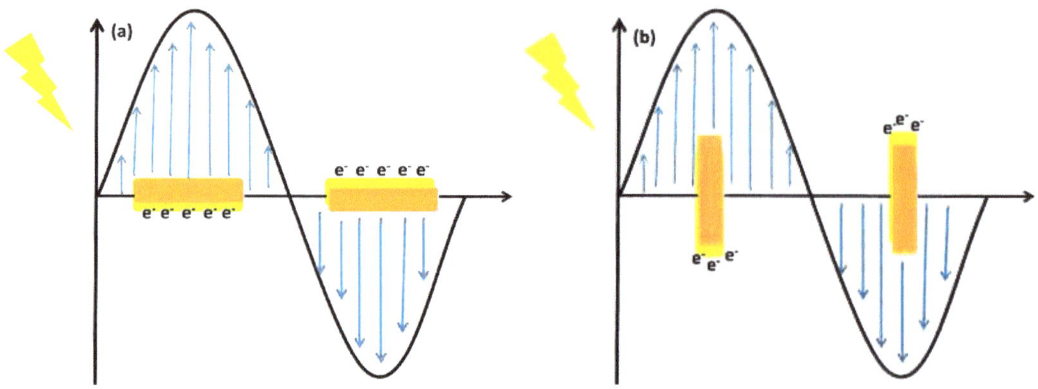

Fig. (2). Show the transverse and longitudinal SPR of the gold nanorod.

The preparation steps of plasmonic photocatalysts include the synthesis of semiconductors by various methods like sol-gel, solvothermal and hydrothermal, *etc*., and integrating the noble metal with the semiconductor with the process like chemical vapor deposition, photo deposition, and wet impregnation method. All the methods are suitable for the synthesis technique. Only the availability of the resources, time taken, and the required temperature play an important role in defining the structure of the prepared composite.

The literature by Man Sig Lee showed the preparation of TiO_2–Ag particles using the sol-gel method involving the reduction technique. Titanium tetra isopropoxide and silver nitrate were used as precursors of titania and silver, respectively. $AgNO_3$ mixed with reducing agent sodium citrate tribasic dihydrate was introduced in the reaction [14].

Work reported by Haolin Li showed that ZnO NWs were prepared by CVD (chemical vapour deposition) method. Au thin films were sputtered and annealed on ZnO NWs. The Au/ZnO nanocomposites were obtained [15, 16].

In the work by Gregor Žerjav, a wet-impregnation synthesis procedure was employed to prepare TiO_2 + Au composites. Fixed stochiometric value of TiO_2 was added into $HAuCl_4 \cdot 3H_2O$ aqueous solution and stirred for 20 h. Before drying for 12 h at room temperature, they obtained material washed several times [17].

According to one of the literature, the preparation of Ag/Bi_2WO_6 plasmonic photocatalyst involved a hydrothermal synthesis of $Bi_2WO_{6,}$ and Ag was photo deposited under xenon lamp irradiation for an hour [18, 19]. Seongwon Jo prepared an Au/TiO_2 plasmonic photocatalyst through TiO_2 pretreatment before depositing Au by light irradiation [20]. The Ag/TiO_2 photocatalyst was made using a hydrothermal process, and Ag was added using a wetness impregnation method, according to another piece of writing by V.I. Simagina [21].

According to Haibo Yin's report, solvothermal synthesis of tungsten and molybdenum oxide hybrid also produced plasmonic results [22]. According to Yasutaka Kuwahara's findings, mesoporous silica combined with the photo deposition of Ag on Ti produced good photocatalytic behavior [23]. According to Hefeng Chengre, the hydrolysis of AB was successful when MoO_{3-x} nanosheets were prepared using a solvothermal process at 160°C for 12 hours [24]. This method also demonstrated plasmonic activity. Platinum with mesoporous silica was synthesized in one pot and had a good catalytic action, according to Maria Irum [25].

Fig. (**3**) shows the pictorial representation of the hydrothermal synthesis technique. X and Y are the chemical compounds mixed and stirred for a certain interval of time before being autoclaved in a water medium under high pressure and temperature. The hydrothermally prepared composite is washed and dried in an oven. The metal particle is photo-deposited on the synthesized composites [2].

All the methods mentioned here are suitable for the synthesis technique. Only the availability of the resources, time taken, and the required temperature play an important role in opting for a particular technique.

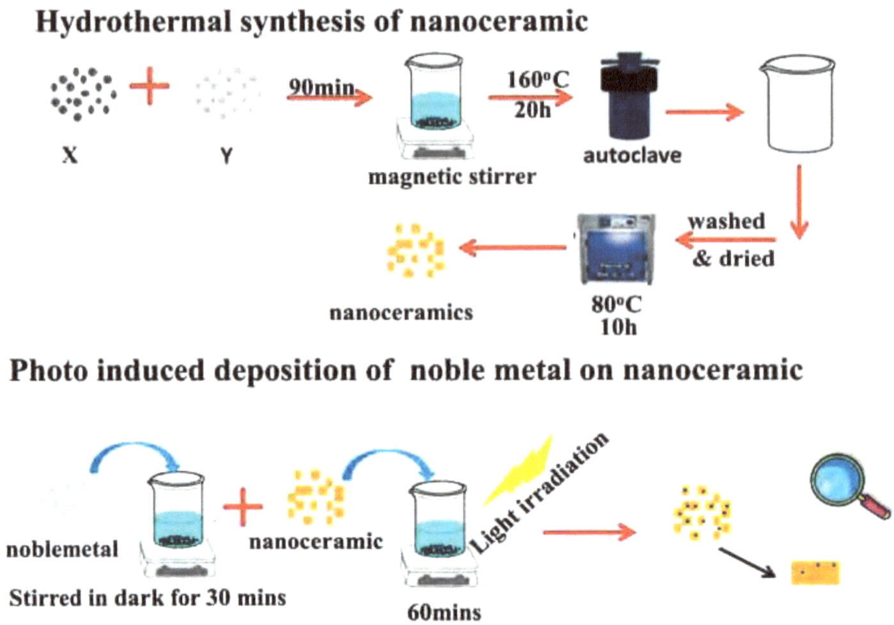

Fig. (3). Synthesis procedure of plasmonic metal-semiconductor.

FACTORS AFFECTING PLASMONIC NANOSTRUCTURE

Geometrical Shape

M. Rycenga *et al.* investigated the LSPR effect of different geometrical shapes of Ag using discrete dipole approximation (DDA) by studying spherical, cube, tetrahedron, octahedron, triangular plate, and cuboid Ag nanoparticles. The LSPR spectra of the 40 nm Ag nanosphere were calculated by Mie theory. The spectra show two resonance peaks, one at 370 nm and another at 410 nm [1]. The weak resonance peak arises due to energy losses. In contrast, the study of silver nanocubes was calculated by the DDA method. The peaks of Ag nanocubes are redshifted as compared to the Ag nanosphere. The red shift in the absorption spectra is due to the accumulation of surface charges at the corners of the silver nanocubes structure. Similarly, the study on tetrahedron and octahedron Ag nanoparticles showed a further red shift in LSPR peaks. These Ag nanostructures have even sharper corners than cubes. When the spectra of the triangular plate were observed, it showed a further redshift and an intense peak. The intense peak is observed due to the increased symmetry of the triangular geometry, and the redshift is due to an increase in charge separation. Further study on Ag nano bars with aspect ratios 2, 3, and 4 showed different absorbance spectra. The red shift and intense peak were observed with an increase in the aspect ratio. The spectra showed the aspect ratio of the nano bars controls the resonance peak. The longitudinal polarization of the nano bars with aspect ratios 2,3 and 4 produced the resonance peak at 700, 900, and 1100 nm.

The non-practical calculation on the gold of three various shapes—cube, sphere, and slab—another piece of literature likewise showed a similar outcome. The three nanocrystals' volume remained constant at 5nm^3 [26]. The slab's effective area was supposed to be 5nm, while the breadth was assumed to be 5nm. The absorption spectra revealed a significant resonance peak for the cube and sphere. Due to an increase in charge separation brought on by the sharpness of the cubic structure, the resonance peak for a cube was red-shifted. At 525 nm for the gold nanosphere and 575 nm for the gold nano cube, respectively, is the resonance peak. In contrast to the slab, the confined nanocrystals exhibit high LSPR. The slab does not provide much plasmonic resonance since the LSPR impact depends on the size quantification factor. The effects of confinement are significantly more noticeable in sphere- and cubic-shaped confined nanocrystals, leading to powerful oscillations in the resonance spectra of photogenerated electrons.

Another intriguing study by Dajian Wu and his team used a core-shell metal nano-shell nanoparticle with a dielectric core covered in a metal shell [27]. The LSPR in metal nanoshells, in contrast to solid metal nanoparticles, is sensitive to the proportions of the core and shell and may be modified over a wide range of wavelengths. For instance, the LSPR of the gold nanoshell can be changed, with decreasing shell thickness, from the visible range (500 nm) to the near-infrared region (800–1200 nm), which contains the transparent window of biological tissues [28]. Gold nanoshells with such near-infrared properties may find use in biomedicine for things like cancer thermal ablation therapy and medical diagnostics [29]. In accordance with this, the core-shell structured metal nanowire ought to have adjustable near-infrared optical characteristics [30]. Recently, Prokes *et al.* [31] discovered that the dielectric core-metal shell structure of Ga2O3-Ag and ZnO-Ag core-shell nanowires accounts for their remarkable SERS sensitivity. However, much research hasn't been conducted on the LSPR characteristics of core-shell gold nanowires, particularly on the impact of incidence angle and polarization direction.

Size of the Plasmonic Nanocrystal

The literature by K. Kluczyk *et al.* explained the effect of the size of the metallic nanoparticles clearly. The numerical modeling of Ag, Au, and Cu nanoparticles has been performed using COMSOL [32]. The result showed that LSPR of different sizes followed a pattern. The absorption spectra of the Au nanosphere from radii 20 nm to 140 nm were observed. The spectra observed a redshift and broadness of the LSPR peak with an increase in the size of the Au nanostructure.

This was mainly because of the generation of photo charge carriers. The separation of charges in plasmonic material depends upon the size of the plasmonic material. A similar pattern was observed in silver and copper nanospheres of different sizes ranging from 30 nm to 100 nm, respectively; as the radii of the nanosphere increased, the redshift was observed in the absorption spectra, showing that polarization of the plasmonic material takes at lower energy with an increase of the size of the nanoparticles.

In their study, Anderson and his team used Ag nanospheres, nanocubes, triangular prisms, and nanowires as model systems to demonstrate the impact of shape [33, 34]. The triangular nano prisms were created using a visible-light-photoinduced conversion process, whereas the nanospheres, nanocubes, and nanowires were created through polyol synthesis [33]. The LSPR band was found to be centered at 379, 407, 438, and 794 nm for the nanowires, nanospheres, nanocubes, and nano prisms, respectively (black, red, green, and blue traces, respectively), proving that the optical properties were significantly shaped-dependent as expected [33, 34].

Material Composition

The strength and intensity of the LSPR peak are strongly influenced by the composition of the plasmonic nanostructures. In order to split water photo-catalytically, a noble metal and semiconductor are often coupled. The separation of charge carriers, which eventually affects the quantity of hydrogen produced, is now significantly influenced by the type of noble metal and semiconductor used. Xinbo Wei *et al.* reported that nanostructures of g-C_3N_4 (nanospheres), g-C_3N_4/TiO_2 (nanofibers), Ag/g-C_3N_4/TiO_2 (nanofibers), Pt/g C_3N_4/TiO_2 (nanofibers), and Au/g-C_3N_4/TiO_2 synthesized through electrospinning reduction and oxidation process is efficient in the photocatalysis of water [35]. Yet, due to the various nanofibers, the amount of hydrogen produced varies. Between Au/g-C_3N_4/TiO_2 and Pt/g-C_3N_4/ TiO_2, Au/g-C_3N_4/TiO_2 had the highest intensity of the LSPR peak at 625 nm. At 410 nm, Ag/g-C_3N_4/TiO_2 exhibits an LSPR peak. Due to the high imaginary component of the dielectric function of Pt, there is hardly any LSPR peak seen for Pt/g-C_3N_4/TiO_2.

The LSPR absorption band of Ag/C at 410nm overlaps with the absorption peak of g-C_3N_4. The H_2 evolution order of these plasmonic nanofibers under simulated sunlight irradiation came out to be Ag/g-C_3N_4/TiO_2 NFs (~1.50 μmol/h) > Au/g-C_3N_4/TiO_2 NFs (~1.19 μmol/h) > Pt/g-C_3N_4/TiO_2 NFs (~0.89 μmol h-1). Whereas the TiO_2 NFs showed H_2 evolution with a rate of ~0.14 μmol/h and the H_2 evolution rate of gC_3N_4 NSs was recorded to be ~0.07 μmol/h.

Ratio of Noble Metal to the Semiconductor

The metal particle on a semiconductor's surface turns out to be a useful plasmonic nanostructure for boosting the photocatalytic activity. By adjusting the plasmonic metal's size, shape, and weight relative to the semiconductor, one can alter its plasmonic properties. The separation and transfer of hole pairs, the oscillation of plasmons, and the improved local electric field development all depend on these tuning variables. LSPR spectra show how the tuning factor differs from one another. Shelton J. P. Varapragasam *et al.*, in the literature provide one such case. By reducing Ag+ with oleyl amine and 1,2-hexadecanediol on TiO_2 nanorods, Ag/TiO_2 nanocrystals were created. Ag to TiO_2 ratios of 1:1 and 4:1 was changed. Ag/TiO_2 nanocomposites generated in ratios of 1:1 and 4:1 revealed one LSPR in their UV visible spectra, with the ratio 4:1 sample exhibiting a stronger LSPR peak than the ratio 1:1 sample [36].

PLASMONIC METAL-SEMICONDUCTOR HETEROSTRUCTURE FOR PHOTOCATALYTIC HYDROLYSIS OF AMMONIUM BORANE

Plasmonic photocatalysts have been developed as a strategy for ensuring high photocatalytic efficacy when exposed to visible light. It is a metal-semiconductor hybrid that is capable of producing photoelectrons, which can drive numerous chemical reactions that separate entities are unable to carry out [2]. When placed together in close proximity, these heterostructured materials mix to develop new properties. Its effectiveness depends on [37] the type of photosensitized catalyst [38], an appropriate photon source for excitation [39], the formation of electron-hole pairs, and [40] the pace at which these pairs recombine. According to the study, AB could hydrolyze and produce H_2 through a variety of molecular processes. In the process of photo-catalytically aided hydrogen evolution, 1 mol of NH_3BH_3 yields 3 mol of H_2. The first occurs when a plasmonic photocatalyst is exposed to radiation, and the metal's excited carriers and holes are pushed to the catalyst's edges. This causes an electric field to be created from the semiconductor towards the metal, creating LSPR charge pairs on the metal NPs, which can then cause AB to degrade. N-B bond elongation is caused by the breakdown process of AB, which drives electrons to the lowest unoccupied molecular orbital (LUMO) [23, 24]. In the second instance, a heterostructure of plasmonic metal and high band gap semiconductors causes a photocatalytic reaction by transferring electrons from the metal to the semiconductor when illuminated [2, 41 - 45]. When materials with low band gap semiconductors, such as noble metal, are investigated, it is discovered that the probability of an electron transfer from semiconductor to noble metal depends on the energy band placement at the intersection of the metal-semiconductor and illumination source. The situation in which a metal and a semiconductor are both light-sensitive is better for causing a photo-catalytic reaction because it can generate more charge carriers than the other two classifications in which a metal or a semiconductor absorbs light. The electron-hole pair's rate of recombination is decreased by the LSPR effect, which also creates a charge separation area. The generated electron (e-) and hole (h+) pair are crucial to the semiconductors' photocatalytic activity, and the hydrogen evolution from AB is benefited by the formation of hydroxyl radicals (•OH), which are produced in the photocatalytic reaction with water. The species involved in the reaction can be identified using scavenger testing. If e-, h+, and •OH are active species in the process, the inclusion of electron, hole, and hydroxyl scavengers in the reaction lowers the rate of hydrogen evolution.

The effectiveness of the response will not be impacted if any of the species do not participate. Some of the many articles reporting on hydrogen evolution with metal-semiconductors are given in Table **1**. Seong Won [46] looked into the impact of gold loading on titanium dioxide (Au/TiO_2) catalysts with varying gold contents (0, 0.5, 1, 2, and 3 wt%) in the evolution of hydrogen from AB when exposed to visible light. The H_2 production peaked at 1 wt% loading (85 mol) in 4 hours, indicating that a good amount of Au loading is necessary for a smooth reaction and that a higher amount of loading decreases the reaction's active sites. Similar findings were reported by Xinhong Qi [47], who used Pt @SiO_2 and PtNi @SiO_2 catalysts to catalyze the hydrolysis of AB for the production of hydrogen. With a release of 4.9 ml/min and 1.1 ml/min hydrogen, the PtNi @SiO_2 catalyst was discovered to be more active than Pt@SiO_2. Fig. (**4**) shows the representation of photocatalytic hydrolysis of hydrogen from ammonia borane.

Table 1. Literature reporting the evolution of hydrogen with metal-semiconductor.

S. No.	Catalysts	Source	Time	H_2 (Evolution)	Refs.
1.	Au/TiO_2	Visible light	4 h	85 μmol	[46]
2.	Ag/TiO_2	LED (365nm)	3 h	330 μmol	[21]
3.	Ag/Ti/SBA	Visible light	1 h	65 μmol	[23]
4.	$Mo_xW_{1-x}O_{3-y}$	LED (420nm)	1 h	60 mol%	[22]
5.	MoO_{3-x}	LED (420nm)	1 h	77 mol%	[24]
6.	Cu/TiO_2	Without light	1 h	90 mol%	[48]
7.	Pt-TiO_2	UV light	4 h	15 μmol	[49]
8.	ZnO-Fe_2O_3-TiO_2	Visible light	50 min	2.7 equivalent of H_2	[50]
9.	PtNi/g-C_3N_4	Visible light	60 min	3 ml/min	[51]
10.	PtNi@SiO_2	Visible light	60 min	4.9 ml/min	[47]
11.	Pt@SiO_2	Visible light	60min	1.1 ml/min	[47]

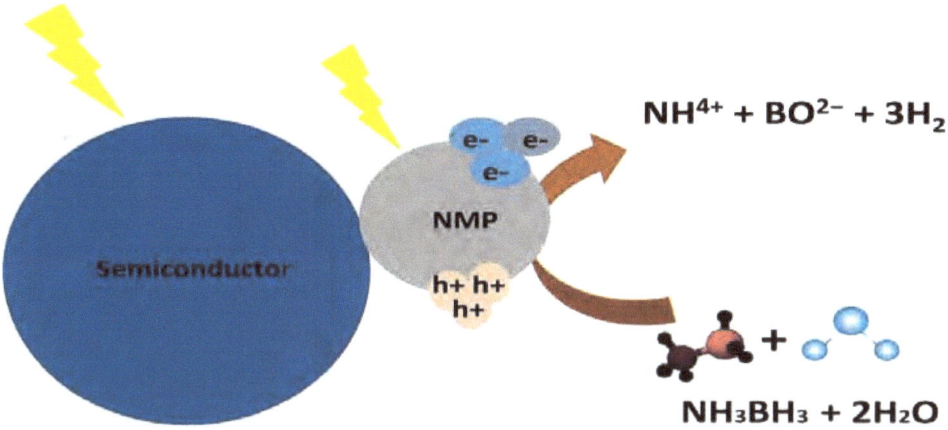

Fig. (4). Pictorial representation of photocatalytic hydrolysis of hydrogen from ammonia borane.

PLASMONIC METAL-SEMICONDUCTOR HETEROSTRUCTURE FOR DYE WATER TREATMENT

The photocatalyst has shown to be effective in a variety of applications, including optical, biosensing, and storage [52 - 55]. Plasmonic-metal-based semiconductor nanostructures have significantly increased the photo-catalytic performance of the photocatalyst. Several researchers have been motivated to create visible-ligh--driven plasmonic photocatalysts as a result of the localized surface plasmon resonance (LSPR) effect of noble-metal nanoparticles (such as Ag and Au) on other semiconductors. Purely visible-light-driven photocatalysts suffer from a number of problems, including poor stability, a low recycling rate, and ineffective photocatalytic degradation of azo dyes. Hence, plasmonic photocatalysts are very sought after in the sectors of energy and the environment. In order to address the issues of the energy crisis, environmental pollution, and global warming, a viable solution is the heterostructure of metal and semiconductor photocatalysis [2, 56 - 58]. Huang *et al.* [59] created an ion-exchange-based Ag@AgCl plasmonic photocatalyst in 2008 and investigated the connection between morphological and photocatalytic effects [60 - 64]. It was discovered that a plasmonic photocatalyst's shape influences its photocatalytic function. Similar to this, the charge separation property of plasmonic photocatalysts is influenced by their geometrical shape. Ag@AgCl plasmonic photocatalysts with cubic shapes have recently been compared to Ag@AgCl catalysts with spheres of comparable size. A superior photocatalyst was found to be a geometric shape with more surface area and active sites [65 - 67]. According to a variety of published studies, plasmonic photocatalysts perform better when it comes to the photocatalytic degradation of organic dyes.

The result of a few papers is listed in Table **2**. The literature by Junki Li *et al.* [2] studied the photocatalytic effect of Bi_2WO_6 and plasmonic Ag/Bi_2WO_6 on RhB. It was reported that bare Bi_2WO_6 had only 52.5% degradation efficiency in 140 mins. At the same time, Ag/Bi_2WO_6 showed a drastic increment in the degradation efficiency and could degrade RhB up to 94%.

Table 2. Literature reporting the photocatalytic degradation of dye molecules.

S. No.	Catalyst	Organic Pollutant	Time	Degradation %	Reference
1.	Ag@AgCl	Phenol	50 min	93	[71]
2.	Ag@AgCl	DCP	60 min	97	[61]
3.	Ag@AgCl	AO7	20 min	97	[61]
4.	Ag/AgBr-CNTs	TBP	50 min	100	[72]
5.	AgBr	TBP	50 min	60	[69]
6.	CNTs	TBP	50 min	45	[69]
7.	Ag/AgBr-CNT	TBP	50 min	100	[69]
8.	Ag/AgCl-CNTs	TBP	50 min	39	[69]
9.	AgCl	TBP	50 min	26	[69]
10.	Ag/AgI-CNTs	TBP	50 min	42	[69]
11.	AgI	TBP	50 min	8	[69]
12.	Ag-$BaTiO_3$	RhB	75 min	83	[73]
13.	TiO_3	RhB	75 min	63	[70]
14.	Ag/Bi_2WO_6	RhB	140 min	94.1	[68]
15.	Au/Bi_2WO_6	Phenol	60 min	93.3	[74]
16.	Bi_2WO_6	Phenol	60 min	65.2	[71]

In the reported work by H. Shi [69], the photocatalytic degradation of TBP by bare CNTs, AgBr, and the produced Ag/AgBr-CNT composites was studied. When Ag/AgBrCNTs were utilized as the photocatalyst under visible light irradiation, 100% of 100 mol L-1 TBP could be photo catalytically destroyed in 50 minutes. When AgBr or CNTs are used as the photocatalyst, however, under identical experimental conditions, only 60 and 45% of TBP are broken down, respectively. Ag/AgCl-CNTs and Ag/AgI-CNTs were used as photocatalysts in the elimination of TBP using photocatalysis. Ag/AgCl-CNTs and Ag/AgI-CNTs can only degrade 39 and 42% of the TBP, respectively. Comparatively speaking, pure AgCl and AgI without CNTs only destroyed 26 and 8% of TBP, respectively. These findings confirm that composite Ag/AgX-CNTs (X = Cl, Br, I) photocatalysts have higher photocatalytic activity than pure AgX for the photocatalytic destruction of TBP. among every catalyst examined in this study.

According to one of Shuya Xu's published studies [70], $BaTiO_3$ degraded by 15% and 63% deterioration was seen in TiO_3 under dark conditions. Ag-$BaTiO_3$ is substantially more photo-catalytically active after being changed by Ag due to the SPR effect. The best photocatalytic activity is demonstrated by 1mAg-$BaTiO_3$, which causes a degradation rate of around 83% in 75 minutes. So, it can be suggested that the generation of OH and the oxidation of RhB are mostly dependent on the plasmonic effect (Fig. **5**).

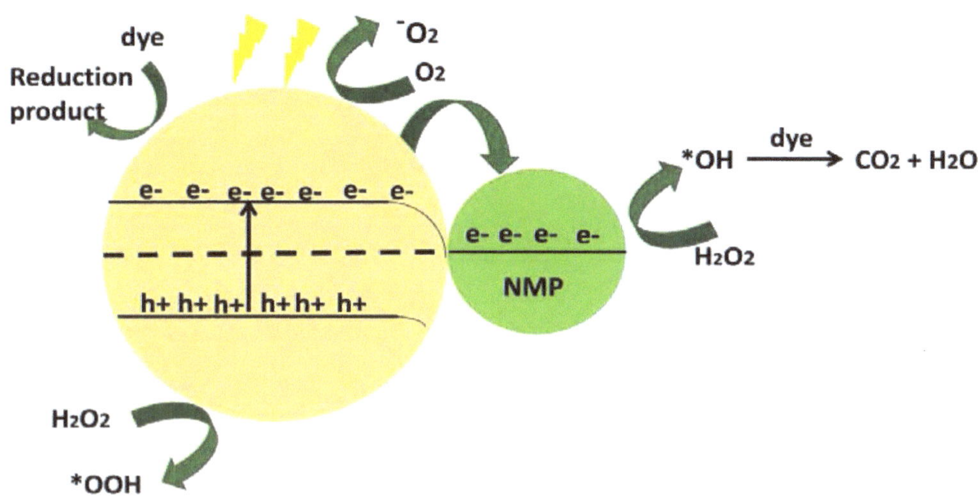

Fig. (5). Representation of photocatalytic degradation of dye molecules.

According to Jian-Yi Liu's research [73], after 60 minutes of exposure to visible light, phenol degradation over pure Au/Bi_2WO_6 is 65.2%. The maximum photocatalytic activity, however, was demonstrated by the Au/Bi_2WO_6. The corresponding phenol degradation rates reached about 93.3% in about 60 minutes, showing that the Au/Bi_2WO_6 displayed improved visible-light-driven photocatalytic performance.

PROPOSED MECHANISM

In the photocatalytic hydrolysis of molecules rich in hydrogen, a plasmonic photocatalyst is effective at treating wastewater. Several processes that satisfy the process have been described in the literature. Generally, the reported literature suggests four mechanisms. The plasmonic catalyst works through four different mechanisms: the Schottky barrier, an elevated local electric field, direct electron transfer, and Plasmon resonant energy transfer.

Schottky Barrier

A typical Schottky junction is created when a noble metal and a semiconductor come into close contact. Noble metals have a larger work function and lower Fermi level than the n-type semiconductor. Fig. (**6**) show the noble metal and an n-type semiconductor before and after the development of a Schottky barrier. A noble metal and semiconductor will have identical Fermi levels when they come into contact. A positively charged area inside the semiconductor and a negatively charged area close by are consequently created as a result of electrons diffusing from the semiconductor to the noble metal. As a result, a zone of space charge is created that is devoid of free charge carriers [2]. As a result, an electric field is created from the semiconductor to the noble metal. In the presence of visible light, the internal electric field will drive photogenerated holes toward noble metal and electrons toward semiconductors, avoiding the recombination of holes and electrons. The created electron-hole pair takes part in the chemical reactions of reduction and oxidation. This increases a photocatalyst's ability to do photocatalysis. The parameter Φ_M represents the metal's work function, and the semiconductor's electron affinity is represented by the variable X_{SM}. The Fermi energy of metal is lower than that of semiconductor. Because of the stability, electrons go from the semiconductor to the metal, and they will keep doing so until the Fermi levels of the two materials are equal. To move from a semiconductor to a metal, electrons must pass this potential barrier, which is created by this deformed band structure. Negative charges build up at the metal interface, and positive charges accumulate below the semiconductor interface as a result. The following equation provides an explanation for how this affects the Schottky barrier (Φ_B), a bending of the conduction band.

$$(\Phi_B) = \Phi_M - X_{SM}$$

An electron is trapped by the Schottky barrier created by the bending of the conduction band, preventing it from returning to the semiconductor and preventing the recombination of an electron-hole pair.

According to the stability theory, the Schottky barrier can effectively act as an electron trap, preventing the electron from flowing back to the semiconductor. As a result, the metal serves as the photo-induced electron's electron sink. The Schottky barrier prolongs the electron's life for the photoreaction and stops the electron-hole from recombining. Finding a correlation between the barrier height and the photoactivity, whether linear or nonlinear, is challenging. It is challenging to identify any linear or nonlinear association between the barrier height and the photoactivity, claims Maksudur R. Khan's review work [75].

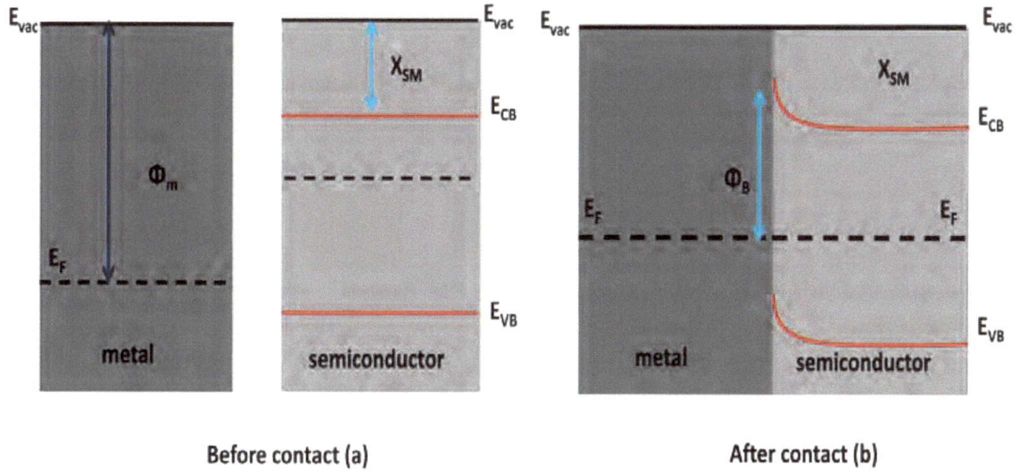

Fig. (6). (**a**) Generation of the Schottky barrier before and (**b**) After the semiconductor and metal make contact.

It's also intriguing to note that while Ru-TiO_2 scored second in hydrogen production, the metal Ru has a lower work function (4.71 eV) than the electron affinity of Bi_2O_3 (4.74 eV). Since the barrier is unable to stop the electron, in this instance, from flowing back to the semiconductor, the work function in this situation is lower than the electron affinity, indicating a tendency for the metal not to capture the electron [76, 77]. It is significant to remember that the arrangement of atoms at a material's surface affects the work function. Additionally, it should be remembered that metals have their own catalytic properties that affect how the semiconductor distributes electrons into the electrolyte. Pt has the potential to significantly enhance hydrogen gas production when considering the water-splitting reaction, as it can make the transfer of electrons to water easier. Rh, however, is discovered to produce the highest yield when the reaction is switched from water splitting to glucose reforming for the same H_2 generation, followed by Cu, Pt, Ni, Au, Ru, and Ir [76, 77].

Direct Electron Transfer

Tian and Tatsuma have shown direct electron transfer, also referred to as the LSPR sensitization effect, using an electrochemical cell and an Au-TiO_2 composite [78, 79]. When a photocatalyst receives enough photon energy to excite the free electrons in the noble metal to a higher Fermi level, hot electron holes are created on the metal. According to Fermi-Dirac non-equilibrium statistics, the energy is redistributed as a result of heated electrons moving above the Fermi level. As shown in Fig. (7), the excited electron is transported to the semiconductor during this redistribution, leaving holes in its way. The LSPR decay effect causes the electron to migrate in the direction of the semiconductor. The metal, in this instance, does not trap electrons; instead, it contributes them to the reaction. With this method, Tian and Tatsuma clarified the role of metal-semiconductors in photocatalysis [78, 79]. Due to its potential to increase the effectiveness and simplicity of electron transfer processes, direct electron transfer has attracted a lot of attention in the fields of bioelectrochemistry and bio-electrocatalysis. The enzyme or biomolecule needs to have a suitable redox-active site that can interact with the electrode surface in order for direct electron transfer to take place. Some scientists disagree with this device since it violates the Schottky barrier. So, whether the metal acts as an electron catcher or as a provider of electrons to the semiconductor's conduction band is still in question.

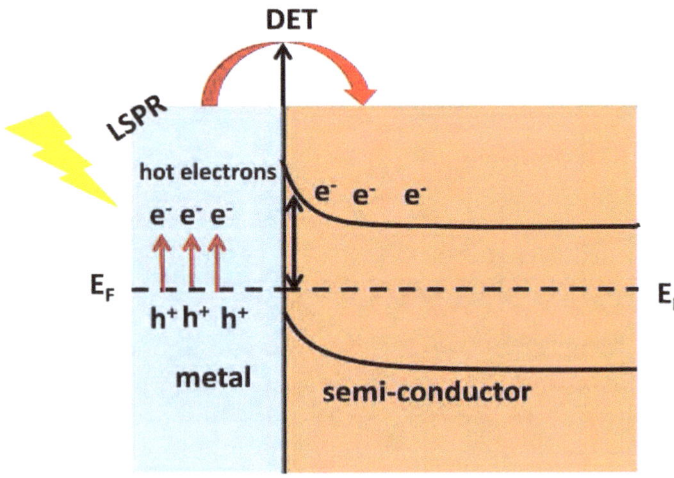

Fig. (7). Mechanism of DET.

Enhanced Local Electric Field

In plasmon dynamics, this phenomenon has been explored in great detail above. The falling photons are absorbed when electromagnetic irradiations irradiate the plasmonic composite with a frequency equal to resonance frequency (ω_p) of the noble metal. As a result, the oscillating electron cloud and the SPR electric field begins to form. The electric field that has formed around the nanocrystal is now stronger since it is a merger of the external field and crystal response field. Due to the enlarged local electric field and accelerated inter-band transition rate, LSPR produces energy that exceeds the band gap of a semiconductor. The photocatalyst's photo activity is improved as a result. Only in the vicinity of the noble metals does the enhanced local electric field exist, and its strength decreases with increasing distance [49]. The noble metal on the semiconductor (Fig. **8**) absorbs the light that strikes it and goes through plasmonic oscillation. This excites electrons and holes. A space charge zone is formed in the n-type semiconductor near the noble metal, creating an electric field from the semiconductor that is directed at the noble metal. This pushes the electrons towards noble metals and the holes towards semiconductors. Moreover, semiconductors absorb incident light and support the synergistic action of photocatalysis.

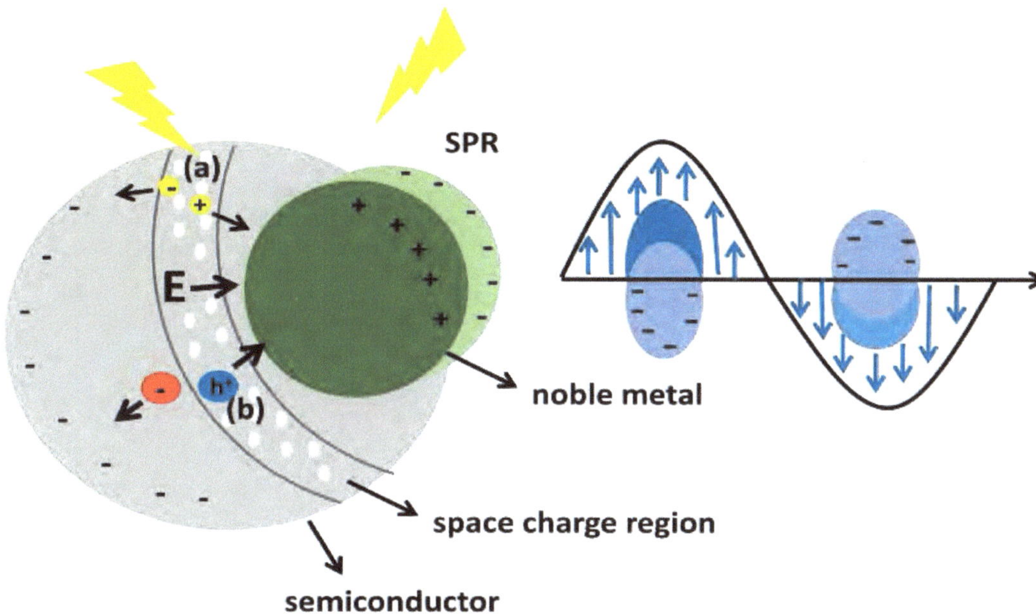

Fig. (8). Shows the formation of enhanced local electric field.

Plasmon Resonant Energy Transfer

When a metal is exposed to light with a Plasmon frequency, an electric field is created in its immediate vicinity. This prompts the heated electron in the metal to ascend above the fermi level and form a hole. Here, the electric field is used to transmit energy from the plasmon to the semiconductor. It alludes to the flow of energy between dipoles. The noble metal's plasmonic dipole moment and the semiconductor's electron-hole pair's dipole moment are coupled by this non-radiative mechanism. The relaxation of the localized surface plasmon dipole creates the electron-hole pair in the semiconductor. Whether the noble metal and semiconductor are in contact or not has no impact on PRET [80]. Mi Gyoung Lee stated that plasmonic Au NPs and an n-type semiconductor must make a Schottky contact in order for the hot electron injection process to occur in Au/$BiVO_4$ [82]. When the PRET mechanism is used, this restriction is lifted. No matter whether there is contact with the semiconductor, the PRET process is always engaged if the spectral overlap with the semiconductor and LSPR peak location is satisfied

[83]. The previous study demonstrated that PRET is dominating even at the sandwich structure's (Au@SiO_2@Cu_2O) LSPR peak's outer range [81]. In the PRET mechanism, an increased electromagnetic field is used to transfer the energy of plasmonic oscillation from a plasmonic material to the semiconductor. Fig. (**9**) shows the process of Plasmon resonant energy transfer.

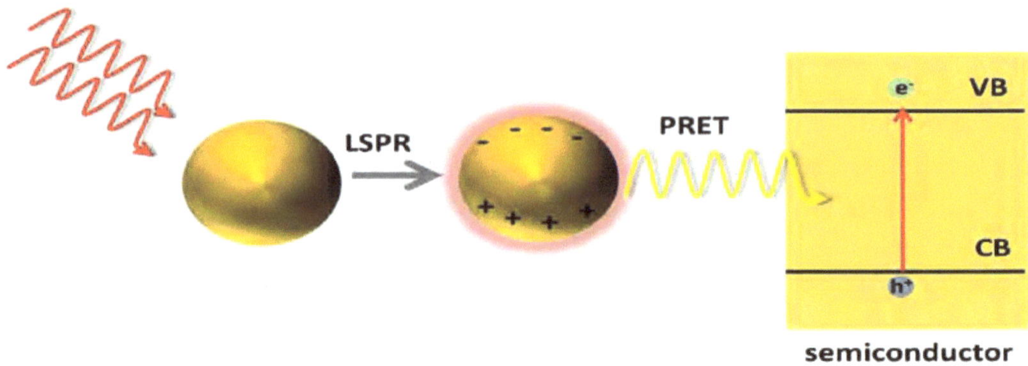

Fig. (9). Liberation of Plasmon energy to the semiconductors.

PROSPECTS AND FUTURE

It has become simpler to create optoelectronic devices, which have the potential to be extraordinarily useful in the future, thanks to research into the plasmonic mechanism that occurs inside a plasmonic material when light interacts with the electrons. The plasmonic material has been helpful in many domains where its electromagnetic field intensity may be regulated due to the tuning of its size and shape. The losses and expense of nanofabrication affect the application of plasmonic. Applications for plasmonics range from extremely sensitive biosensors to invisibility cloaks (a device that can see without being seen). It is possible to create magnetic memory storage on discs, nanoantenna, effective solar cells, and other useful devices by carefully studying the surface plasmon resonance feature of plasmonic materials. In the near future, some amazing work will be launched on plasmonic switches, plasmonic-supported quantum dots, plasmonic nanowires and nanocircuits. In the medical field, plasmonic nanoparticles are being studied to treat cancer. Plasmonic has several applications. Future prospects for plasmonic are promising, as it can be effectively applied to practically all areas of research.

Plasmonic as Energy Inputs

There is a rising desire for inexpensive, sustainable fuels that are derived from solar energy. Therefore, radiation-induced energy sources produced by plasmonic hybrid can be inexpensive. The optical density of states is a key factor in the photoelectric mechanism, which occurs in a radiation-induced energy source

(DOS). The number of channels that will either store or permit the propagation of electromagnetic energy in the medium is defined by DOS. Plasmonic materials are ideal for use as energy inputs because they offer desirable light focusing and DOS modification capabilities. When an external light source strikes a plasmonic nanostructure, the oscillation of electrons is known as plasmons. By adjusting the size, shape, and area occupied by the electrons, it is possible to adjust the collective oscillation dynamics of the plasmons. Heat can be produced from the energy held in the DOS of surface Plasmon oscillation. The development of hot-electron PV cells and photodetectors [89], undercooled boiling [86 - 88], improved frequency up-conversion processes [89], thermoelectric energy conversion [86], smart energy-saving window coatings [90, 91], optical data storage [92], nanoscale heat management [93, 94], and light extraction in solid-state lighting [84, 85] have all been made possible as a result. Plasmonic nanostructures are suitable for photocatalysis because the surface plasmonic effect produced can be exploited to power chemical reactions [95 - 97]. The capacity of plasmonic materials to focus light at the nanoscale results in an increased electric field that can produce local heating and phase transitions and be utilized to control chemical reactions. The transfer of heat is hindered if the plasmonic is very tiny [98]. This enables the particle to be heated locally. This focused heating produces a concentrated area of vapour and may even result in substrate melting [99]. The standard vapour generating systems lack portability, are expensive, depending on high optical concentration, require highly heated working fluids, and have a significant storage issue. The photothermal conversion using plasmonic nanoparticles is the ideal substitute due to the drawbacks of the standard vapour system [100]. Plasmonic has recently demonstrated 24% efficiency in the conversion of solar energy into heat, which is promising evidence that it can outperform rival technologies in the conversion of solar energy into hot vapour.

Quantum Dots and Quantum Dots

The semiconductor nanoparticles known as QDs are created by colloidal chemistry. QDs' optical characteristics can be modified by adjusting their geometry (size and shape). A QD's contact with light determines how it behaves in terms of photophysics. When light strikes QDs, an electron-hole pair is produced. An exciton is a quasi-particle created by this electron-hole pair. The confinement of the electrons and holes caused by the semiconductor's shrinkage raises exciton energy. Several applications are rendered illegible by the QDs' capacity to be tweaked in accordance with form and size [101 - 103]. Plasmonic materials have the potential to revolutionize LEDs and QDs. The plasmonic LEDs are shown in Fig. (**10**). The electric field augmentation of plasmonic materials close to the metal-dielectric interface boosts semiconductor LED intensity and enhances the QD emission rate. The coating of LEDs with plasmonic nanoparticles has been shown to boost their intensity by 14 times. Silicon LEDs that are affordable are now possible thanks to Plasmonic. According to reports, silver coating the silicon quantum dots can increase their emission rate by roughly 10 times. Another benefit of covering with plasmonic material is that it can make silicon brighter by 100 times by adjusting the resonance frequency and distance between the noble metal and semiconductor.

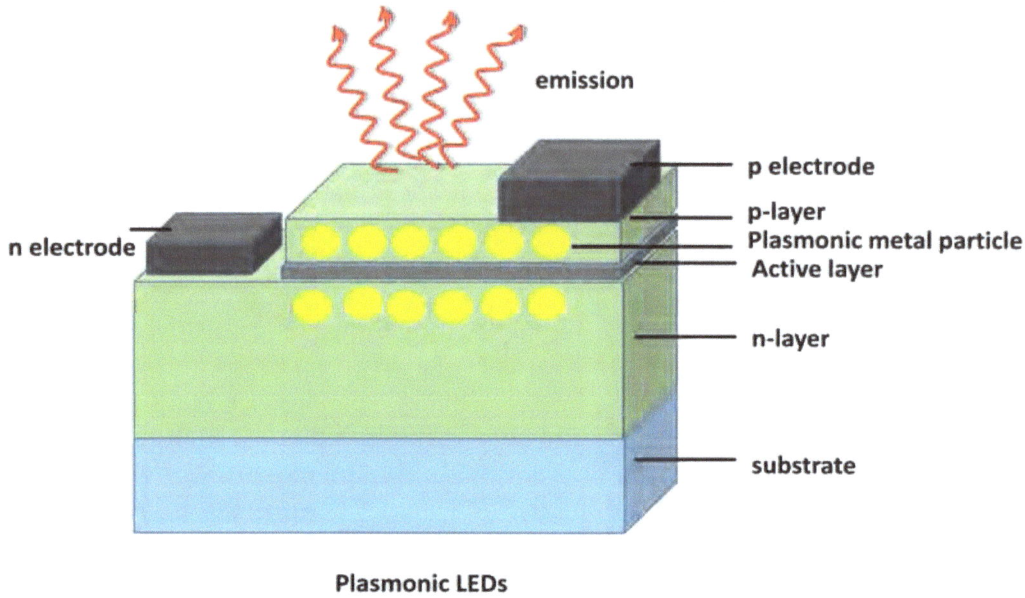

Fig. (10). Plasmonic LEDs.

Communication with Plasmonic

In plasmonic, the electron is excited by photons, and at a specific optical frequency, surface plasmon polaritons are formed. This SPP can oscillate at an optical wavelength and be constrained to a space smaller than that wavelength. These characteristics allow plasmonic materials to communicate effectively since they can transmit data at an optical band gap. It has been discovered that plasmonic metal-dielectric interfaces with the proper design can generate surface Plasmons with wavelengths that are shorter and have the same frequency as electromagnetic waves. This enables the transmission of plasmons, which transport information from one microprocessor to another, in interconnects (nanoscale cables). The movement of SPP within the nanoantenna is depicted in Fig. (**11**). Plasmonic waveguides have attracted a lot of attention lately since they can function in the visible to far-infrared range. Surface Plasmon Polaritons (SPP) are used for the signal processing step [104, 105]. The size disparity between large-scale photonics and small-scale electronics can be overcome using plasmonics. As plasmonic waveguides are small and their transverse dimensions are less than 100 nm, they can speed up optical processing.

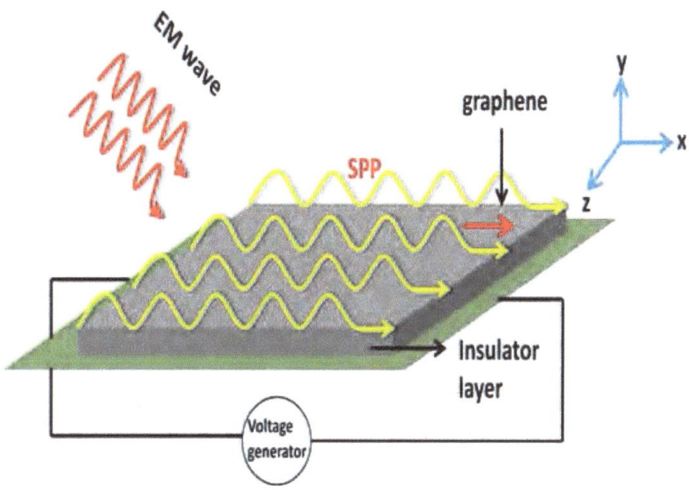

Fig. (11). Traversing of SPP towards nanoantenna.

Plasmonic Nanoparticles in the Treatment of Cancer

Cancer in mice has been reported to be cured using plasmonic nanoshells. Gold-coated silica particles with plasmonic nanoshells measuring about 100 nm have been added to mouse blood. Because more blood flows through the rats' malignant cells, these nanoshells are dispersed all around them. The plasmonic nanoshell experiences surface plasmonic oscillation when infrared light is shone on the mice's body, which finally causes the nanoshell to heat up. The malignant cells next to this plasmonic nanoshell will become hotter due to the heat generated. The malignant cell's temperature will rise from 37°C to 45°C. According to Fig. (**12**), this increase in temperature will destroy the cancer cell, a process known as photothermal death [106 - 108]. Authorities' approval is being sought by researchers in order to conduct a clinical trial on people with head and neck cancer. If it proves successful in treating patients, it will be a time and money-saving strategy.

Plasmonic in Desalination of Water

Global water scarcity is an issue, and the idea that the third world war will be waged over water is alarming. Around the world, there is a lot of study being done on how to desalinate water without harming the environment by using sunshine. Plasmonic also makes an appearance in this area. The Plasmonic Solar Desalination Method's special ability to improve light absorption and localize heating can maximize solar energy transfer. According to Lin Zhou *et al.*, the Plasmon-enhanced solar desalination system was created by adding aluminium nanoparticles to the grooves of a 3D aluminium membrane. The device's design is shown in Fig. (**13**). Water may be desalinated effectively to about 90% by localized heating thanks to the porous aluminium membrane loaded with aluminium nanoparticles, which can float on water and collect more than 96% of solar energy [109]. For up to 25 cycles, the device produced positive results. The plasmonic desalination device's abundant, affordable, and environmentally friendly properties could supply us with freshwater whenever we need it.

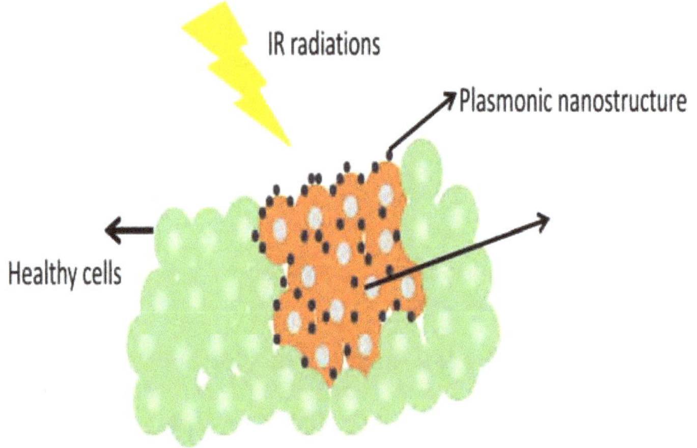

Fig. (12). Action of plasmonic composite on cancerous cells.

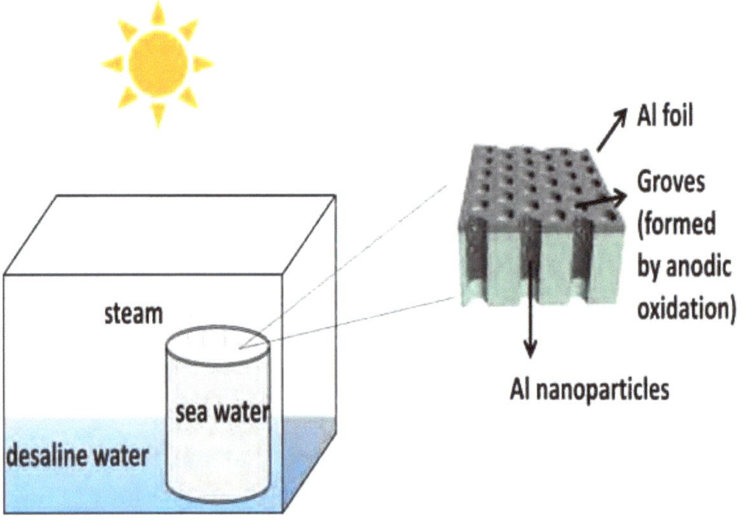

Fig. (13). Desalination of seawater using plasmonic aluminium.

Plasmonic Invisible Photodetector

An invisible device that senses light is called a plasmonic invisible photodetector, also called an invisibility cloak. The object that can be seen without being seen is the light-detecting gadget. It is a brand-new category of devices that can manipulate light at the nanoscale to carry out optical and electrical tasks. The component is made up of gold-plated nanocaps covering silicon nanowires. By adjusting the properties of the metal and semiconductor, the light reflected from each material cancels out the other's reflection, rendering the device invisible, at least to radiations in a specific frequency band. Both the metal and the semiconductor experience an equal strength opposing dipole moment as a result of the light waves. When two dipole moments that are equally strong and opposing collide, they cancel one another out, and the system becomes invisible. Invisibility will be reduced if the alignment of the dipoles is improper [110]. Hence, having the appropriate quantity of material at the nanoscale is crucial. The silicon is covered in gold, which not only makes the wire invisible but allows light to pass through to the silicon, where the current generated is monitored. The other metals, such as copper and aluminium, which reflect light just as gold, can be used in its stead and perform just as well. The arrangement of the metal-covered nanowires, the form of the metal, or the angle of light incidence have no effect on the cloaking effect. All that changes is how metal and semiconductors are tuned.

Electrochromic windows are intelligent windows that may alter their optical properties in response to an applied electrical charge. These windows can thus dynamically control the amount of light and heat that enter a building, providing benefits such as increased comfort, increased energy efficiency, and improved aesthetics.

Electrochromic Smart Window

Electrochromic windows are intelligent windows that may alter their optical properties in response to an applied electrical charge. These windows can thus dynamically control the amount of light and heat that enter a building, providing benefits such as increased comfort, increased energy efficiency, and improved aesthetics. Electrochromic smart windows can be made more functional by using plasmonic composites in their structural design. To interact with and alter the behavior of light entering the window, these composites can be made with tunable optical properties. By using plasmonic nanoparticles, the electrochromic material's ability to regulate light transmission and heat absorption can be further enhanced. Windows can vary their transparency and heat-blocking properties dynamically thanks to plasmonic composites' capacity to selectively absorb or scatter specific light wavelengths. This can improve energy efficiency by reducing the demand for artificial heating or cooling and assisting in managing indoor temperatures. The selective filtering and control of different light colors and wavelengths is possible with electrochromic windows that have plasmonic enhancement. This can produce more even and visually pleasing interior lighting settings while lowering glare and unwanted heat [111, 112].

Plasmonic Sensing

Noble metals are superior plasmonic materials because of their unique characteristics, which make them suitable for a range of plasmonic sensing applications. Due to their ability to magnify electromagnetic fields and their sensitivity to changes in the immediate environment, nanoscale devices are capable of very sensitive and accurate detection of a wide range of compounds and interactions. Two examples of the distinctive optical properties of plasmonic materials are plasmon-enhanced fluorescence and surface-enhanced Raman spectroscopy (SERS), which employs enhanced absorbance and scattering from plasmonic materials. The surface plasmon resonance (SPR) phenomenon, which results from the interaction of free electrons in a metal with an incident light wave to produce collective oscillations of the electron density known as surface plasmons, is used in plasmonic sensing. As a result of this interaction, the material's optical characteristics, such as its reflectance and absorbance, can change dramatically and sensitively [113, 114].

Plasmonic sensing has a wide range of applications, including in biochemistry, medicine, environmental monitoring, and other fields. Due to their ability to intensify the electromagnetic field around their surfaces, plasmonic materials are particularly responsive to changes in the immediate environment.

Antimicrobial Coating

In order to generate surfaces that may actively hinder the growth of germs like bacteria and viruses, antimicrobial coatings with plasmonic components combine the features of antimicrobial agents with the distinctive qualities of plasmonic materials. Due to their resonance characteristic, plasmonic nanoparticles can transform light energy into heat. Microbial cells that are near the nanoparticles can be thermally damaged or destroyed by using this localized heating. When exposed to light, plasmonic nanoparticles can help reactive oxygen species (ROS) form. ROS are very reactive chemicals that can harm DNA, proteins, and microbial cell membranes [115, 116].

LIMITATIONS AND CHALLENGES

To boost the H_2 generation yield and degrading efficiency, the majority of plasmonic metal-semiconductor heterostructures need noble metal co-catalysts, which raises the cost for large-scale commercial applications. Due to the fact that all heterostructures are unable to achieve the LSPR effect, not all metal-semiconductor-based photocatalysts can produce H_2 by breaking down AB. The two mixtures' size, geometrical shape, and composition have a complete bearing on the LSPR intensity. To create the LSPR effect at the intersection of metal and semiconductors, the proper concentration of noble metal is necessary. The active site is diminished by excessive noble metal loading, which decreases the effectiveness of the photocatalysts. Although there is still work to be done in the field of photocatalysis to get a 100% result, further study needs to be done on the plasmonic photocatalyst. The size, content, and synthesis process all affect in enhancing properties of plasmonic photocatalysts constructed of various semiconductors. The greatest plasmonic metal-semiconductor photocatalysts are still far from being developed, and there are still many applications for them to be found.

Enhancing the photoactivity of metal-semiconductor heterostructures depends on their size, composition, and synthesis methods. Nosaka and colleagues changed TiO_2 powder [117], and they discovered that Pt > Rh > Pd > Co > Ni > Ru > V > Cu > Mo had the highest catalytic effectiveness for boosting the photoactivity of metal-TiO_2 composites. According to Atwater *et al.* , hot electrons produced by Cu and Au NCs have higher energies than those produced by Ag NCs, which suggests that metal-semiconductor heterostructures made of Cu or Au metal NCs may have higher hot electron injection efficiency. Au NRs have a significant size influence on the photocatalytic activities of Au NRs-TiO_2 composites, as shown by Majima *et al.* [117].

They revealed that it is simple to create Au NRs-TiO_2 composites with wide band absorption in the UV-vis-NIR region by simply altering the aspect ratio of Au NRs. The H_2 evolution activity is significantly impacted by the aspect ratio of Au NRs. A high aspect ratio is advantageous for increasing NIR light-harvesting and thus promoting photoactivity. By shifting the SPR peak to the low resonance energy, further raising the aspect ratios ultimately results in a poorer efficiency of hot electron generation and injection. The greatest plasmonic metal-semiconductor photocatalysts are still far from being developed, and there are still many applications for them to be found.

CONCLUSION

Due to their limited stability and high rate of electron-hole pair recombination, conventional semiconductor photocatalysts have a low efficiency. The plasmonic metal-semiconductor heterostructure has recently come into its own as the ideal replacement for bare semiconductor photocatalysts. Due to the LSPR energy transfer at the junction of the metal and semiconductor, a well-designed plasmonic metal-semiconductor photocatalyst can efficiently promote charge separation and reduce the charge recombination rate of electron-hole pairs. The use of plasmonic metal-semiconductor nanoparticles as a catalyst for treating dye water and releasing hydrogen in H_2 from ammonia borane is gaining popularity. To create a high-performance photocatalyst, these components must be synthesized using the ideal process. Although just a small number of these plasmonic materials have been created yet, their dependable photocatalytic capabilities drive the development of a wide range of such materials.

ACKNOWLEDGMENTS

One of the authors (SK) is grateful to the Department of Science and Technology, Govt. of India, for providing financial assistance through the WOS-A Fellowship (SR/WOS-A/CS-128/2018) research work.

REFERENCES

[1] Akbayrak, S.; Ozkar, S. ScienceDirect Ammonia borane as hydrogen storage materials. *Int. J. Hydrogen Energy,* **2018**, *43*(40), 1-15.

[2] Khanam, S.; Rout, S.K. A Photocatalytic Hydrolysis and Degradation of Toxic Dyes by Using Plasmonic Metal–Semiconductor Heterostructures: A Review. *Chemistry,* **2022**, *4*(2), 454-479. [http://dx.doi.org/10.3390/chemistry4020034]

[3] Khanam, S.; Rout, S.K. Enhanced photocatalytic oxidation of RhB and MB using plasmonic performance of Ag deposited on Bi_2WO_6. *Chemistry,* **2022**, *4*(2), 272-296. [http://dx.doi.org/10.3390/chemistry4020022]

[4] Denney, M.C.; Pons, V.; Hebden, T.J.; Heinekey, D.M.; Goldberg, K.I. Efficient catalysis of ammonia borane dehydrogenation. *J. Am. Chem. Soc.,* **2006**, *128*(37), 12048-12049. [http://dx.doi.org/10.1021/ja062419g] [PMID: 16967937]

[5] Stephens, F.H.; Baker, R.T.; Matus, M.H.; Grant, D.J.; Dixon, D.A. Acid initiation of ammonia–borane dehydrogenation for hydrogen storage. *Angew. Chem. Int. Ed.,* **2007**, *46*(5), 746-749. [http://dx.doi.org/10.1002/anie.200603285]

[6] Xu, Q.; Chandra, M. A portable hydrogen generation system: Catalytic hydrolysis of ammonia–borane. *J. Alloys Compd.,* **2007**, *446-447*, 729-732. [http://dx.doi.org/10.1016/j.jallcom.2007.01.040]

[7] Jiang, H.L.; Xu, Q. Catalytic hydrolysis of ammonia borane for chemical hydrogen storage. *Catal. Today,* **2011**, *170*(1), 56-63. [http://dx.doi.org/10.1016/j.cattod.2010.09.019]

[8] Kalidindi, S.B.; Indirani, M.; Jagirdar, B.R. First row transition metal ion-assisted ammonia-borane hydrolysis for hydrogen generation. *Inorg. Chem.,* **2008**, *47*(16), 7424-7429.

[http://dx.doi.org/10.1021/ic800805r] [PMID: 18646842]

[9] Frueh, S.; Kellett, R.; Mallery, C.; Molter, T.; Willis, W.S.; King'ondu, C.; Suib, S.L. Pyrolytic decomposition of ammonia borane to boron nitride. *Inorg. Chem.,* **2011**, *50*(3), 783-792. [http://dx.doi.org/10.1021/ic101020k] [PMID: 21182274]

[10] Graham, T.W.; Tsang, C.W.; Chen, X.; Guo, R.; Jia, W.; Lu, S.M.; Sui-Seng, C.; Ewart, C.B.; Lough, A.; Amoroso, D.; Abdur-Rashid, K. Catalytic solvolysis of ammonia borane. *Angew. Chem. Int. Ed.,* **2010**, *49*(46), 8708-8711. [http://dx.doi.org/10.1002/anie.201003074] [PMID: 20922731]

[11] Kawamura, G.; Matsuda, A. Synthesis of plasmonic photocatalysts for water splitting. *Catalysts,* **2019**, *9*(12), 982.

[12] Meng-yu, X.; Kang-yang, S.; Xin-yuan, P.; Ren-jang, W.; Murthy, C.; Wei-chen, C. Hydrogen production by photocatalytic water-splitting on Pt-doped TiO_2-ZnO under visible light. *J. Taiwan Inst. Chem. Eng.,* **2016**, 1-7.

[13] Taylor, P.; Katyal, J.; Soni, R.K. Size- and shape-dependent plasmonic properties of aluminum nanoparticles for nanosensing applications. *J. Mod. Opt.,* **2013**, *60*, 37-41.

[14] Lee, M.S.; Hong, S.S.; Mohseni, M. Synthesis of photocatalytic nanosized TiO_2–Ag particles with sol–gel method using reduction agent. *J. Mol. Catal. Chem.,* **2005**, *242*(1-2), 135-140. [http://dx.doi.org/10.1016/j.molcata.2005.07.038]

[15] Yildiz, T.; Yatmaz, H.C.; Öztürk, K. Anatase TiO_2 powder immobilized on reticulated Al_2O_3 ceramics as a photocatalyst for degradation of RO16 azo dye. *Ceram. Int.,* **2020**, *46*(7), 8651-8657. [http://dx.doi.org/10.1016/j.ceramint.2019.12.098]

[16] Wu, F.; Li, X.; Liu, W.; Zhang, S. Highly enhanced photocatalytic degradation of methylene blue over the indirect all-solid-state Z-scheme g-C_3N_4-RGO-TiO_2 nanoheterojunctions. *Appl. Surf. Sci.,* **2017**, *405*, 60-70. [http://dx.doi.org/10.1016/j.apsusc.2017.01.285]

[17] Wang, C.; Astreuc, D. Nanogold plasmonic photocatalysis for organic synthesis and clean energy conversion. *Chem. Soc. Rev.,* **2014**, *43*, 7188-7216.

[18] Khanam, S.; Rout, S.K. Plasmonic metal/semiconductor heterostructure for visible light-enhanced H2 production. *ACS Omega,* **2022**, *7*(29), 25466-25475.

[19] Bunluesak, T.; Phuruangrat, A.; Thongtem, S.; Thongtem, T. Visible-light-driven heterostructure Ag/Bi_2WO_6 nanocomposites synthesized by photodeposition method and used for photodegradation of rhodamine B dye. *Res. Chem. Intermed.,* **2021**, *47*(7), 3079-3092. [http://dx.doi.org/10.1007/s11164-021-04456-0]

[20] Manuscript, A. *Materials Chemistry A,* **2017**.

[21] Simagina, V.I.; Komoyo, O.; Ozeroya, A.; Netskina, O.; Odegova, G.; Kayl, N.; Filippov, T. TiO_2-based photocatalysts for controllable hydrogen evolution from ammonia borane. *Catal. Today,* **2020**, *379*, 1-10.

[22] Yin, H.; Kuwahara, Y.; Mori, K.; Cheng, H.; Wen, M.; Huo, Y.; Yamashita, H. Localized surface plasmon resonances in plasmonic molybdenum tungsten oxide hybrid for visible-light-enhanced catalytic reaction. *J. Phys. Chem. C,* **2017**, *121*(42), 23531-23540. [http://dx.doi.org/10.1021/acs.jpcc.7b08403]

[23] Verma, P.; Kuwahara, Y.; Mori, K.; Yamashita, H. Enhancement of Ag-based plasmonic photocatalysis in hydrogen production from ammonia borane by the assistance of single-site Ti-oxide moieties within a silica framework. *Chemistry,* **2017**, *23*(15), 3616-3622. [http://dx.doi.org/10.1002/chem.201604712] [PMID: 27933684]

[24] Cheng, H.; Kamegawa, T.; Mori, K.; Yamashita, H. Surfactant-free nonaqueous synthesis of plasmonic molybdenum oxide nanosheets with enhanced catalytic activity for hydrogen generation

from ammonia borane under visible light. *Angew. Chem.,* **2014**, *126*(11), 2954-2958. [http://dx.doi.org/10.1002/ange.201309759]

[25] Irum, M.; Zaheer, M.; Friedrich, M.; Kempe, R. Mesoporous silica nanosphere supported platinum nanoparticles (Pt@MSN): One-pot synthesis and catalytic hydrogen generation. *RSC Advances,* **2016**, *6*(13), 10438-10441. [http://dx.doi.org/10.1039/C5RA25243A]

[26] Zhang, H.; Govorov, A.O. Optical generation of hot plasmonic carriers in metal nanocrystals: The effects of shape and field enhancement. *J. Phys. Chem. C,* **2014**, *118*(14), 7606-7614. [http://dx.doi.org/10.1021/jp500009k]

[27] Wu, D.; Liu, X.; Li, B.; Wu, D.; Liu, X.; Li, B. Localized surface plasmon resonance properties of two-layered gold nanowire : Effects of geometry, incidence angle, and polarization Localized surface plasmon resonance properties of two-layered gold nanowire : Effects of geometry, incidence angle, and polarization. *J. Appl. Phys.,* **2011**, *109.*

[28] Hirsch, L.R.; Jackson, J.B.; Lee, A.; Halas, N.J.; West, J.L. A whole blood immunoassay using gold nanoshells. *Anal. Chem.,* **2003**, *75*(10), 2377-2381. [http://dx.doi.org/10.1021/ac0262210] [PMID: 12918980]

[29] Allain, L.R.; Vo-Dinh, T. Surface-enhanced Raman scattering detection of the breast cancer susceptibility gene BRCA1 using a silver-coated microarray platform. *Anal. Chim. Acta,* **2002**, *469*(1), 149-154. [http://dx.doi.org/10.1016/S0003-2670(01)01537-9]

[30] Limmer, S.J.; Chou, T.P.; Cao, G.; Uni, V.; Science, M.; Hall, R. Formation and optical properties of cylindrical gold nanoshells on silica and titania nanorods. *J. Phys. Chem. B,* **2003**, *107*(48), 13313-13318. [http://dx.doi.org/10.1021/jp034992h]

[31] Prokes, S.M.; Glembocki, O.J.; Rendell, R.W.; Ancona, M.G. Enhanced plasmon coupling in crossed dielectric/metal nanowire composite geometries and applications to surface-enhanced Raman spectroscopy. *Appl. Phys. Lett.,* **2007**, *90*(9), 093105. [http://dx.doi.org/10.1063/1.2709996]

[32] Kluczyk, K.; Jacak, W. Damping-induced size effect in surface plasmon resonance in metallic nano-particles: Comparison of RPA microscopic model with numerical fi nite element simulation (COMSOL) and Mie approach. *J. Quant. Spectrosc. Radiat. Transf.,* **2016**, *168*, 78-88. [http://dx.doi.org/10.1016/j.jqsrt.2015.08.021]

[33] Anderson GM, Thenner S. Rodrigues, Jiale Wang, Liliam K. Yamada, Tiago V. Alves, Fernando R. Ornellas, Romulo A. Ando, and Pedro HC Camargo. The fault in their shapes: investigating the surface-plasmon-resonance-mediated catalytic activities of silver quasi-spheres, cubes, triangular prisms, and wires. *Langmuir,* **2015**, *31*, 10272-10278.

[34] da Silva, A.G.M.; Rodrigues, T.S.; Wang, J.; Camargo, P.H.C. Plasmonic catalysis with designer nanoparticles. *Chem. Commun. (Camb.),* **2022**, *58*(13), 2055-2074. [http://dx.doi.org/10.1039/D1CC03779J]

[35] Sajid, M. Toxicity of nanoscale metal organic frameworks: a perspective. *Environ. Sci. Pollut. Res. Int.,* **2016**, *23*(15), 14805-14807. [http://dx.doi.org/10.1007/s11356-016-7053-y] [PMID: 27300166]

[36] Varapragasam, S.J.P.; Mia, S.; Wieting, C.; Balasanthiran, C.; Hossan, M.Y.; Baride, A.; Rioux, R.M.; Hoefelmeyer, J.D. Ag–TiO_2 Hybrid Nanocrystal Photocatalyst: Hydrogen Evolution under UV Irradiation but Not under Visible-Light Irradiation. *ACS Appl. Energy Mater.,* **2019**, *2*(11), 8274-8282. [http://dx.doi.org/10.1021/acsaem.9b01730]

[37] Maeda, K.; Domen, K. Photocatalytic water splitting: Recent progress and future challenges. *J. Phys. Chem. Lett.,* **2010**, *1*(18), 2655-2661. [http://dx.doi.org/10.1021/jz1007966]

[38] Zhao, J.; Osterloh, F.E. Photochemical charge separation in nanocrystal photocatalyst films: insights from surface photovoltage spectroscopy. *J. Phys. Chem. Lett.,* **2014**, *5*(5), 782-786. [http://dx.doi.org/10.1021/jz500136h] [PMID: 26274067]

[39] Osterloh, F.E. Boosting the efficiency of suspended photocatalysts for overall water splitting. *J. Phys. Chem. Lett.,* **2014**, *5*(15), 2510-2511. [http://dx.doi.org/10.1021/jz501342j] [PMID: 26277937]

[40] Liao, L.; Zhang, Q.; Su, Z.; Zhao, Z.; Wang, Y.; Li, Y.; Lu, X.; Wei, D.; Feng, G.; Yu, Q.; Cai, X.; Zhao, J.; Ren, Z.; Fang, H.; Robles-Hernandez, F.; Baldelli, S.; Bao, J. Efficient solar water-splitting using a nanocrystalline CoO photocatalyst. *Nat. Nanotechnol.,* **2014**, *9*(1), 69-73. [http://dx.doi.org/10.1038/nnano.2013.272] [PMID: 24336404]

[41] Linic, S.; Christopher, P.; Ingram, D.B. Plasmonic-metal nanostructures for efficient conversion of solar to chemical energy. *Nat. Mater.,* **2011**, *10*(12), 911-921. [http://dx.doi.org/10.1038/nmat3151] [PMID: 22109608]

[42] DuChene, J.S.; Sweeny, B.C.; Johnston-Peck, A.C.; Su, D.; Stach, E.A.; Wei, W.D. Prolonged hot electron dynamics in plasmonic-metal/semiconductor heterostructures with implications for solar photocatalysis. *Angew. Chem. Int. Ed.,* **2014**, *53*(30), 7887-7891. [http://dx.doi.org/10.1002/anie.201404259] [PMID: 24920227]

[43] Wang, F.; Jiang, Y.; Gautam, A.; Li, Y.; Amal, R. Exploring the Origin of Enhanced Activity and Reaction Pathway for Photocatalytic H_2 Production on Au/B-TiO_2 Catalysts. *ACS Catal.,* **2014**, *4*(5), 1451-1457. [http://dx.doi.org/10.1021/cs5002948]

[44] Xing, M.Y.; Yang, B.X.; Yu, H.; Tian, B.Z.; Bagwasi, S.; Zhang, J.L.; Gong, X.Q. X.Q., Enhanced photocatalysis by Au nanoparticle loading on TiO_2 single-crystal (001) and (110) facets. *J. Phys. Chem. Lett.,* **2013**, *4*(22), 3910-3917. [http://dx.doi.org/10.1021/jz4021102]

[45] Bian, Z.; Tachikawa, T.; Zhang, P.; Fujitsuka, M.; Majima, T. Au/TiO_2 superstructure-based plasmonic photocatalysts exhibiting efficient charge separation and unprecedented activity. *J. Am. Chem. Soc.,* **2014**, *136*(1), 458-465. [http://dx.doi.org/10.1021/ja410994f] [PMID: 24308587]

[46] Calvo, M.E. Materials chemistry approaches to the control of the optical features of perovskite solar cells. *J. Mat. Chem.,* **2017**, *5*, 20561-20578. [http://dx.doi.org/10.1039/C7TA05666D]

[47] Qi, X.; Li, X.; Chen, B.; Lu, H.; Wang, L.; He, G. Highly active nanoreactors: Patchlike or thick ni coating on pt nanoparticles based on confined catalysis. *ACS Applied Materials & Interfaces,* **2016**, *8*, 1922-1928.

[48] Yousef, A.; Barakat, N.A.M.; EL-Newehy, M.H.; Ahmed, M.M.; Kim, H.Y. Catalytic hydrolysis of ammonia borane for hydrogen generation using Cu(0) nanoparticles supported on TiO_2 nanofibers. *Colloids Surf. A Physicochem. Eng. Asp.,* **2015**, *470*, 194-201. [http://dx.doi.org/10.1016/j.colsurfa.2015.02.004]

[49] Nasalevich, M.A.; Goesten, M.G.; Savenije, T.J.; Kapteijn, F.; Gascon, J. Correction: Enhancing optical absorption metal–organic frameworks for improved visible light photocatalysis. *Chem. Comm.,* **2015**, *51*, 961-962.

[50] Barakat, N.A.M.; Motlak, M.; Taha, A.; Nassar, M.M.; Mahmoud, M.S.; Fouad, H. Super effective Zn-Fe-doped TiO_2 nanofibers as photocatalyst for ammonia borane hydrolysis. *Int. J. Green Energy,* **2016**, *13*(7), 642-649. [http://dx.doi.org/10.1080/15435075.2015.1004575]

[51] Zhang, M.; Xiao, X.; Wu, Y.; An, Y.; Xu, L.; Wan, C. Hydrogen Production from Ammonia Borane over PtNi Alloy Nanoparticles Immobilized on Graphite Carbon Nitride. *Catalysts,* **2019**, *9*(12), 1009.

[http://dx.doi.org/10.3390/catal9121009]

[52] Cai, W.; Wang, J.; Liu, H.; Chen, W.; Wang, J.; Du, L.; Hu, J.; Wu, C. Gold nanorods@metal-organic framework core-shell nanostructure as contrast agent for photoacoustic imaging and its biocompatibility. *J. Alloys Compd.,* **2018**, *748*, 193-198. [http://dx.doi.org/10.1016/j.jallcom.2018.03.133]

[53] Lee, Y.; Schade, N.B.; Sun, L.; Fan, J.A.; Bae, D.R.; Mariscal, M.M.; Lee, G.; Capasso, F.; Sacanna, S.; Manoharan, V.N.; Yi, G.R. Ultrasmooth, Highly Spherical Monocrystalline Gold Particles for Precision Plasmonics. *ACS Nano.,* **2013**, *7*(12), 11064-11070. [http://dx.doi.org/10.1021/nn404765w]

[54] Pham, T.T.; Le, L.N.; Nguyen, H.N.; Luong, T.T.K.; Pham, T.N.; Nguyen, H.L.; Nguyen, T.K. Encapsulating gold nanoparticles in zeolitic imidazolate framework crystal for novel optical response. *Polyhedron,* **2018**, *148*, 171-177. [http://dx.doi.org/10.1016/j.poly.2018.04.002]

[55] Shang, W.; Zeng, C.; Du, Y.; Hui, H.; Liang, X.; Chi, C.; Wang, K.; Wang, Z.; Tian, J. Core – Shell Gold Nanorod @ Metal – Organic Framework Nanoprobes for Multimodality Diagnosis of Glioma. *Adv. Mat.,* **2016**, *29*(3)

[56] Zhang, Q.; Lima, D.Q.; Lee, I.; Zaera, F.; Chi, M.; Yin, Y. A highly active titanium dioxide based visible-light photocatalyst with nonmetal doping and plasmonic metal decoration. *Angew. Chem. Int. Ed.,* **2011**, *50*(31), 7088-7092. [http://dx.doi.org/10.1002/anie.201101969] [PMID: 21710514]

[57] Bavykin, D.V.; Friedrich, J.M.; Walsh, F.C. Protonated titanates and tio_2 nanostructured materials: Synthesis, properties, and applications. *Adv. Mater.,* **2006**, *18*(21), 2807-2824. [http://dx.doi.org/10.1002/adma.200502696]

[58] Mo, J.; Zhang, Y.; Xu, Q.; Lamson, J.J.; Zhao, R. Photocatalytic purification of volatile organic compounds in indoor air: A literature review. *Atmos. Environ.,* **2009**, *43*(14), 2229-2246. [http://dx.doi.org/10.1016/j.atmosenv.2009.01.034]

[59] Wang, P.; Huang, B.; Qin, X.; Zhang, X.; Dai, Y.; Wei, J.; Whangbo, M.H. Ag@AgCl: A highly efficient and stable photocatalyst active under visible light. *Angew Chem. Int. Ed. Engl.,* **2008**, *47*(41), 7931-7933.

[60] Wang, P.; Huang, B.; Lou, Z.; Zhang, X.; Qin, X.; Dai, Y.; Zheng, Z.; Wang, X. Synthesis of Highly Efficient Ag @ AgCl Plasmonic Photocatalysts with various structures. *Chem. Euro. J.,* **2010**, *16*, 538-544.

[61] Dong, R.; Tian, B.; Zeng, C.; Li, T.; Wang, T.; Zhang, J. Ecofriendly Synthesis and Photocatalytic Activity of Uniform Cubic Ag@AgCl Plasmonic Photocatalyst. *J. Phys. Chem. C,* **2013**, *117*(1), 213-220. [http://dx.doi.org/10.1021/jp311970k]

[62] Li, Y.; Ding, Y. Porous AgCl/Ag nanocomposites with enhanced visible light photocatalytic properties. *J. Phys. Chem. C,* **2010**, *114*(7), 3175-3179. [http://dx.doi.org/10.1021/jp911678q]

[63] Lou, Z.; Huang, B.; Wang, P.; Wang, Z.; Qin, X.; Zhang, X.; Cheng, H.; Zheng, Z.; Dai, Y. The synthesis of the near-spherical AgCl crystal for visible light photocatalytic applications. *Dalton Trans.,* **2011**, *40*(16), 4104-4110. [http://dx.doi.org/10.1039/c0dt01795g] [PMID: 21384009]

[64] Lou, Z.; Huang, B.; Qin, X.; Zhang, X.; Cheng, H.; Liu, Y.; Wang, S.; Wang, J.; Dai, Y. One-step synthesis of AgCl concave cubes by preferential overgrowth along and directions. *Chem. Comm.,* **2012**, *48*, 3488-3490.

[65] Xia, Y.; Xiong, Y.; Lim, B.; Skrabalak, S.E. Shape-controlled synthesis of metal nanocrystals: simple chemistry meets complex physics? *Angew. Chem. Int. Ed.,* **2009**, *48*(1), 60-103.

[http://dx.doi.org/10.1002/anie.200802248] [PMID: 19053095]

[66] Chen, Z.; Yu, A.; Higgins, D.; Li, H.; Wang, H.; Chen, Z. Highly active and durable core – Corona structured bifunctional catalyst for rechargeable metal – Air battery application. *Nano Lett.*, **2012**, *12*, 1946-1952.
[http://dx.doi.org/10.1021/nl2044327]

[67] Zhu, M.; Chen, P.; Liu, M. Sunlight-driven plasmonic photocatalysts based on Ag/AgCl nanostructures synthesized *via* an oil-in-water medium: enhanced catalytic performance by morphology selection. *J. Mater. Chem.*, **2011**, *21*(41), 16413-16419.
[http://dx.doi.org/10.1039/c1jm13326h]

[68] Li, J.; Guo, Z.; Zhu, Z. Ag/Bi_2WO_6 plasmonic composites with enhanced visible photocatalytic activity. *Ceram. Int.*, **2014**, *40*(5), 6495-6501.
[http://dx.doi.org/10.1016/j.ceramint.2013.11.102]

[69] Liu, J.Y.; Bai, Y.; Wang, P.Q. Photocatalytic degradation of phenol using Au/Bi_2WO_6 composite microspheres under visible-light irradiation. *Micro & Nano Lett.*, **2013**, *8*(2), 90-93.
[http://dx.doi.org/10.1049/mnl.2012.0759]

[70] Xu, S.; Liu, Z.; Zhang, M.; Guo, L. Piezotronics enhanced photocatalytic activities of Ag-$BaTiO_3$ plasmonic photocatalysts. *J. Alloys Compd.*, **2019**, *801*, 483-488.
[http://dx.doi.org/10.1016/j.jallcom.2019.06.115]

[71] Dong, R.; Tian, B.; Zeng, C.; Li, T.; Wang, T.; Zhang, J. Ecofriendly Synthesis and Photocatalytic Activity of Uniform Cubic Ag@AgCl Plasmonic Photocatalyst. *J. Phys. Chem. C*, **2013**, *117*(1), 213-220.
[http://dx.doi.org/10.1021/jp311970k]

[72] Shi, H.; Chen, J.; Li, G.; Nie, X.; Zhao, H.; Wong, P.K.; An, T. Synthesis and Characterization of Novel Plasmonic Ag/AgX-CNTs (X = Cl, Br, I) Nanocomposite Photocatalysts and Synergetic Degradation of Organic Pollutant under Visible Light. *ACS Appl. Mater. Interfaces.*, **2013**, *5*(15), 6959-6967.
[http://dx.doi.org/10.1021/am401459c]

[73] Xu, S.; Liu, Z.; Zhang, M.; Guo, L. Piezotronics enhanced photocatalytic activities of Ag-$BaTiO_3$ plasmonic photocatalysts. *J. Alloys Compd.*, **2019**, *801*, 483-488.
[http://dx.doi.org/10.1016/j.jallcom.2019.06.115]

[74] Liu, J.Y.; Bai, Y.; Wang, P.Q. Photocatalytic degradation of phenol using Au/Bi_2WO_6 composite microspheres under visible-light irradiation. *Micro & Nano Lett.*, **2013**, *8*(2), 90-93.
[http://dx.doi.org/10.1049/mnl.2012.0759]

[75] Khan, M.R.; Chuan, T.W.; Yousuf, A.; Chowdhury, M.N.K.; Cheng, C.K. Schottky barrier and surface plasmonic resonance phenomena towards the photocatalytic reaction: study of their mechanisms to enhance photocatalytic activity. *Catal. Sci. Technol.*, **2015**, *5*(5), 2522-2531.
[http://dx.doi.org/10.1039/C4CY01545B]

[76] Einaga, H.; Ogata, A.; Futamura, S.; Ibusuki, T. The stabilization of active oxygen species by Pt supported on TiO_2. *Chem. Phys. Lett.*, **2001**, *338*(4-6), 303-307.
[http://dx.doi.org/10.1016/S0009-2614(01)00296-2]

[77] Li, X.; Liu, C.; Wu, D.; Li, J.Z.; Huo, P.; Wang, H. Improved charge transfer by size-dependent plasmonic Au on C_3N_4 for efficient photocatalytic oxidation of RhB and CO_2 reduction. *Chin. J. Catal.*, **2019**, *40*(6), 928-939.
[http://dx.doi.org/10.1016/S1872-2067(19)63347-4]

[78] Tian, Y.; Tatsuma, T. Plasmon-induced photoelectrochemistry at metal nanoparticles supported on nanoporous TiO_2. *Chem. Commun. (Camb.)*, **2004**, *16*(16), 1810-1811.
[http://dx.doi.org/10.1039/b405061d]

[79] Tian, Y.; Tatsuma, T. Mechanisms and applications of plasmon-induced charge separation at TiO_2

films loaded with gold nanoparticles. *J. Am. Chem. Soc.,* **2005**, *127*(20), 7632-7637. [http://dx.doi.org/10.1021/ja042192u] [PMID: 15898815]

[80] Cushing, S.K.; Li, J.; Meng, F.; Senty, T.R.; Suri, S.; Zhi, M.; Li, M.; Bristow, A.D.; Wu, N. Photocatalytic activity enhanced by plasmonic resonant energy transfer from metal to semiconductor. *J. Am. Chem. Soc.,* **2012**, *134*(36), 15033-15041. [http://dx.doi.org/10.1021/ja305603t] [PMID: 22891916]

[81] Li, J.; Cushing, S.K.; Meng, F.; Senty, T.R.; Bristow, A.D.; Wu, N.; Wu, N. Plasmon-induced resonance energy transfer for solar energy conversion. *Nat. Photonics,* **2015**, *9*(9), 601-607. [http://dx.doi.org/10.1038/nphoton.2015.142]

[82] Zhang, X.; Chen, Y.L.; Liu, R.S.; Tsai, D.P. Plasmonic photocatalysis. *Rep. Prog. Phys.,* **2013**, *76*(4), 046401. [http://dx.doi.org/10.1088/0034-4885/76/4/046401] [PMID: 23455654]

[83] Cushing, S.K.; Li, J.; Bright, J.; Yost, B.T.; Zheng, P.; Bristow, A.D.; Wu, N. Controlling plasmon-induced resonance energy transfer and hot electron injection processes in Metal@ TiO_2 core–shell nanoparticles. *J. Phys. Chem. C,* **2015**, *119*(28), 16239-16244. [http://dx.doi.org/10.1021/acs.jpcc.5b03955]

[84] Boriskina, S.V.; Ghasemi, H.; Chen, G. Plasmonic materials for energy: From physics to applications. *Biochem. Pharmacol.,* **2013**, *16*, 375-386.

[85] Pillai, S.; Green, M.A. Plasmonics for photovoltaic applications. *Sol. Energy Mater. Sol. Cells,* **2010**, *94*(9), 1481-1486. [http://dx.doi.org/10.1016/j.solmat.2010.02.046]

[86] Neumann, O.; Urban, A.S.; Day, J.; Lal, S.; Nordlander, P.; Halas, N.J. Solar vapor generation enabled by nanoparticles. *ACS Nano,* **2013**, *7*(1), 42-49. [http://dx.doi.org/10.1021/nn304948h] [PMID: 23157159]

[87] Fang, Z.; Zhen, Y.; Neumann, O.; Polman, A.; Garc, F.J. Evolution of light-induced vapor generation at a liquid-immersed metallic nanoparticle. *Nano Lett.,* **2013**, *13*(4), 1736-1742. [http://dx.doi.org/10.1021/nl4003238]

[88] Polman, A. Solar steam nanobubbles. *ACS Nano,* **2013**, *7*(1), 15-18. [http://dx.doi.org/10.1021/nn305869y] [PMID: 23282196]

[89] Atre, A.C.; García-Etxarri, A.; Alaeian, H.; Dionne, J.A. Toward high-efficiency solar upconversion with plasmonic nanostructures. *J. Opt.,* **2012**, *14*(2), 024008. [http://dx.doi.org/10.1088/2040-8978/14/2/024008]

[90] Garcia, G.; Buonsanti, R.; Runnerstrom, E.L.; Mendelsberg, R.J.; Llordes, A.; Anders, A.; Richardson, T.J.; Milliron, D.J. Dynamically modulating the surface plasmon resonance of doped semiconductor nanocrystals. *Nano Lett.,* **2011**, *11*(10), 4415-4420. [http://dx.doi.org/10.1021/nl202597n] [PMID: 21859093]

[91] Flatté, M.E.; Kornyshev, A.A.; Urbakh, M. Electrovariable nanoplasmonics and self-assembling smart mirrors. *J. Phys. Chem. C,* **2010**, *114*(4), 1735-1747. [http://dx.doi.org/10.1021/jp9083234]

[92] Zijlstra, P.; Chon, J.W.M.; Gu, M. Five-dimensional optical recording mediated by surface plasmons in gold nanorods. *Nature,* **2009**, *459*(7245), 410-413. [http://dx.doi.org/10.1038/nature08053] [PMID: 19458719]

[93] Joulain, K.; Mulet, J.P.; Marquier, F.; Carminati, R.; Greffet, J-J. Surface electromagnetic waves thermally excited: Radiative heat transfer, coherence properties and Casimir forces revisited in the near field. *Surf. Sci. Rep.,* **2005**, *57*(3-4), 59-112. [http://dx.doi.org/10.1016/j.surfrep.2004.12.002]

[94] Bermel, P.; Boriskina, S.V.; Yu, Z.; Joulain, K. Control of radiative processes for energy conversion and harvesting. *Opt. Express,* **2015**, *23*(24), A1533-A1540.

[http://dx.doi.org/10.1364/OE.23.0A1533] [PMID: 26698801]

[95] Mukherjee, S.; Libisch, F.; Large, N.; Neumann, O.; Brown, L.V.; Cheng, J.; Lassiter, J.B.; Carter, E.A.; Nordlander, P.; Halas, N.J. Hot electrons do the impossible: plasmon-induced dissociation of H_2 on Au. *Nano Lett.,* **2013**, *13*(1), 240-247. [http://dx.doi.org/10.1021/nl303940z] [PMID: 23194158]

[96] Hung, W.H.; Aykol, M.; Valley, D.; Hou, W.; Cronin, S.B. Plasmon resonant enhancement of carbon monoxide catalysis. *Nano Lett.,* **2010**, *10*(4), 1314-1318. [http://dx.doi.org/10.1021/nl9041214] [PMID: 20350008]

[97] Adleman, J.R.; Boyd, D.A.; Goodwin, D.G.; Psaltis, D. Heterogenous catalysis mediated by plasmon heating. *Nano Lett.,* **2009**, *9*(12), 4417-4423. [http://dx.doi.org/10.1021/nl902711n]

[98] Chen, G. Nonlocal and nonequilibrium heat conduction in the vicinity of nanoparticles. *J. Heat Transfer.,* **1996**, *118*(3), 539-545.

[99] Atanasov, P.A.; Nedyalkov, N.; Imamova, S.E.; Miyanishi, T.; Obara, M. Substrate nanomodification based on heating and near field properties of gold nanoparticles. *International Journal of Nanoparticles,* **2010**, *3*(3), 206-219. [http://dx.doi.org/10.1504/IJNP.2010.035876]

[100] Richardson, H.H.; Hickman, Z.N.; Govorov, A.O.; Thomas, A.C.; Zhang, W.; Kordesch, M.E. Thermooptical properties of gold nanoparticles embedded in ice: characterization of heat generation and melting. *Nano Lett.,* **2006**, *6*(4), 783-788. [http://dx.doi.org/10.1021/nl060105l] [PMID: 16608284]

[101] Gao, X.; Cui, Y.; Levenson, R.M.; Chung, L.W.K.; Nie, S. *In vivo* cancer targeting and imaging with semiconductor quantum dots. *Nat. Biotechnol.,* **2004**, *22*(8), 969-976. [http://dx.doi.org/10.1038/nbt994] [PMID: 15258594]

[102] Jamieson, T.; Bakhshi, R.; Petrova, D.; Pocock, R.; Imani, M.; Seifalian, A.M. Biological applications of quantum dots. *Biomaterials,* **2007**, *28*(31), 4717-4732. [http://dx.doi.org/10.1016/j.biomaterials.2007.07.014] [PMID: 17686516]

[103] Tian, J.; Shen, T.; Liu, X.; Fei, C.; Lv, L.; Cao, G. Enhanced Performance of PbS- quantum-do--sensitized Solar Cells *via* Optimizing Precursor Solution and Electrolytes. *Nat. Publ. Gr.,* **2016**, 1-9.

[104] Jornet, J.M.; Akyildiz, I.F. Graphene-based Plasmonic Nano-Antenna for Terahertz Band Communication in Nanonetworks. *IEEE J. Sel. Areas Comm.,* **2013**, *31*(12), 685-694. [http://dx.doi.org/10.1109/JSAC.2013.SUP2.1213001]

[105] Jornet, J.M.; Akyildiz, I.F.; York, N. Graphene-based Plasmonic Nano-transceiver for Terahertz Band Communication *Proceedings of the 8th European Conference on Antennas and Propagation (EuCAP),* **2014**, 492-496. [http://dx.doi.org/10.1109/EuCAP.2014.6901799]

[106] Huang, X.; Jain, P.K.; El-sayed, I.H. Plasmonic photothermal therapy (PPTT) using gold nanoparticles. *Laser Med. Sci,* **2008**, 217-228.

[107] Bao, T.; Yin, W.; Zheng, X.; Yu, J. One pot synthesis of PEGylated plasmonic MoO3-x hollow nanosphere for photoacoustic imaging guided chemo-photothermal combinational therapy of cancer. pp. 11-24. *Biomaterials,* **2015**. [PMID: 26517561]

[108] He, W.; Ai, K.; Jiang, C.; Li, Y.; Song, X.; Lu, L. *State Key Laboratory of Electroanalytical Chemistry, Changchun Institute of Applied.,* **2017**,

[109] Zhou, L.; Tan, Y.; Wang, J.; Xu, W.; Yuan, Y.; Cai, W.; Zhu, S.; Zhu, J. 3D self-assembly of aluminium nanoparticles for plasmon-enhanced solar desalination. *Nat. Photonics,* **2016**, *10*(6), 393-398. [http://dx.doi.org/10.1038/nphoton.2016.75]

[110] Fleury, R.; Alù, A. Cloaking and invisibility: A review. *FERMAT,* **2014**, *1*.

[111] Runnerstrom, E.L.; Llordés, A.; Lounis, S.D.; Milliron, D.J. Nanostructured electrochromic smart windows: traditional materials and NIR-selective plasmonic nanocrystals. *Chem. Commun. (Camb.),* **2014**, *50*(73), 10555-10572. [http://dx.doi.org/10.1039/C4CC03109A] [PMID: 24935022]

[112] Wang, K.; Meng, Q.; Wang, Q.; Zhang, W.; Guo, J.; Cao, S.; Elezzabi, A.Y.; Yu, W.W.; Liu, L.; Li, H. Advances in energy-efficient plasmonic electrochromic smart windows based on metal oxide nanocrystals. *Adv. Energy Sustain. Res.,* **2021**, *2*(2100117)

[113] Ferhan, A.R.; Kim, D.H. Nanoparticle polymer composites on solid substrates for plasmonic sensing applications. *Nano Today,* **2016**, *11*(4), 415-434. [http://dx.doi.org/10.1016/j.nantod.2016.07.001]

[114] Lim, J.K.; Majetich, S.A. Composite magnetic–plasmonic nanoparticles for biomedicine: Manipulation and imaging. *Nano Today,* **2013**, *8*(1), 98-113. [http://dx.doi.org/10.1016/j.nantod.2012.12.010]

[115] Faupel, F.; Zaporojtchenko, V.; Strunskus, T.; Elbahri, M. Metal-polymer nanocomposites for functional applications. *Adv. Eng. Mater.,* **2010**, *12*(12), 1177-1190. [http://dx.doi.org/10.1002/adem.201000231]

[116] Alissawi, N. *Ion Release from Silver, Polymer Nanocomposites.,* **2013**.

[117] Yuan, L.; Geng, Z.; Xu, J.; Guo, F.; Han, C. Metal-semiconductor heterostructures for photoredox catalysis: Where are we now and where do we go? *Adv. Funct. Mater.,* **2021**, *31*(27), 2101103. [http://dx.doi.org/10.1002/adfm.202101103]

CHAPTER 2

$BaZrO_3$-Based Ceramics and Ceramic Composites as Smart Materials for Advanced Applications

D. Vignesh[1] and **Ela Rout**[1,*]

[1] *Birla Institute of Technology, Mesra, Ranchi, Jharkhand, India*

Abstract: Ancient human history invites significant learning with unknowing facts and fascinations. Significant development and transitions in the human lifestyle are visualized from the capitalized materials. "Ceramics", as antique as it sounds, is frequently used for innumerable applications. From pottery to pellets, ammunition to antennas, electrolytes to electronics, all exist under the radar of ceramic materials. The dominant trait of ceramic materials for advanced applications is constantly replenished to extract peerless products for future utilization. Ionic or covalent bonding in ceramic microstructures administers their suitable mechanical, electrical and chemical characteristics. Pristine ceramics display low conductivity and chemical stability, while doped ceramics *via* implanted impurities empower their characteristics. The nature of dopants and defect substitution differs on the target application. The vastly introspected energy sector is permeated with acceptor-doped perovskite ceramics, while the defense sector inquests over piezoelectric ceramics and ceramic composites. The trivial facet amongst all is the use of Barium Zirconate ($BaZrO_3$) based ceramic compositions. It has been substantially contemplated to visualize the role played by $BaZrO_3$ in multiple domains. Either as a parent material or as an additive, $BaZrO_3$ attracts research groups from diverse sectors. Compiled with innumerable advantages, it accompanies a few limitations. The vital thing is the high sintering temperature along with the trade-off between proton conductivity and chemical stability. However, $BaZrO_3$-based ceramics are keenly monitored and tailored in an attempt to subsidize the maximum possible drawback with a simultaneous improvement in their properties. In the following chapter, we emphasize $BaZrO_3$-based ceramic and ceramic composites as smart materials for advanced applications. The extended applications in the energy sector, photocatalysts for hydrogen production, smart bullet systems in defense and microwave dielectric resonators for wireless communications are elaborately introspected with key insights.

Keywords: $BaZrO_3$, Dielectric resonator, Hydrogen production, Photocatalyst, Proton conductor, Smart bullets.

[*] **Corresponding author Ela Rout:** Birla Institute of Technology, Mesra, Ranchi, Jharkhand, India; E-mail: elasinha@bitmesra.ac.in

Dibya Prakash Rai, Kingsley O. Obodo, & Jitendra Pal Singh (Eds.)

INTRODUCTION

Material Science, as it is much heard of, is concerned with the study of contemporary materials for diverse applications which pave as an eye-opener for future technologies. Of the frontline nature of materials known to date, ceramics are undoubtedly influential and demanding. It wouldn't be erroneous to affirm us all besieged by ceramics. The root for ceramics arrives from the Greek word 'keramikos' for clay products. Ceramics are non-metallic, inorganic and often crystalline materials, with clay being traditionally recognized. Ceramic artifacts date long back to ancient human history as a source of evolution and survival [1, 2]. Archaeologists and multiple research groups consider 'ceramography' as an effective strategy to examine and unveil numerous unanswered questions about human history and development. While fundamental ceramics remain as handy products, ceramic matrix composites, on the other hand include advanced domestic and industrial products over monolithic ceramics [3, 4]. The spectrum of material architecture using ceramic materials has flourished with a subsequent revelation of multiple domains. The physical and chemical properties of ceramics inevitably rely upon their crystalline nature. The underlying study of solid-state chemistry reveals an interrelation between the properties and microstructure of ceramics. The elementary idea of material assessment for future advancement lies in an appropriate choice with requisite characteristics. Analysis and characterizations of ceramic microstructure occur at similar microscopic scales that vary from a few Angstroms (Å) to micrometers (μm). This invites ceramics and their composite counterparts for utilization in optics.

From the material's perspective, Barium Zirconate ($BaZrO_3$) and their allied compositions establish suitable bulk, mechanical, optical and dielectric properties due to supportive microstructures [5 - 7]. $BaZrO_3$ is used in distinct sectors either as a host material or as an additional reagent. This illustrates our former claims of their growing demand in the energy sector [8, 9]. Population growth and rapid decline in the intermittent and usable form of eco-friendly energy sources impels researchers and practitioners to introspect efficient energy storage technology. Solid oxide fuel cells (SOFC) and proton-conducting solid oxide fuel cells (PC-SOFC) showcase promising results as a spark for future technological advancements [10]. The study of Bi, Lei *et al.* [11] gauges the proton conductivity and power density of 20%Yttrium (Y) doped $BaZrO_3$. the outcome reveals considerable proton conductivity of 0.53 x 10^{-3} S-cm^{-1} under ambient conditions. The following observation attributes to the large grain size that enables unimpeded proton mobility. In support of the former literature, the works of Satapathy [12], showcase a rise in proton conductivity in Yb-doped $BaZrO_3$ under the N_2 atmosphere from 10^{-5} to 10^{-3} between $x = 0.05 - 0.20$, respectively. The conductivity, however decreased after that with a subsequent increase in dopant concentration due to the proton trapping. In compliance with former discussions, Draber *et al.* [13] certified the proton trapping effect *via* a comparative difference in the activation energy for proton conduction between pure and doped $BaZrO_3$ compositions. While the activation energy in pure $BaZrO_3$ varies between 0 - 0.27 eV, the activation energy in the latter case with high doping concentrations ($x > 0.25$) elevates close to ~ 0.5 eV. This affirms the inability of long-range proton diffusion due to dopant-proton interaction. Recent studies by Vignesh [14, 15] as a continuum, highlights the optimal acceptor substitutions to desirable outcomes and eliminate the trapping effect. The material characteristics is preserved at lower doping concentrations ($x < 0.25$) while the predominant and early instance of proton trapping effect emerges at elevated acceptor substitutions ($x > 0.25$). While intermediate doping concentrations showcases multiphase co-existence and refrains adequate proton conductivity due to compositional induced structural phase transitions. This provides an optimal doping range between $x = 0.125 – 0.25$ which varies among different acceptor substituents.

Following the applications of $BaZrO_3$-based ceramics in different domains, the works of Fu *et al.*, Zuo *et al.* and Wang *et al.* [16 - 18] depict the use of $BaZrO_3$ and Sb^{5+} as additives too $_{(0.97-x)}(K_{0.48}Na_{0.52})Nb_{0.965}Sb_{0.035}$ (Sb-KNN) which leads to the dispersion and broadening of dielectric peaks and enables relaxor like ferroelectric characteristics with low hysteresis. This grants their implication as multilayer capacitors, dielectric resonators and also in advanced defense technologies. Probing an eco-friendly and alternate source of energy, hydrogen fuel is preferably indicated as the best possible option among all available sources [19, 20]. Production of hydrogen through electrolysis of water depicts slow chemical kinetics. A recent study conducted by Tao *et al.* [21] in 2021 concerns hydrogen production using photocatalytic activity *via* perovskite-based ceramics. The machine learning models were proposed to predict the accurate bandgap (E_g) of the resultant ceramic. The predicted outcomes suggest an active hydrogen production rate (R_H) and extend the scope of ceramics as a potential photocatalyst with trivial modulations. Kolodiazhnyi *et al.* [22] on the other hand investigated the plausibility of $BaZrO_3$-based ceramics as a microwave dielectric resonator. Established studies over resonators demand and operate materials with high dielectric constant, quality factor and high absorption coefficient with large bandwidth. The following requirements allow the material to serve as an accomplished sensor and actuator. Pure $BaZrO_3$ manifests high dielectric loss at room temperature to serve as a microwave dielectric resonator [22, 23]. The product of quality factor (Q) and resonator frequency (f) at f = 10 GHz is evaluated to lie in the range of 5 – 20 THz, which goes in agreement with the works of Stetson *et al.*, Yamaguchi *et al.* and Sivasubramanian *et al.* [24 - 26] respectively. To overcome the high dielectric loss, $BaZrO_3$ is used as an alloy with a host ceramic material to obtain an overall ceramic composite

$BaZr_{0.96}Ga_{0.02}Nd_{0.02}O_3$ with an improved microwave dielectric constant and Q x f values. A sudden decline in the microwave dielectric loss and enhanced dielectric constant (ε) of ε = 36.7 attributed to the role of donor ions over the random electric field and anti-ferro-distortive instability. As a result, high ε and large positive value of temperature coefficient of resonating frequency make $BaZrO_3$ befitting as microwave dielectric resonator composites.

State-of-the-art high-end technological advancements in terms of ceramic implants have intrigued significant attention over prior technologies. Illuminating the recent progress, the high mechanical strength of $BaZrO_3$-based ceramic composites is exercised as bio-ceramics for artificial bone, dental implants and phosphors for radiology shielding. According to Borja Urby *et al.* [27], the source of photoluminescence in $BaZrO_3$ is devoted to structural disorder in the resultant BaO_{12} and ZrO_6 polyhedra due to lattice defects (VO..) during the ceramic synthesis. While distinct fabrication techniques are implemented to preserve compositional homogeneity, uniform morphology and physiochemical stability, impurity substitution on the other hand constitutes distinct crystalline structure to vary the native characteristics respectively. As a result, highlighting the material aspect of ceramics natural $BaZrO_3$ possesses a high melting point (3193 K), dielectric constant, mechanical strength and wide bandgap respectively [28 - 30]. However, material engineering and processing to skilfully utilize $BaZrO_3$-based hybrid framework as propitious materials for future advancements. In the following chapter, we capitalize to elucidate $BaZrO_3$-based ceramics and ceramic matrices in crucial domains. From a broader perspective, the chapter is framed for the readers to avail clear perspective of the involvement of $BaZrO_3$-based ceramic and their composite counterparts to emphasize the role of distinct requisite characteristics to serve diverse advanced applications as a smart material.

SYNTHESIS STRATEGY OF CERAMICS

The synthesis of single crystal, polycrystalline, composite materials and thin films is one of the critical processes to reveal the particulars of examined specimens in material science. In contrast, distinct synthesis procedures are practiced to accomplish desired outcomes with each process targeting specific attributes. Copious synthesis techniques such as solid-state reaction, sol-gel, and wet-chemical procedures, to name a few, are well established and adopted over the years to obtain physiochemical and morphological compatibility. However, each synthesis technique demands to follow unique procedures to resolve the experimental and compositional complexity. While standalone ceramics for various applications are largely synthesized using solid states reactions, sol-gel, on the other hand, has captivated the attention of researchers globally on account of compositional phase purity of nano-particles and ceramic composites with minimal residue [31]. However, the growing popularity of wet chemical procedures occurs due to their promising appliance in optoelectronics and photochemical stability. The synthesized product is a synergetic combination of a gel-like network involving solid and liquid phases, which eases the demand for technological (harsh experimental conditions and machinery) and economic facets. As a result, we summarize the significance of various synthesis strategies over a diverse class of materials for advanced applications in Table **2**.

According to Table **1**, the synthesis of bulk and ceramic composites is well suited by the solid-state reaction route to accomplish polycrystalline materials with multiphase coexistence involving various intermediate steps. However, the afore synthesis limits their utility due to slow reaction kinetics *via* solid diffusion of ions in the primary mixture. As a result, the solid-state reaction principle invites major shortcomings in the form of stoichiometric disorder, heterogeneous particle size distribution and high sintering conditions [37]. The combustion and wet chemical procedures, on the contrary, provide promising hope for the synthesis of ceramic composites and nanomaterials for diverse applications with compositional homogeneity. However, improvisations to former synthesis procedures such as modified sol-gel (Pechini method), molten salt synthesis, solvothermal synthesis and many others are consistently practiced to finely tune the material characteristics. While conventional techniques forbid the physiochemical stability of thin films, modified routes are adopted to combat the material limitations for compositional homogeneity. As a result, $BaZrO_3$-based thin films are synthesized *via* numerous techniques involving the vapor phase or liquid phase deposition [38, 39]. However, high-temperature treatment of the powdered precursors limits the technological ease with the demand of high-temperature autoclaves. As a result, the advanced synthesis route, such as hydrothermal–galvanic couple, according to the works of Teng *et al.* [40] assists the synthesis of highly crystalline thin films at very low temperatures (< 100 °C). The cumulative outlook of the synthesis route relies upon the end product and the target application.

CERAMICS AND CERAMIC COMPOSITES IN PC-SOFC TECHNOLOGY

Perovskite-based ceramic oxides have often fascinated researchers with their high electrical conductivity. One of the principal leverages of perovskite ceramics in the context of fuel cell application is their high catalytic activity and ionic mobility [41, 42]. ABO_3 perovskite crystal structure, the inherent flexibility of bonds and large cationic sites establish the essentiality of such ceramics and ceramic composites for electrochemical energy conversion. High proton conductivity due to rapid proton diffusion across the electrode-electrolyte interface in distorted proton conductors (PCs) serves as a suitable electrolyte material for PC-SOFC [43, 44]. Fuel cells are distinguished based on the operating temperature regimes. The eco-friendly fuel flexibility in PC-SOFC offers adequate electrochemical kinetics, which, following multiple surveys, is ill-suited for proton exchange membrane fuel cells (PEMFC) [45]. Electrochemistry between H_2 and O_2 endures at the molecular scale. The fundamentals of which result due to O—H bond evolution *via* successive H—H and O—O bond dissociation. The resultant H_2O by-product occurs due to oxidized H_2 molecules. Exothermic electrochemistry is attributed to the gradient in chemical bonding. The external circuit is powered as a consequence of electron transfer between the molecular constituents. The appreciable spatial separation of reactants develops as a requisite criterion to accomplish electron transfer at enhanced length scales and harness ionic mobility from fuel to oxidant as power. The combined heat and power can be utilized to escalate the operational efficiency by close to 80%. The fluctuations in operational efficiency between 60 – 80% emerge due to material-oriented limitations [46].

Table 1. Outlook of different synthesis strategies with their experimental requisites (Vignesh D Paper [31], 2023, Pg -10).

Experimental Technique	Nature of Synthesized Materials	Synthesis Procedure	Advantages of the Synthesis Route	Limitations of the Synthesis Route	Refs.
Solid state reaction (SSR)	• Bulk proton conducting (PC) electrolytes	• Grinding/ball milling of powdered precursors	• High oxygen permeability of the mixed ceramic membrane	• Variable sample's surface area	[32]
		• Calcination *via* resistive heating in a furnace	• Economical and eco-friendly large-scale synthesis	• Free energy variations	
		• Re-grinding for phase purity (optional)	• Synthesis of heterogeneous polycrystalline materials	• Precursor's reactivity	
		• Sintering *via* resistive heating in a furnace (~1000–1500 □)	• Fine grain size distribution (~ 0.1 – 10 µm)	• Harsh sintering conditions to overpower lattice energy for stable atomic arrangement	
		• Pelletizing the end product *via* hydraulic press	-	• Multi-phase coexistence at elevated temperatures	

(Table 1) cont.....

Experimental Technique	Nature of Synthesized Materials	Synthesis Procedure	Advantages of the Synthesis Route	Limitations of the Synthesis Route	Refs.
Sol-gel	• Thin film PC-electrolytes	• Hydrolysis and polycondensation of precursors in H_2O/ organic solvent	• Large sample surface area and controlled dopant distribution	• Expensive precursors	[33]
	• Organic-inorganic hybrid polymer	• Gelation (gel formation) and vaporization of excess solvent	• Reduced thermodynamical conditions (temperature, pressure) (200 – 800)	• Tedious synthesis process with extended processing duration	
	• Dense nanoparticles	• Aging (chemical reactivity of sol particles to form gel-network, pore size enhancement and compression of gel)	• High chemical homogeneity and phase purity	• Aging and rapid dehydration of gel under heat treatment	
-	• Solid oxide electrolytes	• Drying, densification and crystallization to form solid oxide materials	• Ease of organic-inorganic hybrid materials	• Increases sample thickness with decreased porosity	-
	-		• Minimized vaporization loss and precipitate agglomeration with crystallized microphases	• Soluble and reactive precursors for gel formation	
	-		• Higher dissolution rate with negative surface charge	• Possible OH^-/hydrocarbon residues	
	-		• Suspended colloidal particle size (~ 1 -100 nm)	-	

(Table 1) cont.....

Experimental Technique	Nature of Synthesized Materials	Synthesis Procedure	Advantages of the Synthesis Route	Limitations of the Synthesis Route	Refs.
Wet chemical precipitation	• Homogeneous nanoparticles	• Blending of precursors in H_2O solvent	• Homogeneous and porous morphology with small grain size (10 -100 nm)	• Segregation of sludge accommodated toxic constituents	[34]
	• Thin film PC-electrolytes	• Imply reflex condition (simultaneous heating and condensation of a chemical reaction) for 1 – 2 h at a temperature (~ 5 - 10°) > boiling point of the lower solvent	• High phase purity of crystalline materials with a simpler synthesis procedure at low temperatures	• Filtration of products post-precipitate formation	
	• Nano-powdered proton conductors	• Aging of precipitate at room temperature for ~ 24 h	• Feasible for large-scale production with reduced synthesis duration	• Tedious to regulate the deposition parameters	
-	-	• Washing and filtering of precipitates	-	• Expensive synthesis route	
	-	• Drying of precipitates in an oven < 100 °C	-		
	-	• Crushing of dry precipitates to obtain powered product	-		
Microwave synthesis	• Nano-powdered proton conductors	• Mixing of weighed precursors in aqueous deionized H_2O	• Faster reaction kinetics	• Inhomogeneous sample shrinkage	[35]
	• Organic materials	• Preparation of homogeneous aqueous solution *via* stirring	• Adequate yield and phase purity	• Distorted palette structure	
	• Nanoparticle synthesis	• Combustion reaction *via* microwave heating for a few minutes (~5 – 15 min)	• Controlled and selective heating of precursors	• Non-uniform particle size distribution with density variations	

(Table 1) cont.....

Experimental Technique	Nature of Synthesized Materials	Synthesis Procedure	Advantages of the Synthesis Route	Limitations of the Synthesis Route	Refs.
	-	• Vaporization of the solvent yields the requisite solid with desired phase.	• Uniform heat distribution to the sample during the synthesis procedure	• Plausible inhomogeneous structures	
	-		• Utilization of microwave power to regulate the reaction kinetics	-	
	• Powdered proton conductors	• Stoichiometric proportions of precursors in a deionized aqueous medium (H_2O) are prepared	• Variable crystallinity and morphology *via* a change in the reaction parameters	• The density gradient of the sample affects the solvation process	
Hydrothermal synthesis	• Fibers	• Transfer of solution into a mineralizing agent with constant stirring (~ 1 h)	• Economical and eco-friendly synthesis route	• Requirement of expensive autoclave	[36]
	• Single crystals	• Subsequent stirring of the solution *via* magnetic stirrer post-heating in a water bath under a constant temperature < 100 □ for ~ 30 min	• Low-temperature processing of volatile solutions	• Inability to regulate the crystal growth rate and introduction of seed crystals	
-	• Nanoparticle	• Transfer of solution into hydrothermal kettle followed by drying of solution	• Direct precipitate of powders from the solution	• Long heating duration (~ several hours)	-
-	-	• Filtration of precipitates *via* washing and centrifugation to yield powdered oxides.	• Highly homogeneous nanoparticles with a particle size between (~ 6 nm – 10 μm)	-	-
-	-		• Reduced synthesis duration	-	-

In a relative comparison between SOFC and PC-SOFC, the high operating temperature of the former limits their commercialization due to the reduced start-up time, material degradation and thermal instability [47 - 49]. PC-SOFC counterbalances vital limitations offered by conventional SOFC due to chemical, mechanical, thermal and electrical characteristics. The effective mass of the active diffusing charge carriers additionally alters the electrochemical kinetics. While oxide ions are heavier, protons, on the contrary, achieve rapid bulk diffusion through the host electrolyte, as observed in PCs. Proton conductivity in PC-SOFC thus follows the Grotthuss mechanism for long-range proton transport. It is the process by which an excess proton diffuses through a hydrogen bond network of water molecules *via* hydrogen bonding with neighbouring oxygen atoms. Proton migration between two neighbouring oxygen atoms can either occur within the same perovskite octahedra (intra-octahedra) or between two neighbouring octahedra (inter-octahedra) based on structural distortions caused by defects and impurities [50 - 54]. Amidst numerous existing ceramic materials, pure and rare-earth-doped $BaCeO_3$ and $BaZrO_3$ PCs have been under scrutiny for a longer duration. While acceptor-doped $BaCeO_3$ illustrates high proton conductivity, $BaZrO_3$, on the contrary, offers high chemical stability. Assimilated works of literature and outcomes epitomize a trade-off between proton conductivity and chemical stability besides the proton trapping effect [55 - 59].

Grotthuss Mechanism in Proton Conductors

A fuel cell is an energy conversion device that generates electricity by virtue of electrochemical redox reactions across the electrode-electrolyte interface in the presence of a fuel and oxidant [60]. Numerous types of fuel cells are distinguished by the nature of charge carriers (protons (H^+)/ oxide ions (O^{2-})), operating temperatures and the employed electrolyte material. Amongst them, solid oxide fuel cells (SOFC) and proton-conducting solid oxide fuel cells (PC-SOFC) are drastically examined owing to their high functional efficiency, fuel flexibility and low pollutant emission [61]. The proton conducting mechanism, also referred to as the "Grotthuss mechanism" or proton jumping, is the process of long-range proton diffusion *via* rotation and translation of excess proton around the oxygen atom of the water molecule, forming bonds with neighbouring atoms amongst octahedra of perovskite proton conductor [62 - 64]. The proton transfer can either occur within the same octahedra or between two different octahedra within the same material. The former occurrence is termed intra-octahedra proton transfer, while the latter is called inter-octahedra proton transfer. Strong and extensive hydrogen-bonded clusters reduce the neighbouring O—O bond distance eases the approach of rapid and long-range proton diffusion. Fig. (**1**), shown below, schematically demonstrates the Grotthuss mechanism with induced oxygen vacancy due to acceptor impurities within the host composition.

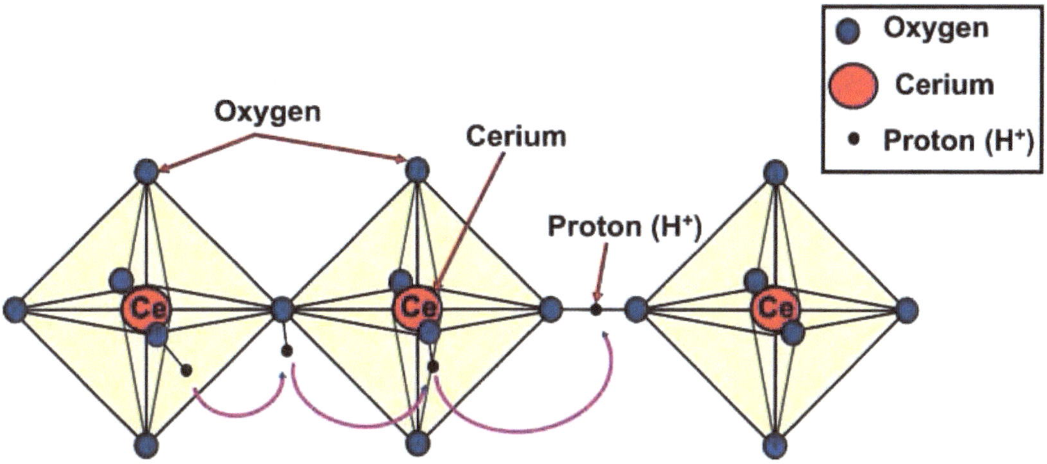

Fig. (1). Grotthuss mechanism within proton conductor.

The elemental pre-requisite for proton incorporation is the presence of a considerable amount of oxygen vacancies which is implanted *via* trivalent acceptor substituents. From the perspective of electroneutrality, the replacement of two B-site tetravalent cations with two trivalent impurities constitutes oxygen vacancies due to the resultant charge imbalance. Consequent hydration allows the H_2O molecules to occupy the available oxygen vacancies within the overall composition. The operating temperature of the fuel cell technology offers adequate activation energy for proton migration. It is to be noted that the gradient in the activation energy allows only the loosely bonded hydrogen atom in the H_2O molecule to be detached as a free proton for bulk diffusion following the Grotthuss mechanism. The favourable proton transfer pathway thus follows the path with minimal activation energy. The ionic conductivity is proportionally associated with the number of oxygen vacancies. As a result, rare-earth substituents become essential to amplify the resultant ionic conductivity in diverse PCs. In compliance with prior discussions, the proton conduction of chemically stable $BaZrO_3$ PCs renders profound significance to innumerable advanced applications of standalone and their derived energy materials. The high thermal conductivity of zirconates poses a potential candidate for proton-conducting electrolytes in PC-SOFC.

$BaZrO_3$ Based Proton Conductors (PCs) for Fuel Cells

Optimization of PCs for lucrative applications acts as the central motive among diverse researcher communities. Multiple permutations of dopant substitutions and compositions are experimented with avail fruitful results. The electrical, mechanical and chemical characteristics besides suitable electrochemical activity avows $BaZrO_3$ as a legitimate preference. Prior works of literature emphasize Y^{3+}, Yb^{3+}, Gd^{3+}, Sc^{3+}, Nd^{3+} and In^{3+} as favourable rare-earth substituents to enhance the proton conductivity under diverse atmospheres. Additionally, single, co-doped and other complex ceramics are examined lately. Each synthesized composition yields a weighted characteristic that compromises over the other. In compliance, Han *et al.* [65] investigated the inter-relations between bulk conductivity and structural characteristics of trivalent doped $BaZrO_3$. The outcomes demonstrate a linear relationship between structural distortions and bulk conductivity. The experimental results illustrate an increase in bulk proton conductivity with hydrated Thulium (Tm) doped $BaZrO_3$ and a decrease with larger acceptor dopants having an ionic radius larger than Dysprosium (Dy).

The study of Gilardi *et al.* [66] emphasizes the influence of dopant-host ionic radius upon the electrophysical properties of the host PCs. The study acquaints the choice of appropriate dopants based on the dopant-host compatibility to uplift proton conductivity at the expense of feeble structural distortions. The comparative analysis by Gilardi *et al.* [66] discloses a moderate proton conductivity with smaller acceptor substituents (In^{3+} and Sc^{3+}) with a homogeneous and large average grain size distribution. While large acceptor dopants, on the contrary (Y^{3+}, Yb^{3+}, Tu^{3+}, Er^{3+}, Ho^{3+}), constitute non-uniform and smaller average grain size and high proton conductivity. Substantial surveys suggest non-uniform proton conduction in $BaZrO_3$-based ceramics. In support of the afore-outcomes, the works of Bi *et al.* [11] accomplish considerable proton conductivity and power density with 20% Y-$BaZrO_3$. The aftermath of electrochemical impedance spectroscopy advocates adequate proton conductivity due to large grain size distribution and minimum grain boundary interactions. Such a proposition contradicts the former microstructural claims. The rising ambiguities demand an elaborate investigation.

Weighing over recent advancements exercising $BaZrO_3$ PCs, Hossain *et al.* [67], summarized the variation of proton conductivity of single and co-doped $BaZrO_3$ under varying parameters (atmosphere, temperature, dopant concentration). The following analysis justifies a dominance of Y^{3+} over electrical characteristics of the overall composition among other acceptor substituents at an optimal doping concentration of ($x = 0.20$). The consequence of such an attribute is associated with large structural deviations *via* octahedral distortions. Amplified dopant-host ionic radius mismatch induces lattice strains and reduces the O—O bond distance between the neighbouring octahedra to aid inter-octahedral proton transfer. However, an inverse effect in terms of chemical instability accompanies it and justifies the prevailing trade-off. Besides the implicit compositional characteristics, distinct atmospheric constraints additionally dictate the chemical and electrical response of the host material. Table **2**, shown below, indicates a variable proton conductivity between Y and (Y + Co) doped $BaZrO_3$ under differing environments:

Table 2. Proton conductivity of Y-$BaZrO_3$ and (Y + Co)- $BaZrO_3$ under varying atmospheres (Hossain *et al.* [67], 2022, Pg – 8).

Composition	Operating Temperature	Conductivity in Differing Atmospheres ($S\text{-}cm^{-1}$)			
		Ar	H_2	H_2O	O_2
Y-$BaZrO_3$	500	2.76×10^{-5}	6.59×10^{-5}	1.34×10^{-4}	5.90×10^{-3}
	600	1.10×10^{-4}	3.29×10^{-4}	6.42×10^{-4}	2.76×10^{-2}
	700	2.34×10^{-4}	1.23×10^{-3}	1.23×10^{-3}	9.54×10^{-2}
	800	1.18×10^{-3}	4.18×10^{-3}	5.18×10^{-3}	2.29×10^{-1}
	900	4.36×10^{-3}	1.18×10^{-2}	1.42×10^{-2}	3.92×10^{-1}
	1000	1.48×10^{-2}	3.02×10^{-2}	3.58×10^{-2}	4.46×10^{-1}
(Y + Co)-$BaZrO_3$	500	1.20×10^{-4}	2.02×10^{-4}	3.88×10^{-4}	7.10×10^{-4}
	600	3.74×10^{-4}	9.35×10^{-4}	1.66×10^{-3}	3.21×10^{-4}
	700	1.25×10^{-3}	2.86×10^{-3}	4.92×10^{-3}	1.06×10^{-3}
	800	3.31×10^{-3}	6.83×10^{-3}	1.04×10^{-2}	2.35×10^{-3}
	900	7.67×10^{-3}	1.43×10^{-2}	2.02×10^{-2}	4.26×10^{-3}
	1000	1.64×10^{-2}	2.82×10^{-2}	3.55×10^{-2}	6.32×10^{-2}

The statistical results in Table **2** dictate a prominent ionic conductivity under dry and wet H_2 atmosphere argon atmosphere for both the compositions due to isotropic effect. The outcome of the study also displays a promising difference in the proton conductivity of Y-$BaZrO_3$ and (Y + Co)-$BaZrO_3$ at the same temperature. The difference emerges due to the role of supplementary substituents in co-doped compositions over the singly doped counterparts. As a result, Co, in the former case, bridges the grain boundary region *via* agglomeration for unimpeded proton diffusion. Additionally, the temperature-dependent surge in proton conductivity attributes to the enhanced electrochemical activity across the electrode-electrolyte interface. The secondary reason also accounts for the proton de-trapping effects from the dopant's sphere of influence due to high inductive effects. However, the mixed ionic conductivity among PCs activates at higher temperatures with a dominant oxide ion conductivity over proton conduction. As a consequence, the study also presents a significant oxide ion conductivity under the O_2 atmosphere at higher temperature regimes (> 700 °C). A dominant proton conductivity at low and intermediate temperatures in (Y + Co)-$BaZrO_3$ under selective atmospheres (dry H_2, H_2O) illustrates a larger protonation in co-doped compositions. This elevates the hope and scope of complex substitutions to achieve optimum proton conductivity.

While acceptor substitutions are preferred to accomplish desirable proton conductivity *via* explicit protonation, a prominent effect around the dopant's close coordination is detected to hinder proton mobility as a function of structural instability. Multiple literatures have reported the influence of acceptor substituents over reduced proton diffusion due to defect scattering events and charge redistribution. According to Vignesh *et al.* [31], a smaller acceptor substituent alters the crystal symmetry and reorients the lattice defects and proton-coordinated oxygen atoms due to dopant charge screening effects. As a result, the coupling of the afore-effects with the rigid hydrogen bonds activates the proton trapping effect around the proton neighbourhood. However, contrary behaviour is observed with large acceptor substituents, which constitute shallow proton traps relative to their former counterparts and decreases the degree of hindered proton mobility. Meanwhile, a recent study by Mielewczyk-Gryń *et al.* [68] and Fujii *et al.* [69] reported the impact of dopant ionic radii over the hydration tendency of distinct proton conductors. The outcomes illustrate a divergent behaviour between the dopant ionic radii and the hydration enthalpy of the host material. While large trivalent impurities aid improved hydration, smaller acceptors, on the other hand, lower the saturation limit to material hydration due to physiochemical instability. As a result, larger acceptor substituents such as Y^{3+}, Gd^{3+}, Yb^{3+} and many others have gained immense importance with physiochemical compatibility.

From prior studies, it is evident that large protonation *via* optimum hydration boosts ionic conductivity, the extent of which depends on diverse parameters such as the nature of charge carriers (H^+/O^{2-}), operating temperature and the oxygen partial pressure in the host octahedra and the functional atmospheric conditions respectively. Saini *et al.* [70] attained a high electrical conductivity of 5.8 x 10^{-3} S-cm^{-1} at 600 °C for 20% Ho-doped $BaZrO_3$ composition. The varying doping concentrations (x = 0.05, 0.1, 0.15, 0.20) within the overall compositions synthesized *via* the flash pyrolysis technique provide a robust method to fabricate pure and nanosized samples. The high conductivity obtained under a 3% humidified O_2 atmosphere accomplishes at the expense of high activation energy. The following disparity appears owing to the contribution of both protons and oxide ions. Table **3** below indicates the compiled outcomes of proton conductivity of differing acceptor dopants substituted in the host $BaZrO_3$ composition.

Table 3. Assimilated outcomes of proton conductivity of various acceptor-doped $BaZrO_3$ compositions.

Reference	Acceptor Dopant	Doping Concentration (x)	Proton Conductivity ($S\text{-}cm^{-1}$)	Sintering Condition	Measuring Atmosphere
[66]	Sc	0.20	2.9×10^{-4} at 500 °C	1600 °C - 8 h - air	Humidified Atmosphere
	In	0.20	6.9×10^{-4} at 600 °C	1600 °C - 8 h - air	Humidified Atmosphere
	Lu	0.20	2.8×10^{-5} at 400 °C	1600 °C - 8 h - air	Humidified Atmosphere
	Y	0.20	5.8×10^{-4} at 550 °C	1600 °C - 8 h - air	Humidified Atmosphere
	Gd	0.20	1.8×10^{-4} at 600 °C	1600 °C - 8 h - air	Humidified Atmosphere
	Sm	0.20	8.9×10^{-5} at 600 °C	1600 °C - 8 h - air	Humidified Atmosphere
[71]	Yb	0.10	5.6×10^{-4} at 500 °C	1500 °C - 48 h - air	Dry Air
[72]	In	0.75	1.0×10^{-4} at 600 °C	1325 °C - 48 h - air	Dry Ar Atmosphere
[73]	Y	-	2.29×10^{-2} at 600 °C	1600 °C - 24 h - air	H_2O Saturated N_2 Atmosphere
[74]	Yb	0.20	1.10×10^{-2} at 600 °C	1600 °C - 24 h – O_2	5% H_2O- H_2 Atmosphere
	Tm	0.20	1.13×10^{-2} at 600 °C	1600 °C - 24 h – O_2	5% H_2O- H_2 Atmosphere
	Er	0.20	1.36×10^{-2} at 600 °C	1600 °C - 24 h – O_2	5% H_2O- H_2 Atmosphere
	Y	0.20	1.39×10^{-2} at 600 °C	1600 °C - 24 h – O_2	5% H_2O- H_2 Atmosphere
[75]	Nd	0.07	2.01×10^{-5} at 628 °C	1700 °C - 1 h – Vacuum	Humid Air
-		0.05	2.07×10^{-4} at 600 °C	1600 °C - 24 h - air	Wet N_2 Atmosphere
[76]	Dy	0.10	3.79×10^{-4} at 600 °C	1600 °C - 24 h - air	Wet N_2 Atmosphere
-		0.15	1.93×10^{-3} at 600 °C	1600 °C - 24 h - air	Wet N_2 Atmosphere
-		0.20	4.15×10^{-3} at 600 °C	1600 °C - 24 h - air	Wet N_2 Atmosphere

(Table 3) cont.....

Reference	Acceptor Dopant	Doping Concentration (x)	Proton Conductivity (S-cm^{-1})	Sintering Condition	Measuring Atmosphere
-		0.05	6.06 x 10^{-5} at 628 °C	1600 °C - 24 h - air	Wet N_2 Atmosphere
-	Sm	0.10	9.07 x 10^{-5} at 600 °C	1600 °C - 24 h - air	Wet N_2 Atmosphere
-		0.15	2.42 x 10^{-4} at 600 °C	1600 °C - 24 h - air	Wet N_2 Atmosphere
-		0.20	3.96 x 10^{-4} at 600 °C	1600 °C - 24 h - air	Wet N_2 Atmosphere
[70]	Ho	0.05	2.01 x 10^{-5} at 650 °C	1600 °C - 8 h - air	3% O_2 Humidified Atmosphere
		0.10	1.43 x 10^{-4} at 650 °C	1600 °C - 8 h - air	-
		0.15	1.42 x 10^{-3} at 650 °C	1600 °C - 8 h - air	-
		0.20	1.11 x 10^{-2} at 650 °C	1600 °C - 8 h - air	-

From Table **3**, it can be observed that proton conductivity estimated corresponding to the dopants Y^{3+}, Yb^{3+}, Tm^{3+}, Eu^{3+} and Ho^{3+} exhibits the highest conductivity among other acceptor substituents. The plausible explanation for the high electrical efficiency of such acceptor-doped $BaZrO_3$ compositions originates from the significant dopant-host ionic radius mismatch and structural deviations with reduced O—O bond distance. This pacifies the resultant translational activation energy required for subsequent proton hopping mechanisms through the bulk.

Table **3** also reflects the critical role of sintering temperature upon the proton conductivity besides different atmospheres. High sintering temperature assists high electrochemical catalysis across the electrode-electrolyte interface with the absence of an external catalyst. The afore-arguments go in agreement with prior predictions and experimental results of high sintering temperatures associated with $BaZrO_3$-based compositions. Additionally, high sintering temperatures become a requisite criterion with $BaZrO_3$-based compositions for a homogeneous phase formation with a definite crystal structure. Despite discrete positive influence, high sintering temperatures limit the commercialization of $BaZrO_3$-based energy materials for large-scale production due to high-end technological demands. Nevertheless, material architecture is conceptualized and implemented to improve the existing drawback with a supplementary motive to decrease the operating temperature of PC-SOFC. The accomplished process opens the gateway for typical and modern technological devices and applications.

$BaZrO_3$-based Thin Film Composites for Fuel Cell Technology

One of the significant milestones in the modern-day machinery is the implication of thin film technology. The use of thin films has successfully bridged numerous energy applications for future sustainability, including their growing appliance in the domain of fuel cell technology. Thin film electrolytes among low and intermediate-temperature fuel cells have profoundly increased the cell performance relative to bulk counterparts due to distinct thermionic responses. The thickness of a thin film electrolyte largely varies between 1μm -10Å respectively, which in principle is fused between two thin parallel plates extended infinitely upon a substrate along the mutually perpendicular directions as shown in Fig. (**2**).

The difference in the thin film operation to their bulk counterpart arises due to such difference in the dimensions where the length scales along the third axes are always unique from the latter, which increases the finite probability of quantum effects within the thin films. As a result, the pre-requisites for a thin film electrolyte are as follows:

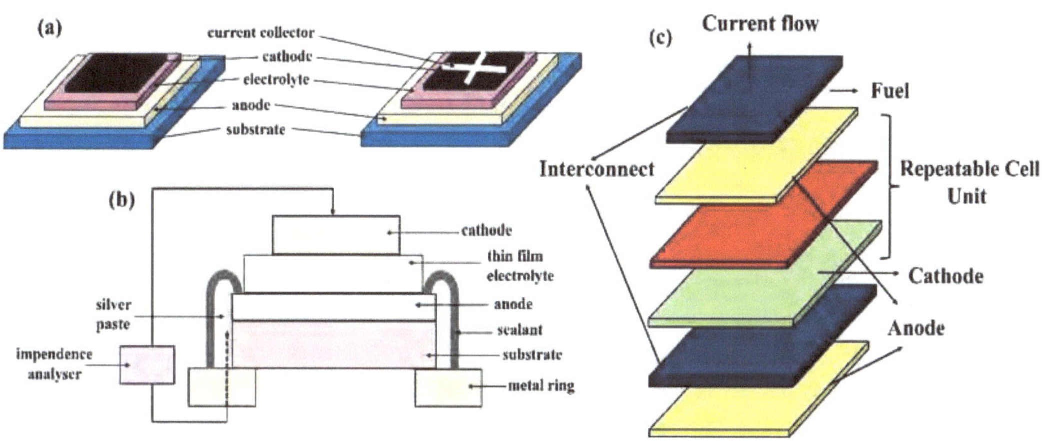

Fig. (2). Utilization of thin film electrolyte for fuel cell technology, **(a)** thin film electrolyte with and devoid of current collector, **(b)** fuel cell system with thin film electrolyte, **(c)** electrode-electrolyte assembly.

• Chemical inertness: thin film electrolytes must establish a requisite contact between the electrode-electrolyte assembly with the presence of a suitable and compatible sealant to illustrate physiochemical inertness against reactive atmospheres.

• Electrophysical landscape: The thin film electrolyte must avow high hydration enthalpy with desirable protonation for high ionic conductivity.

• Fuel flexibility: The thin film electrolyte must display an electrochemical response against distinct hydrogen and hydrocarbon fuels.

• Technological and Economical facet: thin film electrolytes must illustrate thermal resistance against high operating temperatures for high-performance efficiency with low experimental cost [77].

As a result, Hossain *et al.* [78] coupled a proton conducting ($BaZr_{0.9}Y_{0.1}O_{3-\delta}$) nanoparticle within a polymer nanocomposite membrane to lower the operating temperature and enhance the performance efficiency of the intermediate temperature fuel cell technology. The fallout of the study indicates a consistent surge in the proton conductivity with a proportional rise in the volume fraction of the nanoparticles by 1.8%, which declines steeply thereafter due to agglomeration of nanoparticles across the microstructural boundary with reduced proton conductivity landscape. The consequent effect also emerges due to H—H rearrangement between the $-SO_3H$ and -OH groups within the sulfonated poly(ether ketone) polymer. However, the synthesized thin film electrolyte showcases promising thermal, physiochemical and mechanical stability due to the heterogeneous distribution of acceptor substituents within the host material. The trivalent impurity serves as an H_2O intake site to enhance the degree of protonation *via* explicit oxygen vacancy with a consequent hike in the power density (0.16 -0.43 Wm^{-1}) at lower operating temperature (~80 °C) respectively. As a result, the ceramic composite within the polymer matrix illustrates advancement relative to their standalone counterpart with a demand for fine optimizations. According to the reports of Campos Covarrubias *et al.* [79], the strain developed in the $BaZr_{0.8}Y_{0.2}O_{3-\delta}$ thin film epitaxially grown over an MgO substrate occurs due to phase instability and lattice disorder. As a result, a $BaZr_{0.6}Ce_{0.4}O_3$ buffer layer is sandwiched to neutralize the induced strain. The buffer layer aids lattice reorganization and escalates the proton conductivity due to a decline in the localized energy barrier. Pergolesi *et al.* [80] illustrated an improved protonic conduction using Y-doped $BaZrO_3$ thin film electrolyte synthesized *via* pulsed layer deposition at intermediate temperatures. However, the high reactivity of Ni in the anode hinders electrolyte operations due to the formation of an impure phase across the electrode-electrolyte assembly, resulting in poor operational performance. However, the chemical inertness of Yb^{3+} doped $BaZrO_3$ with NiO substrate aids electrochemical stability with high power density. As a result, the recent study by Niwa *et al.* [81] illustrates the synthesis of high proton conducting acceptor doped $BaZrO_3$-based thin film electrolyte using a pulsed layer deposition technique. Two targets synthesized using Pechini and the solid-state reaction route were adapted for the deposition of $BaZr_{0.8}Yb_{0.2}O_{3-\delta}$ on the electrode substrate. The outcome reflects the cumulative performance of the thin film electrolyte upon the reliability and the response synthesized targets. As a result, the $BaZrO_3$-based thin film electrolyte display considerable protonic conduction with the target synthesized using the Pechini method over the solid-state reaction counterpart due to high density and homogeneous morphology. The study highlights the salient feature of synthesized targets over a physiochemical and morphological landscape of the thin film electrolyte to achieve desirable functioning for fuel cell technology. The study projects the analogous protonic conduction among bulk and $BaZrO_3$-based thin film electrolytes with a scope to enhance the overall conductivity *via* increasing crystallinity and regulated orientation of the resultant thin film.

ADVANCED CERAMIC COMPOSITES IN DEFENCE

National security is offered significant priority considering the nation's pride and integrity. Various defence organizations are involved in highly advanced and classified research to strengthen surveillance and maintain peace to avoid intended security breaches. From a conventional perspective, arms and ammunition play a critical role to counterbalance undesirable combats. Post the era of World War 1, the domain of defense was rendered much priority to encounter future disasters due to constant confrontation amongst developed and developing countries in greed for power. Armour and bullets lay the foundation for the primary defense system. Meeting the growth in technologies, appropriate materialistic modulations are executed to former weaponry. The significance of ceramic armour grew steeply after the Vietnam conflict. The lightweight ceramic and ceramic composites are utilized for bulletproof vests and shields [82 - 86]. Advanced ceramics and ceramic composites, on the other hand, lay the foundation for modern-day artillery to obstruct major ballistic impacts. Ceramic derivatives are employed as the functional materials for multi-layer composite armours. The front face of it consists of tough ceramic material with an energy-absorbing backplate to enforce mechanical strength to the armour depending on the strength of the projectile. The distribution of weights of ceramic materials varies for differing shields [87]. Bulletproof automobiles like cars are built with heavyweight multilayer ceramic composites, while bulletproof aircraft are designed distinctly to combat catastrophes due to large ballistic missiles [88]. The incoming projected artillery towards a target is attenuated by the hard ceramic shards, which mollifies the bulk mass of the projectile besides the projectile's velocity. As a consequence, the kinetic energy of the projectile is reduced substantially by the backplate, which causes plastic deformation of the armour. The principal origin of smart systems and materials is concentrated on emulating the human being's functionality in the aspect of prediction and improvisation. Smart systems developed are foreseen to exhibit extreme potential with significant advantages to cope with distressing situations and counterattack the combat with immediate effect. Smart Skin, Smart aircraft, Smart bullets and many more are some of the smart systems widely introspected to overpower conventional defense systems.

Recent and novel innovation in weapons was brought in by a Defence Advanced Research Project Agency (DARPA) that invented a 0.50 calibre course-changing bullet, also described as smart bullets [89]. The bullet acquires the ability to change its track mid-flight and overcomes the limitations of conventional bullets. Additionally, the front cap of the bullet or the ballistic missile is equipped with hard ceramic protective caps or plates to improve the aerodynamics that shield the internal electronics involving actuators and sensors in the bullets. Adequate protection is provided to the core electronics of the smart bullet system. The internal electronics with sensors and actuators form the core of the controlled combat of the bullet-target system. Piezoelectric ceramic materials become the elemental component for actuators along with optical sensors to regulate the aerodynamics of the mid-flight course-changing bullets [90 - 93]. Comprehending the characteristics of smart materials and different fabrication strategies for suitable compositions is essential to developing state-of-the-art smart systems for technological advancements. Large emphasis on the behaviour and properties of smart materials is given to minimize the extended weight and unintended physical dimensions of the fabricated shield or defense system. It is in amendment with prior developed armour, which focuses on regulating the impact of the projectile with subsequent increase in the layers of composite materials. However, with substantial exploration and understanding over percussions imparted by various ammunitions, smart materials and systems are thus drastically encouraged and tested by multiple defense organizations for commercialization. Although, ambiguities persist in the response of such materials to the generated and transmitted stress waves. This demands accurate estimations while designing to accommodate suitable aspect ratios and weight distribution of the armour. As an overview, ceramic and ceramic composites have gained connotation as one of advisable materials for smart defense systems.

Mechanism of Course-Changing Bullets

The mid-way course-changing bullets are equipped with optical sensors mounted on their nose along with a control system and moveable tail fin to be deployed over the laser-driven target. Additionally, the bullets also accommodate a pressure-activated battery system that powers the actuators and other chips present in them. The mid-way course change of the bullet is taken care of by the actuators attached to them, as shown in Fig. (**3**) [94].

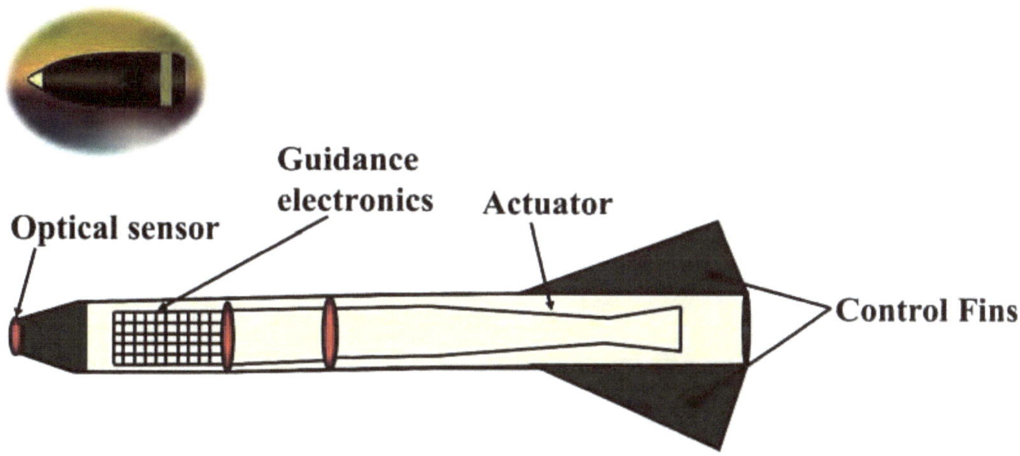

Fig. (3). Schematic representation of smart bullet system.

Unlike conventional bullets, where the number of rotations of the bullet becomes a critical factor to encounter the targets at variable distances, smart bullets, on the other hand, preserve rotational invariance to strike the target. The optical sensor at the nose of the bullets acquires the tendency to detect the target and estimate their distance. The materialistic architecture was motivated to serve as a multipurpose bullet to improvise their appliance according to the ongoing situation on the battlefield. The bullet is mobilized as the target is fixed by the laser. The optical sensor detects the infrared waves emitted by the object (target), and the actuators across the fins follow the signal of the microcontroller. The bullet's positioning is altered with the displacement of the target following it. The sensors in the bullet evaluate the distance of the target during their approach. The bullet explodes utilizing the generated shock waves and destroys the entire system very close to the target, causing serious damage [95]. The accuracy of target detection and distance estimation by the bullet depends on the use of precise optical sensors and actuators.

$BaZrO_3$ Based Ceramic Composites in Defence

The performance of smart bullets relies heavily upon the internal electronics and the accuracy of the optical sensors. To enhance the quality of existing actuators, $BaZrO_3$-based ceramics composites with high piezoelectricity are extensively introspected. However, large strain hysteresis and driving electric field are two major limitations that obstruct their commercialization as high piezoelectric actuators. Zhang *et al.* [96] laid the study over $(_{0.97-x})(K_{0.48}Na_{0.52})Nb_{0.965}Sb_{0.035-0.03}Bi_{0.5}(K_{0.18}Na_{0.82})_{0.5}ZrO_{3-x}BaZrO_3$ for ($x = 0 - 0.06$), as a ternary $BaZrO_3$-based complex ceramic composite fabricated by solid-state reaction (SSR) method. The affiliation of microstructure, phase transition and electrical properties of the complex ceramic reveals a high bipolar strain (S_{max}) of about 29% with low hysteresis of 13.8% at a low electric field (4 kV/mm) and promises a prominent piezoelectric coefficient. The high piezoelectric properties of the former compositionattributes to the rhombohedral-orthorhombic-tetragonal (R-O-T) and relaxor-ferroelectric phase transition [97 - 99]. From *x* = 0.02, the composition is found to exist between ferroelectric-relaxor states. As a result, the long-range ferroelectric domains annihilate due to the evolution of micro-domains and augmentation of piezoelectric properties, which goes in agreement with the works of Wu *et al.* [100], Fu *et al.* [101], Huan *et al.* [102] and Koruza *et al.* [103] respectively. The variation of bi-polar strain with doping concentration indicates a rise in the piezoelectric properties due to induced R-O-T phase transition is shown in Table **4**.

Table 4. Altering bipolar strain as a function of dopant concentration in $(_{0.97-x})(K_{0.48}Na_{0.52})Nb_{0.965}Sb_{0.035-0.03}Bi_{0.5}(K_{0.18}Na_{0.82})_{0.5}ZrO_{3-x}BaZrO_3$ (*x* = 0 - 0.06) composition.

Doping Concentration (x)	Bipolar-strain (S_{max}) %
0	0.24 (1)
0.01	0.25 (1)
0.02	0.29 (1)
0.03	0.26 (1)
0.04	0.15 (1)
0.06	0.05 (1)

In support of prior literature, Zhou *et al.* [104] illustrated the significance of R-O-T phase coexistence to achieve high piezoelectric properties using $BaZrO_3$ additives. The volume ratios of orthorhombic (V_O), rhombohedral (V_R) and tetragonal (V_T) are estimated using XRD profile fitting, which demonstrates only O-T phase coexistence in the absence of $BaZrO_3$ with V_O = 54.2% and V_T = 45.8% respectively. On the other hand, $BaZrO_3$, as an additive in host composition between *x* = 0.01 - 0.02, is found to contain a variable amount of rhombohedral phase owing to R-O-T phase transition at room temperature [105, 106]. Table **5**, shown below, illustrates d_{33} and dielectric values as a function of variable doping concentration.

Table 5. Dielectric properties of $BaZrO_3$ based KNN ceramic $(_{0.96-x})(K_{0.48}Na_{0.52})(Nb_{0.96}Sb_{0.04})O_{3-0.04}(Bi_{0.50}Na_{0.50})ZrO_{3-x}BaZrO_3)$ with varying doping concentration (Zhou *et al.*[104], 2020, Pg – 2).

Dopant Concentration	d_{33} (pC/N)	ε at 100 kHz	tan δ
x = 0	535	2660	0.049
x = 0.01	610	3460	0.055
x = 0.015	550	3720	0.062
x = 0.02	340	3060	0.076
x = 0.03	120	2730	0.078

While the R-T phase transition of complex KNN-based ceramics exhibits d_{33} of 120 pC/N, the introduction of the R-O-T phase causes a sudden elevation in d_{33} to 610 pC/N. This reflects the significance of the R-O-T phase transition upon the increase in the piezoelectric coefficient (d_{33}). Additionally, standalone $BaZrO_3$ exhibits considerable mechanical and thermal durability, making it an ideal candidate to be incorporated with other ceramic complexes for fabricating armour and weapons.

CERAMIC COMPOSITES AS PHOTOCATALYSTS FOR HYDROGEN PRODUCTION

Undoubtedly, water as a resource is adequately available throughout the globe. However, the accessible water resources are shadowed by certain limitations to be exploited as suitable energy sources. Urbanization, together with rapid population growth, has impelled researchers and practitioners to counterbalance the demand-supply ratio. Such stringent constraints arguably invoke deeper concerns to introspect compatible, renewable and eco-friendly energy source and storage technology. A gradual shift from conventional to renewable energy sources aims to accomplish sustainable development in the upcoming years. However, among all available renewable energy sources, solar energy is considered to be imminent over others. In the following context, hydrogen is endowed to constitute a compelling, economic and pollutant-free energy source. The advantageous factor lies in terms of their abundant natural and anthropogenic productions. Hydrogen as a gaseous constituent is available in abundance among other naturally available gaseous elements. Additionally, several chemical reactions also liberate H_2 gas as one of the by-products that can be preserved for futuristic applications. Several techniques involving the thermochemical process, bio-hydrogen production using anaerobic respiration of algae and bacteria and the conventional electrolysis process have been substantially explored since the 1970s [107]. Modern-day energy conversion devices and green energy materials invite H_2 as one of the most favourable fuels to activate the electrochemical energy supply [108 - 111]. The performance parameters of hydrogen fuel, as utilized in several automobiles and industries, impel researchers to speculate a feasible way of large-scale hydrogen production.

Conventional H_2 production *via* water splitting reaction involves the utilization of renewable visible solar radiation with the abundantly available water resource. The photocatalytic water splitting using reusable photocatalysts forbids the generation of unwanted by-products and the threat of toxic pollutants for global warming. Diverse photocatalysts include phosphides, sulphides, nitrides, solid-state oxides and many more [112 - 114]. An optimal photocatalyst must initiate the redox reaction involving water oxidation and proton reduction reactions. The photocatalytic activity in solid-state oxide occurs through photogenerated electrons and holes that accommodate the redox reactions. The liberation of photogenerated electrons, as a result, relies upon the electronic band structure and the redox potential of water of the opted material. This summons that the material must possess a structure with the valance band (VB) edge greater than the redox potential of water and the conduction band edge less than the proton reduction level. In synergy, the resultant bandgap must be greater than 1.23 eV and have a high charge recombination rate, respectively [115]. The requisite bandgap demand is to utilize the ultraviolet and visible regions of the electromagnetic spectrum. The apprehension of the chemical properties of H_2 is a mandatory requirement for their legitimate use as an alternate fuel or energy source. The speculated chemical properties establish the pressure and temperature sensitivity of the H_2 molecule. The temperature dependency introduces a variation in the spin direction of the nucleus, which imposes energy difference between H_2 molecules [116]. As a result, the demand for appropriate ambiance and thermodynamic conditions becomes a vital aspect to consider for maximum H_2 production. The flammability of H_2 gas is equivalently a prime factor to overcome the vulnerability of H_2 gas at elevated temperatures. The elemental H_2 is compatible at room temperature; however, it becomes vulnerable at high temperatures, driving the way for disastrous explosions [117]. Liquification of gaseous H_2 at reduced temperatures is a befit route for long-term storage. However, pressure sensitivity awaits the technological facets for felicitous compression. The effectiveness of liquid hydrogen thus manifests as a best-fit fuel for rocket propulsion with considerable thrust. Besides H_2 as an alternate fuel, they also exhibit strong reducing behaviour and provide an alternate choice as a strong reducing agent over other available reagents. Such characteristics ensure feasible photo-induced redox reactions across the electrode-electrolyte interface.

Photocatalytic Mechanism for Hydrogen Production

Familiarised with the growing demands of energy sources over supply and, a gradual shift of attention towards harnessing solar energy has imposed a significant challenge of solar power extraction during cloudy weather conditions. With significant advantages of solar energy sources, scientists have discovered hydrogen to be an alternative eco-friendly fuel to power devices and automobiles *via* fuel cells. As a result, photocatalytic splitting of water into hydrogen and oxygen proves to be a handy technique exploiting solar radiation to store eco-friendly fuel [118 - 121]. On the other hand, material engineering for photocatalysis provides a complete set of challenges. Ceramics and ceramic composites involving $BaZrO_3$ prove their worth as an effective material for photocatalytic water splitting with improved chemical kinetics [122 - 124]. The mechanism of water splitting begins with the absorption of solar radiation (of a specific wavelength) by the electrodes to generate electron-hole charge carriers. These charge carriers are further transported to the H_2O molecule, which knocks off the electrons to form H_2 gas. The holes, on the other hand, contribute to generating O_2 gas. The water-splitting electrodes, as a result, must be suitable light absorbers, charge separators and charge conductors. The individual reactions across each electrode-electrolyte interface are shown in Eqs. 1- 3 below:

$$H_2O \rightarrow 2H_2 + O_2 \quad (1)$$

$$H_2O \rightarrow 4H^+ + 4e^- + O_2 \quad (2)$$

$$4H^+ + 4e^- \rightarrow 2H_2 \quad (3)$$

Fig. (**4**) shown below depicts the photocatalytic activity of H_2O-producing H_2 gas.

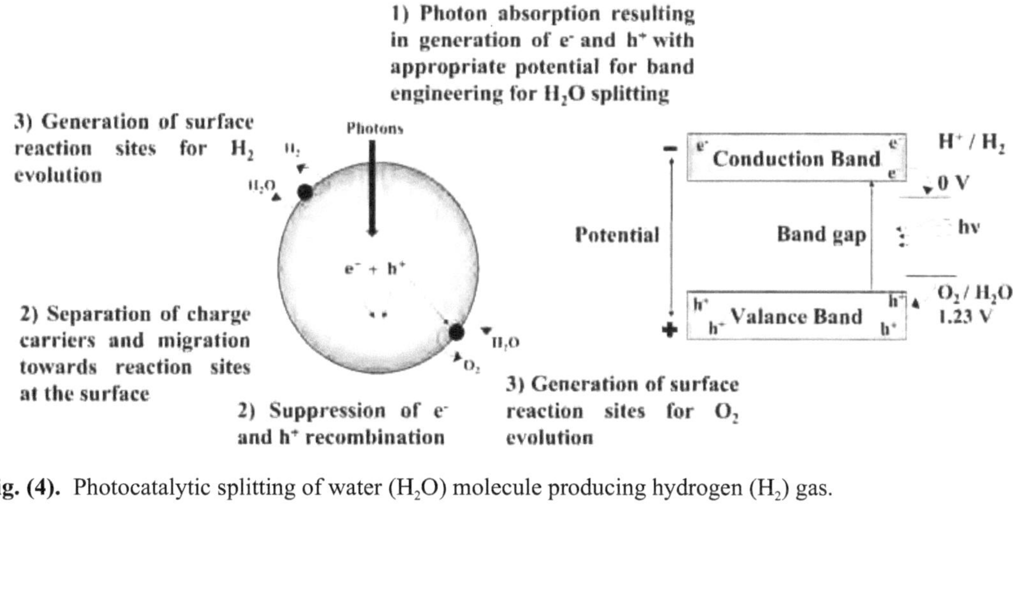

Fig. (4). Photocatalytic splitting of water (H_2O) molecule producing hydrogen (H_2) gas.

The cathode is illuminated with solar radiations, mostly in the UV region of the electromagnetic spectrum, such that electrons are excited from the valance band to the conduction band, crossing over the energy barrier to equally generate holes in the valance band. The holes so generated react with the H_2O molecule acting as the aqueous electrolyte, which results in the generation of O_2 gas. On the other hand, the electrons cross over to the anode *via* an external circuit reacting with the H_2O molecule to release H_2 gas [125 - 127]. The electronic structure of electrodes becomes the essential criterion for water splitting to yield adequate H_2 gas. The optimal bandgap of the electrodes must be close to 1.23 eV, while materials with wide bandgaps require additional energy for the excitation of charge carriers, which curtails the quantum efficiency of the photocatalytic process. This is attributed to the fact that the conduction band of the electrode material must be less than the reduction potential of H_2O or H_2 while the valance band must be at a position greater than the reduction potential of H_2O or $H_{2,}$ respectively [128 - 131]. Tailoring bandgaps of perovskite-based ceramics are implemented owing to their tuneable electronic behaviour to bridge diverse optoelectronics applications. $BaZrO_3$-based ceramic composites act as a potential candidate with satisfactory prerequisites as electrodes for the photocatalysis of H_2O.

$BaZrO_3$ Based Ceramic Composites for Hydrogen Production

The chemical and mechanical stability of $BaZrO_3$ perovskites offer numerous applications as energy materials yet they lag in critical electrophysical characteristics. $BaZrO_3$ fails to combat the limitations due to its electronic configuration. The O-2p and Zr-4d states with a large bandgap restrict their absorption spectrum. The extension of the resultant absorption spectral response is plausible by suitable material engineering. The modified design strategy with a suitable non-metallic dopant composed of lower valence electrons substituted at one of the oxygen sites in $BaZrO_3$ aids in manipulating the valence and conduction band edges. This would allow a shift in the valence band edge over the water redox potential.

In an initiative to achieve a suitable photocatalyst, Yuan *et al.* [28] examined the photocatalytic activity of $BaZrO_3$ with UV illumination. The chemical activity in terms of successive excitations and de-excitations escalated with a non-metallic Sn^{4+} modified composition ($BaZr_{0.7}Sn_{0.3}O_3$). Khan *et al.* [132], on the other hand, examined the photocatalytic ability of Tantalum (Ta^{4+}) doped $BaZrO_3$ ($BaZr_{1-x}Ta_xO_3$). The outcomes of the study suggest a progressive surge in the hydrogen and oxygen production rate with a subsequent rise inTa^{4+} doping concentration, as shown in Table **6**.

Table 6. Hydrogen and Oxygen production rate with differing doping concentrations in doped $BaZrO_3$.

References	Photocatalyst	H_2 production (μmol/h/g)	O_2 production (μmol/h)
	$BaZrO_3$	65	33
	$BaZr_{0.99}Ta_{0.01}O_3$	80	40
[132]	$BaZr_{0.98}Ta_{0.02}O_3$	128	64
	$BaZr_{0.97}Ta_{0.03}O_3$	152	76
	$BaZr_{0.96}Ta_{0.04}O_3$	180	90
	$BaZr_{0.95}Ta_{0.05}O_3$	116	58
	$BaZrO_3$	405	-
[133]	$BaZr_{0.99}Ce_{0.01}O_3$	482	-
	$BaZr_{0.98}Ce_{0.02}O_3$	570	-
	$BaZr_{0.97}Ce_{0.03}O_3$	823	-
	$BaZr_{0.96}Ce_{0.04}O_3$	684	-

From Table **6**, an increase in hydrogen production is obtained with a consequent rise in the doping concentration. The following trend is preserved up to a dopant concentration of $x = 0.04$. However, a downfall in the production rate occurs with a feeble surge in the doping concentration ($x > 0.04$). This is attributed to the bandgap variation with a shift in the Zr-Ta cations within the perovskite octahedra. The efficiency of photocatalytic activity is evaluated based on the number of photons absorbed by the photocatalyst upon illumination. The UV–diffuse reflectance spectroscopy replicates and supports the former claims. Besides bandgap variations, a curtail in the surface area of the photocatalyst constricts the number of incident photons and diminishes the resultant hydrogen production [134 - 136]. However, Gundeboina and co-workers [137], in 2020, established the drawbacks of photocatalytic activity using visible radiations. Limited absorption of visible spectra, rapid electron-hole recombination, and lower utilization of visible-light generated photoelectrons and holes, to name a few, enforce the essentiality to tailor the electronic structure of the host composition [138 - 140]. The most favourable strategy includes a suitable metal/non-metal substitution within the host composition. Metals/non-metals as a substituents generate impurity energy levels either above or below the valence or conduction band and constitute a modified electronic structure. An overview of the impurity energy level is demonstrated schematically in Fig. (**5**).

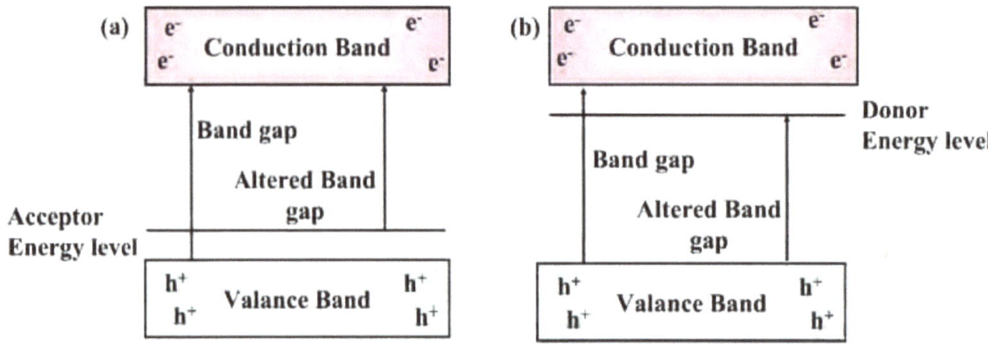

Fig. (5). Impurity energy levels formed above and below valence and conduction band respectively due to metal/non-metal substitution **(a)** acceptor level, **(b)** donor level.

Akhtar *et al.* [141] in 2020 further took into account the afore-disclosed characteristics and experimented with Vanadium (V) and Phosphorus (P) dopants, respectively. The analysis of the pentavalent doped $BaZrO_3$ compositions illustrates a variation in the electronic structure under atomic relaxations. Besides, the density of states (DOS) of V-doped $BaZrO_3$ examined *via* density functional theory (DFT) simulations reveals a bandgap of 3.102 eV for the spin-up configuration. The electronic structure for the spin-down configuration varies due to the introduction of new states in the distorted V-doped $BaZrO_3$ composition. Analogical behaviour is visualized with P-doped $BaZrO_3$, where the spin-up configuration constitutes a bandgap of 2.929 eV. The evaluated bandgaps of individually doped compositions are found to exceed immensely relative to the standard statistical requirement. Meanwhile, a combination of (V+P) doped composition showcases optimum results. The induced mid-gap electronic states by V and P lie beneath the bulk conduction band due to V-3d and P-3p hybridized states, which minimizes the bandgap sufficiently. The simulated results reveal an estimated bandgap of about 1.49 eV, which agrees positively with the requirement. Realizing the elementary requirement for optimum hydrogen production, extensive research continues to accomplish the maximum harvest of hydrogen yield with enhanced chemical kinetics.

The source of future sustainability and technological advancement lies in the energy materials. While bulk ceramics and ceramic composites have been much explored for hydrogen production, a transit to 2D materials has recently captivated the attention of researchers and practitioners globally. According to the recent works of Saeed *et al.* [142], heterogeneous doping through nanostructured thin films profoundly escalates the transport and mass storage property of halides due to decoupled charge compensating species. The outcomes of the study reveal an increase in lattice defects (VO..) and protonic defects (OHO.) across the

$BaZrO_3$ interface due to flux in the electrostatic potential at intermediate temperature which aids charge separation and hence an accelerated H_2 production respectively. Meanwhile, according to the reports of Maeda *et al.* [143], $BaZrO_3$-$BaTaO_2N$ solid solutions find their application in photocatalytic oxidation and reduction under ultraviolet irradiance (~ 660 nm), respectively. The outcomes reveal the potential application of $BaTaO_2N$ in photocatalytic H_2O oxidation and reduction in the existence of sacrificial agents under a photo-illumination of 740 nm. In synchrony with former analysis, the study conducted by Tasleem *et al.* [144] underlines an increased photocatalytic H_2 production by $BaZrO_3$-$BaTaO_2N$ solid solutions under the visible light spectrum. In contrast, prior investigations illustrate the shortcomings in the synthesis of $BaZrO_3$ ceramics and ceramic composites due to high sintering temperature *via* solid-state reaction synthesis and long-range lattice disorder, among other synthesis procedures. However, the synthesis of $CaTaO_2$-$CaZrO_3$ and $BaZrO_3$-$BaTaO_2N$ solid solution according to the polymerized complex route indicates low anionic vacancies and Ta_3N_5 defect density with improved crystallinity. The poor photocatalytic activity of Ta_3N_5 is decreased during the synthesis of the former solid solution. However, the heterostructure due to the coupling of Ta_3N_5-$BaTaO_2N$ encourages the charge separation *via* the transfer of excited electrons from $BaTaO_2N$ to Ta_3N_5 and the holes in the contrary direction.

CERAMIC PHOSPHORS FOR MEDICAL APPLICATIONS

Ceramic phosphors are functional materials with multifarious applications ranging from light-emitting diodes to radiology shielding. However, they find their extensive implications in modern-day sensors and security systems. Ceramic phosphors are designed to display luminescence along the visible spectrum of electromagnetic radiations due to subsequent optical activity. However, unlike typical phosphors, ceramic phosphors offer fundamental benefits due to high thermal stability and chemical inertness, which escalates their practical utility in harsh environments. As a result, ceramic phosphors have achieved immense attention from global researchers and practitioners for the testing of durable ceramic and ceramic composite phosphors for medical applications. The utilization of ceramic phosphors in the medical domain explains their versatility and range of physiochemical characteristics. Recent events have visualized the use of rigid $BaZrO_3$-based ceramic phosphors and their resultant composites for radiation shielding [145, 146]. An outlook of ceramic phosphor for medical application is illustrated in Fig. (**6**).

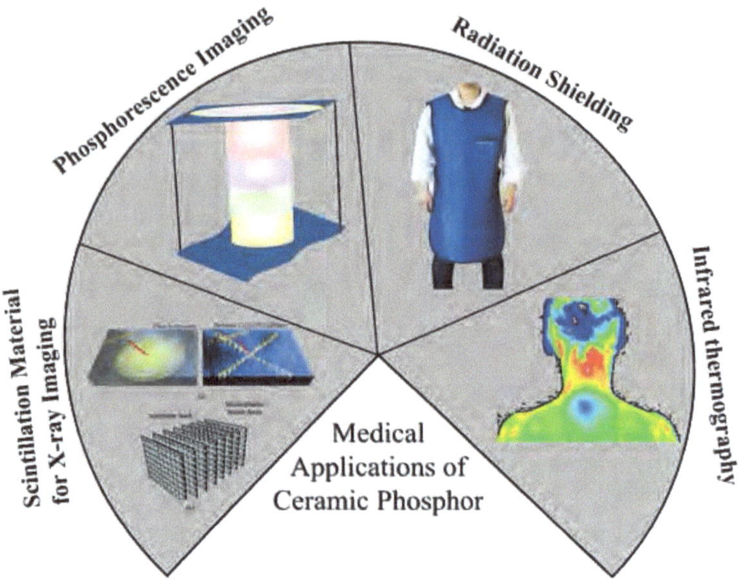

Fig. (6). Medical applications of ceramic phosphor.

High-intensity radiation exposure, such as X-rays and gamma rays are, universally experienced across diverse sectors ranging from nuclear reactors, industries and medical diagnostic centers. However, demanding study involving nuclear waste management, biological examinations, defect engineering of materials and space exploration [147] requires the constant use of high-intensity radiations. While traditional shields are adopted by researchers and technicians as a protective layer to limit the degree of exposure, recent study illustrates the advent of medical conditions due to the leakage of high-intensity radiation for long exposure time. Inadvertent exposure to X-rays and gamma rays with high penetration depth raises great concerns due to detrimental impacts on human health such as cancer, organ damage and radiation sickness, respectively [148]. As a result, stability from the inimical influence of high-intensity radiations for the global population and future sustainability is critical, which relies upon 4 significant factors such as location, duration, shielding and operation. Increase in distance from the radiation source with a subsequent decrease in exposure duration, the degree of exposure can be minimized marginally. A competent means to attenuate the influence of pernicious radiations in the use of suitable shields. Radiation shielding is in consistent debate due to their increased utilization among modern technological advancements. As a result, the search for a new material is required to combat the radiation effects. Diverse material characteristics such as the optical response, structural landscape and durability hinder the researchers from designing and fabricating appropriate radiology shields. A relevant radiation shield is one that can attenuate and eliminate a paramount of gamma radiations [147]. As a result, the nature of radiation (gamma/X-rays) and the underlying principle of interacting radiation with a durable material is essential for a practical and economic radiation shield. According to the works of Abulyazied *et al.* [149], radiation shields used to combat a gamma source must be composed of compositional constituents with high atomic number and density to allow sizeable energy transfer between the host material and the interacting particles, respectively. However, Khozemy *et al.* [150], on the other hand, report the possibility of low atomic number-based ceramic materials and composites to gain appreciable thickness (analogous to highly dense materials) to combat the penetrating influence of high-frequency radiations. Multiple pieces of literature indicate an extensive preference and utilization of lead, barium, copper and concrete as the leading raw materials used to construct a ceramic composite for radiation shielding. However, despite extensive investigations, lead and concrete are abandoned due to their corrosion and chemical instability against reactive atmospheres *via* prolonged usage, respectively. The influence of moisture on the former raw materials serves as a troublesome measure to predict the obnoxious impact of high frequency radiations over the human health hazards.

Optical Activity of Ceramic Phosphors

Ceramic phosphors are recognized due to their luminescence characteristics, where they capture energy and re-emit in the visible spectrum. The following mechanism is vital for diverse applications, as discussed in the former section. In compliance with radiation shielding, the utility of traditional materials involves a definite interaction of optically active chiral molecules with the polarized light due to molecular asymmetry. The light interaction of ceramic phosphor, on the contrary, occurs *via* scattering, absorption and emission over the optical rotation of the polarization plane. The optical response of ceramic phosphor is devoted to the fundamental light-matter interactions. The interaction of photons occurs distinctly with the distinct atomic structures of standalone and ceramic composites, respectively. The unique response of the host material is responsible for the absorption and scattering of excess energy (phosphorescence) in the visible spectrum. The genesis of the outcome is associated with the electronic excitations among various allowed energy levels with the host material. The breakdown of the optical response of ceramic radiation shields is as follows:

• **Absorption:** Ceramic phosphor and their composites absorb the energy from high-frequency radiations such as X-ray or gamma ray photons to activate the electrons from the stable ground state to excited states equal to the energy difference between the former states, respectively. The shielding materials are chosen based on the ability of the atoms and molecules to absorb high-frequency radiation. A durable and rigid ceramic complex curtails the penetration depth of the high-frequency radiations due to an energy transfer *via* inelastic collision between the compositional constituents. The excitons dissipate excess energy in the form of heat. As a result, a thicker ceramic shield is a suitable radiation absorber, reducing the penetration depth of the high frequency photons.

• **Relaxation and Emission:** The exciton process is, in principle, known as the relaxation with the release of excess energy in the form of light in the visible spectrum. However, the wavelength and intensity of the resultant emission rely upon the allowed transition states, which is a characteristic feature of distinct materials. However, ceramic phosphors as radiation shields are generally an effective scattering type material that alters the direction of incident photons due to the subsequent collision with the host's compositional constituents. As a result, the gradient in the trajectory of high-frequency photons forces them to travel prolonged distances and time within the material, thereby reducing the intensity of impact by providing requisite shielding. The afore-influence is illustrated according to Beer Lambart's law according to Eq.4 below:

$$I = I_0 e^{-(\mu l)} \quad \textbf{(4)}$$

Where,

I = Intensity of transmitted through the shield

I_0 = Intensity of light before entering the shield

μ = absorption coefficient

l = optical path length within the material

The study also describes the degree of transmittance of high-frequency photons through the radiation shield.

• **Luminescence Response:** Ceramic phosphors illustrate various luminescence characteristics based on the relaxation time scales. The delay in phosphorescence of distinct ceramic phosphors solely relies upon the nature (radiative/non-radiative transitions) and the allowed energy states for consistent excitations. While the incident radiation is attenuated as it progresses through the ceramic shield, the degree of impedance is dictated by the material properties (photoinduced transitions) and the energy of the incoming photons. While low-frequency radiations such as microwave and radiowaves are shielded using traditional shields, x-rays and gamma rays, on the contrary, demand dense ceramics and ceramic composites to hinder their impact from excessive exposure. As a result, the material's density and atomic number are two critical components that decide the resistive character of the radiation shield.

However, despite the awareness of prerequisites, the ongoing search for an optimum radiation shield occurs due to unintended energy loss within the material during the relaxation process. In contrast, the optical response of ceramic phosphors is highly tuneable, allowing the researchers and engineers to improvise the optical properties (emission and absorption spectra) of the host material. An effective means to execute the afore-requisites is based on dopant-induced impure energy levels. According to multiple pieces of literature, ceramic phosphors and their composites are doped with p-type/n-type impurities to engineer explicit optical characteristics. The versatility of impurity-contained ceramic phosphors has rapidly increased their widespread utility across diverse domains. In compliance with former discussions, $BaZrO_3$ agrees duly with the requisite characteristics for the efficient performance of the radiation shield. As a result, smart and advanced ceramic phosphors are synthesized and examined consistently to bridge the research gap in the medical domain.

$BaZrO_3$-based Ceramic Phosphor for Radiation Shielding

A critical revolution in the domain of medical application is the advent of ceramic phosphors for their utilization in nuclear medical facilities such as positron emission tomography, diagnostic radiology, radiation shield, oncology and many others respectively [151]. The United Nations Scientific Committee on the Effects of Atomic Radiation (UNSCEAR) offers and renews stringent guidelines annually regarding the radiation dosage and exposure for technologists and civilians to regulate global radiation dispersion [152]. Table **7** indicates the average exposure of noxious radiations from distinct technological devices.

According to Table **7**, the statistical reports indicate the maximum toxicity from high-frequency radiations emitted *via* medical X-rays and gamma rays while relatively lower effects from other technological devices such as televisions, respectively. While prolonged exposure and a constant absorption of high-frequency radiations by electronic devices degrades their quality and lifetime [155], the attenuation of nuclear radiation is still essential to control the degree of exposure and ensure requisite safety and protection from health hazards [156]. The synergetic effects interest the need for a suitable radiation shield, which has been explored over the years; while traditional radiation shields composed of lead ingots have been employed as a temporary solution to the detrimental influence of X-rays and gamma radiations, the bulky and toxic with lower chemical inertness against reactive and humid atmospheres increases their risk of rapid aging and accelerated corrosion. The prolonged utilization of lead-based radiation shields avows due to economic facets. In recent years, multiple pieces of literature illustrate concretes, ceramics, alloys, and glasses as alternative radiation shielding materials [157 - 160]. However, significant attention is devoted to standalone ceramic and ceramic composites due to their higher durability and chemical stability against diverse atmospheres. A sizeable significance is offered to ceramics and ceramic composites due to the tuneability of material characteristics and the determination of toxicity, density, fabrication cost and durability during the production phase. Meanwhile, the lighter weight of ceramic composites offers additional advantages over concrete and alloy counterparts. As a result, an upcoming decade will visualize a significant growth in the chemical filler-doped and lead-free radiation shields. According to the outcomes of Malekzadeh *et al.* [161], the presence of micro-sized grains and filers containing Ba, Bi, W and other heavy elements impedes the proton transport, thus upgrading the mechanical and structural response of the host composite. As a result, $BaZrO_3$-based ceramic phosphors are valuable assets for X-ray imaging and radiation shielding due to competent scintillation response, which amends the X-ray into visible spectra. The afore-conversion aids the generation of high-resolution images for authentic diagnosis. The ceramic composites composed of $BaZrO_3$ phosphors offer enhanced light output and sizeable stopping potential for high-frequency radiation shielding. The synergetic property makes them a potential candidate for their utilization in global medical facilities.

Table 7. Radiation dosage and target area distinct source type.

Source Type	Reference Area	Exposure Rate	Ref
Ionizing Radiation	0.1 m from the surface of the apparatus	Max < 0.1 mR/h	[153]
Medical X-ray	1m from source except diaphragm region	Max < 1R/h and dose rate of 1m from the source	[154]
Industrial X-ray and Gamma ray	0.05 m from the surface	Max < 25mR/h	

Borja Urby *et al.* [27] claimed a direct correlation of the photo-luminescence intensity of $BaZrO_3$ with the magnitude of order-disorder crystallinity of the host lattice. The outcomes of the study display the bandgap narrowing of defect-induced lattice due to a consistent rise in the oxygen vacancy concentration, which reflects in a subsequent hike in the degree of lattice disorder and distorted crystal lattice. In relevance to the phosphorescence property of distinct acceptor-doped $BaZrO_3$-based perovskites with various ionic radii and their increasing utility with other well-known nanomaterials for medical applications has been addressed using down-conversion mechanisms, respectively. A recent study by

Ghrib *et al.* [162] experimentally reveals the shielding of gamma-ray, protons, alpha-particles and other charged particles by $BaZrO_3$-based ceramic composites using physiochemical and optical evaluations. The study was motivated to increase the density and shielding response of the host ternary oxide *via* TiO_2 substitutions. The mass attenuation of the resultant zirconia-based ceramic composite was experimentally analyzed against neutral and charged particles and eventually compared with the XCOM database, respectively. The experimentally observed shielding parameters were in agreement with the XCOM data. According to the experimental outcomes, the degree of gamma ray shielding of the resultant complex increases linearly with increase in TiO_2 concentration which is unlikely with the charged particles respectively. However contrary impact occurs with the charged particles where the extent of attenuation inversely varies with TiO_2 concentration and the nature of charged particles due to a heterogeneous contribution among Compton scattering, photoelectric effect and pair production to the total linear attenuation, respectively. The ternary samples with distinct TiO_2 concentrations were synthesized using the traditional solid-state reaction route with single-phase purity for different physiochemical and optical investigations. From the optical landscape, the mass absorption coefficient (MAC) reveals a subsequent decrease with high-energy photons due to the large penetration depth of the incoming radiations. However, a sufficiently high MAC corresponding to $x = 0.25$ in $_{0.5}ZrO_{2\text{-}(0.5\text{-}x)}TiO_{2\text{-}x}Tl_2O_3$ with 3.84 cm^2/g at 0.06 Mev respectively. The following study agrees with the former results of narrowing bandgap within the defect-induced lattice. The Urbach energy evaluated using the slope of the absorbance function (F(R)) from the Tauc plot indicates a dwell in the disorder due to a consistent rise in the oxygen vacancies and Tl_2O_3 mixing. The consequent attribute emerges due to the evolved charge imbalance between the Tl^{3+} and Ti^{4+} cations in the nanocrystal, which is neutralized by the generated oxygen vacancy as the charge trapping sites. As a result, the surge in the Urbach energy increases the lattice defects and tails the dual band edge, leading to reduced bandgap, respectively.

The evaluation of half value layer (HVL) from the morphological landscape illustrates their consistent decrease with increasing Tl_2O_3 concentration and material density, which escalates the radiation shielding response. The synergetic outcomes indicate a positive correlation of Tl_2O_3 concentration with the degree of attenuation, irrespective of TiO_2 concentration, respectively. A more critical parameter associated with radiation shielding is the mean free path (MFP) due to light-matter interaction. According to the outcomes, the MPF increases with a surge in photon energy due to the curtailed probability of light-matter interactions (number of collisions) of the high-frequency photons. The consequent attribute emerges due to a sizeable contribution of photoelectric effect among others (Compton scattering and pair production) at a high energy range, which increases the degree of penetration. The outcomes of the study also reveal a rise and dwell in the radiation absorption dosage and MFP, respectively, with increasing Tl_2O_3 and TiO_2 ($x = 0.25$) and therefore dictates the potential application of zirconia-based ceramic composites as an improved radiation shield. In compliance with the former study, the works of Marí *et al.* [163] illustrate a profound rise in the photoluminescence intensity of Eu^{3+} and Li^{3+} doped $CaZrO_3$ and $BaZrO_3$ respectively, due to a decline in the radiation-less energy transfer due to the presence of impurity bands in the defect induced lattice.

A novel gateway towards radiation shielding was implemented using ceramic-based polymer composites. The light weight and ease of synthesis escalate their demand as ideal radiation absorbers for shielding applications. As a result, $Ba_2P_2O_7$ and $BaZrO_3$-based polymer composites examined by Ozel *et al.* [164] reported competent shielding against high-energy ionized radiations. The outcome of the study reveals an effective shielding with improved performance using 20% $BaZrO_3$ and $Ba_2P_2O_7$ as fillers in the unsaturated polyester. While the MAC illustrates an increase with an increase in filler concentration, contrary trends with MFP and HVL indicate their effectiveness against gamma-ray attenuation. However, the optimization of polymer nanocomposites involving $BaZrO_3$ draws research interest due to their low-temperature synthesis limitations. A homogeneous crystallinity among $BaZrO_3$ demands high sintering temperatures *via* heat treatment methods while different other conditions with wet chemical methods. As a result, a need for a suitable synthesis route is essential to stabilize the compositional performance. Therefore, in search of a compatible synthesis route, More *et al.* [165] established a rapidly growing and adopted in-situ technique for the synthesis of polymer matrix using nanoparticles as the precursor, as shown in Fig. (7).

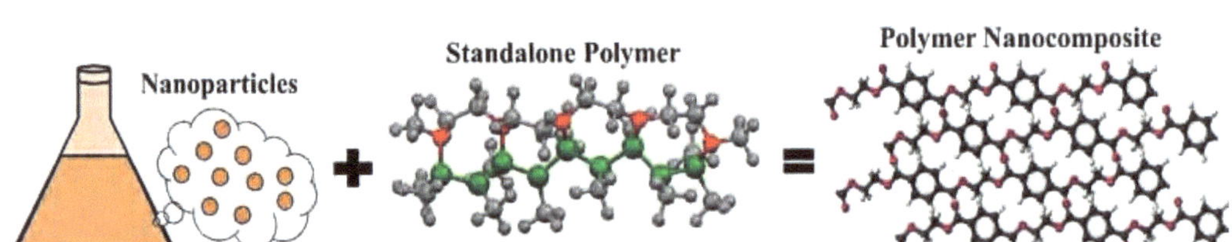

Fig. (7). Schematic representation for the synthesis of polymer nanocomposite.

According to Fig. (7), the early stage of synthesis begins with the nucleation of nanoparticles and their growth within the matrix to hinder the aggregation of nanoparticles. The following technique is advantageous to achieve stabilization of functional groups within the host complex. According to the synthesis procedure, the particle size of the nanoparticle can be regulated, and the agglomeration of nanoparticles is impeded with a preserved spatial distribution. As a result, the in-situ polymerization is a highly effective route based on the dispersion of the filler to fabricate nanocomposites based on thermoset polymers. The filler nanoparticles are maximized to constitute intercalated sheets to commence polymerization *via* caption exchange post-nanoparticle swelling. The in-situ intercalative polymerization follows a solvent-free route, which unifies the polymerization and intercalative procedures to accommodate compositional uniformity. As a result, the works of Maksoud *et al.* [166] over gamma-ray shielding using poly(vinyl butyral)/ Bi_2O_3@$BaZrO_3$ nanocomposite. An increase in Bi_2O_3@$BaZrO_3$ content reduces the optical bandgap of the nanocomposite with a consequent rise in the Urbach energy (E_u). The durability test of the polymer nanocomposite also demonstrates a rise in the mechanical toughness with a surge in the Bi_2O_3@$BaZrO_3$ concentration. The study also reveals the lowest possible HVL and MFP corresponding to a high ($x = 0.50$) Bi_2O_3@$BaZrO_3$ filler content, respectively. As a result, the synergetic outlook of the former discussions reflects the promising characteristics of standalone and defect-induced $BaZrO_3$ based ceramic and ceramic composites with an additional contribution from the novel polymer matrix as suitable radiation shields.

CERAMIC MATERIALS FOR MICROWAVE DIELECTRIC RESONATORS

The size, shape and orientation of materials are significant topological characteristics to discern the direction of scattered energy. The dielectric constant is one such property used to analyze the amount of energy scattered along a specific direction. It can otherwise be stated as a property to measure the extent of electromagnetic interaction with a given material [167]. A resonator, in general, is a device that exhibits higher oscillations at certain specific frequencies called the resonant frequency. The significant use of resonators is to distinctly separate a single frequency in a signal. A dielectric resonator, on the other hand, is a dielectric material, conventionally ceramic materials are engaged to operate in the microwave range of the electromagnetic spectrum [168, 169]. They contain 'puck' of ceramics with high dielectric constant and low dielectric loss. The former property paves its cruciality towards wireless communication operating in microwave wavelengths. The resonance frequency for any communication is estimated by the dimensions of the resonator and the dielectric constant of the material employed. Microwave dielectric resonator finds their applications as filters and oscillators in addition to communication systems and antennas. Metal cavities or surfaces exhibit lossy behaviour at microwave frequencies and appeal for appropriate modifications. The replacement of dielectric resonators over resonating cavities in microwave frequencies opens a new gateway for researchers. They provide cost-effective materials with high-quality factor (Q) to be engaged for fabricating compact devices. Isolated microwave dielectric resonators are identified by their resonating frequency, which resembles their minimum dielectric loss. The corresponding resonant frequency is estimated with the material's dielectric constant (ε_r) and the volume of the resonator (V), as shown in Eq. 5:

$$f = \frac{233}{V^{\frac{1}{3}} \sqrt{\varepsilon_r}} \quad \mathbf{(5)}$$

The dielectric resonators, unlike waveguides, can be compiled, providing leverages of integrated circuits in the aspect of dimensions. The functionality of dielectric resonators resembles that of cavity resonators at microwave frequency regimes. The basic distinction between metallic and dielectric resonators lies in the material aspect. The tendency of dielectric resonators to store energy in the form of electric and magnetic fields relative to the power dissipated allows their quality factor to be assessed accordingly. Additionally, the tendency of dielectric resonators to reflect the radio waves as a result of the change in permittivity over conductivity in metals adds a crucial distinction between their functionality. The high dielectric properties provided by various ceramics and ceramic composites are used to design suitable antennas with a high Q factor. Such antennas tend to display high efficiency and stability. The dimensional dependency of resonating frequency, along with the design flexibility offered by dielectric resonators based on their compact aspect ratios, allows the fabrication of efficient dielectric resonating antennas. Conventional fabrication of dielectric resonator disks involves mixed oxide processing with distinct ε_r and temperature coefficient of resonant frequency (τ_f) for different compositions. As a result, a demand for a suitable composition is vital to accommodate future advancements.

Functioning of Microwave Dielectric Resonator

A small disc of high dielectric constant as a dielectric resonator is sandwiched between two microstrips. The one with a broad input band and the output with a low, narrow band are respectively incorporated within the microwave integrated circuits. Three prominent resonant modes (transverse electric (TE), transverse magnetic (TM) and hybrid electromagnetic (HEM)) are more likely excited within the resonators [170]. However, multiple resonant modes are available, and the relevant choice is determined based on the target application. The input signal fed by microstrip 1 causes an excitation of the dielectric resonator upon gaining it, which subsequently couples the obtained input broadband signal to microstrip 2 at its corresponding resonating frequency. Fig. (**8**), shown below, indicates a dielectric resonator mounted between two microstrips.

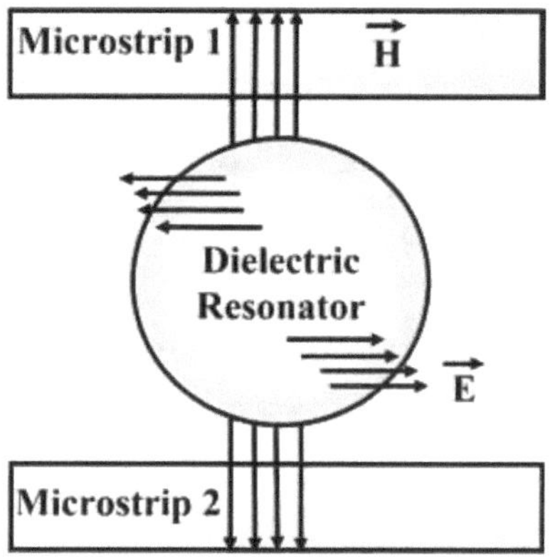

Fig. (8). Schematic representation of dielectric resonator mounted between two microstrips.

The dielectric resonator, as shown in Fig. (5), traps the energy within the microwave frequency limit confined to a small resonating volume, which is later reflected back to the resonator. The consequent effect arises due to the difference in the permittivity between the surface and resonator boundary that constitutes standing waves with large amplitude corresponding to their resonating frequency. However, colossal dielectric properties of microwave dielectric resonators enclosed in conducting cavities often exhibit sensitivity towards temperature variations and mechanical vibrations. Recent progress in material science has brought in various advancements in material architecture and material processing techniques to counter conventional and vital limitations.

$BaZrO_3$-Based Ceramic Materials as Microwave Dielectric Resonators

In compliance with research progression in the domain of hydrogen production, the extended absorption spectrum of $BaZrO_3$-based ceramic compositions motivates researchers to extract the relation between the dielectric constant and the absorption spectrum of ceramic materials. Zang *et al.* [171] reported the sensitivity of Born effective charge, dielectric and spontaneous polarization of $BaZrO_3$ towards strain. The outcomes additionally report a tendency for piezoelectric-ferroelectric phase transition with compressive and tensile strain to complement the tuneable dielectric constant of the synthesized compositions. The investigation by Al Azar *et al.* [169] upon the dielectric properties of $BaZrO_3$ demonstrates the influence of the imaginary part of the dielectric constant (ε_2 (ω)) over the absorption spectrum. An increment in the values of ε_2 (ω) indicates an increase in the absorption spectra as a consequence of an increased number of transitions between the available valance and conduction bands. In an attempt to reform the microwave dielectric properties, Khirade *et al.* [172] investigated the dielectric properties of iron (Fe^{2+}) doped $BaZrO_3$ as a function of frequency. A pattern of dielectric fluctuations with frequency was put forth from the experimental analysis. The results reveal a high dielectric constant at lower frequencies and a contrasting trend at higher frequencies. The surge in ε_r emerges due to the presence of impurities in the form of oxygen vacancies, voids and other defects. Meanwhile, the decline in ε_r arises due to the contraction at high frequencies as a consequence of ineffective polarization of dipoles with reduced relaxation time. Additionally, the grain contribution, besides electron hopping within the material perturbed by the short electric field fluctuations, results in the downfall of the dielectric constant at higher frequencies. While pure $BaZrO_3$ displays a reduction in dielectric loss from 3.04 x 10^{-3} at 50 Hz to 0.507 x 10^{-3} at 5 MHz, the dielectric constant and dielectric loss curbs steeply with rising Fe^{2+} dopant concentration.

A steep gradient in the dielectric constant and loss at higher concentrations of Fe^{2+} reflect the donor behaviour of Fe^{2+} replacing $Ti^{2+,}$ respectively. This escalates the carrier mobility with a consequent rise in the electrical conductivity and curtailing dielectric constant. A supplementary reason accounts for the high porosity of Fe-doped $BaZrO_3$ composition. This cautions the careful use of additives at appropriate doping concentrations to maximize the yield. A comparative study by Charoonsuk *et al.* [173] features a novel sample fabrication technique and displays the extent of divergence in the dielectric properties of $BaZrO_3$-based ceramic. The ceramics synthesized *via* the salt flux dissolution (SFD) technique advocate a considerable rise in the dielectric characteristics of the host material relative to frequently used techniques. The resultant attributes arise due to the compact microstructure with homogeneous grain distribution. The investigated dielectric properties of the synthesized ceramic composite using the SFD technique validate their preference over other available techniques. Powder $BaZrO_{3,}$ in the context of their study, was sintered at 1400 ºC for about 10 h, yielding a density and average grain size of about 98% and 4.45 3.50 μm, respectively. Table **8**, shown below, indicates the difference in the dielectric properties estimated using samples fabricated with different synthesis methods.

From Table **8**, it can be speculated that the latter-adopted SFD technique exhibits optimum results relative to the other sample fabrication techniques. A microstructural least density is featured for the composites fabricated using the SSR method. However, sintering aids are frequently used additives to enhance the density of the sample. It is noticed that the powder obtained from the SFD technique can be utilized to enhance the ceramic density by close to 98%. Concentrating over dielectric properties such as the resonance frequency, dielectric constant, dielectric loss and product Q x f in addition to the diameter for microwave measurements are found to be around 8974.9 MHz, 37.3, 2.045 x 10^{-3}, 4388 GHz and 5.54 mm respectively. The dielectric constant obtained from the following study is closely in agreement with those reported by Parida *et al.* [174], Sivasubramanian *et al.* [26] and Kumar *et al.* [175], respectively.

Table 8. Comparative analysis of dielectric properties of $BaZrO_3$-based ceramics fabricated using differing techniques.

Reference	Sample fabrication technique	Sintering temperature	Sintering time	Dielectric constant (ε_r)	Q- factor
[130]	Microwave-assisted hydrothermal	1670 °C	4 h	38.40	5731
[28]	Solid state reaction (SSR)	1650 °C	4 h	35.0	8800
[137]	Combustion	1650 °C	2 h	32.23	-
[138]	Soft-mechanochemical	1450 °C	6 h	37.42	5096.91
[129]	Salt flux dissolution (SDF)	1400 °C	10 h	37.26	4387.8
[174]	Solid state reaction (SSR)	1670	4 h	38.40	5731

Suhaib *et al.* [176] explored the temperature and frequency dependence of colossal dielectric properties of Ce and Y-doped $BaZrO_3$ compositions in 300 – 1000 K temperature and 10^2 – 10^6 Hz frequency range, respectively. They proposed the colossal behavior attributed to the virtue of two different relaxation behaviour at low and high-temperature limits, respectively. Such distinctions emerge due to the presence of oxygen vacancy, -Y and OH-Y defects and dipole complex pairs, respectively. The following compositions go in agreement with the prior discussions over defect-induced (complex dipole) requisite dielectric properties. The colossal dielectric constant (CDC) for $BaZr_{0.1}Ce_{0.7}Y_{0.2}O_{3-\delta}$ elevates in a stepwise rise in temperature and frequency limit. The CDC gradually increases from 300 K with a first stepwise elevation at 500 K in the low-frequency range of 100 Hz. The consequent increase of about 2 x 10^4 is higher than the prior reported magnitude of 2.6 x10^3. Thus, high CDC at low temperature (< 600 K) is probed to occur due to pseudo-relaxor behaviour caused by the ion hopping mechanism within the perovskite material. Meanwhile, the stepwise rise in the CDC at higher temperatures (> 600K), is observed to occur due to the negative capacitance effect. The resultant downfall in the voltage (dV) with a subsequent rise in the charge (dQ) establishes a negative capacitance effect. In the case of $BaZr_{0.4}Ce_{0.4}Y_{0.2}O_{3-\delta}$, the stepwise increase in CDC occurs around the intermediate temperature regime of 400-500 K, respectively. On the other hand, an opposing attribute prevails with a consequent rise in the frequency beyond 10^6 Hz. Conventional concepts propose the influence of thermal relaxations for a steep rise in dielectric properties at higher temperatures. However, a contrasting effect of an analogical rise in the dielectric characteristics at low and intermediate temperatures occurs due to the implanted defect-dipole complexes. The former findings still demand additional exploration with suitable evidence to support the microscopic viewpoint. Further investigations concerning $BaZrO_3$-based ceramics and ceramic composites in the upcoming era would allow rapid growth in the technology industry with smart materials for advanced applications in diverse domains.

Critical Shortcomings and Future Scope

In concurrence with prior discussion over diverse applications of $BaZrO_3$-based ceramic and ceramic composites, the critical shortcomings occur due to variable physiochemical responses *via* numerous synthesis techniques. A key concern with $BaZrO_3$-based ceramics is the demand for high sintering temperatures to attain phase purity with homogeneous morphology. However, such high temperatures constitute partial Ba and vaporization with a resultant A-site impurity substitution, which consumes the oxygen vacancy to hinder the degree of protonation and cumulative performance of the proton-conducting electrolyte within the fuel cell technology. Meanwhile, a major drawback of $BaZrO_3$-based ceramic photocatalyst for H2 production lies with the large fundamental bandgap (> 3eV) of the host material, which constraints the use of a marginal region of the electromagnetic spectra (UV radiation (~ 5% of the em-spectrum)) for charge recombination. The poor energy absorbance of certain $BaZrO_3$-based ceramics forbids their utilization as suitable radiation shields due to atomic displacements and lattice damage *via* high energy light-matter interaction. Meanwhile, inconsistent polarization with small relaxation time at high frequencies and perturbed charge transport due to heterogeneous morphology escalates the dielectric loss and decreases the Q-factor for their utility in microwave resonating antennas.

• $BaZrO_3$ based thin film and polymer electrolyte with appropriate synthesis techniques devoid of heat treatment must be used to avoid stoichiometric deviation and random impurity distribution.

• A higher doping strategy with a suitable acceptor dopant must be used to alter the fundamental electrophysical characteristics of the $BaZrO_3$ polymorph to aid accelerated charge recombination for improved photocatalytic hydrogen production.

• Defect engineering and the use of appropriate additives can regulate the random grain growth to preserve morphological homogeneity to lower the dielectric loss.

• The high mechanical strength, thermal conductivity and chemical stability of $BaZrO_3$ must be ensured to achieve stable and economical ceramic implants and radiation shields with non-corrosive behaviour.

• Implanted mechanical flexibility in $BaZrO_3$-based complexes can be utilized as dental ceramics and ceramic implants for knees and hips. They can further be tested as an alternative artificial structure to support defective joins and bone functionality.

• The high thermal conductivity of advanced $BaZrO_3$-based ceramic composites must be adopted to constitute their application as favourable thermal protection system.

CONCLUSION

Ceramic materials and artifacts have been in use over an elongated period in human history. Pragmatic characteristics have attracted the dominant use of ceramics and ceramic composites as the backbone for technological advancements in multiple domains. The outcome of the following introspection expands the scope of $BaZrO_3$-based ceramics and ceramic composites in the forthcoming era to play a vital role in lucrative applications. The predominant focus hovers over the energy, defense, communication and medical sectors, respectively. The non-equilibrium supply-demand ratio has enforced the need for an effective and efficient energy conversion device in terms of fuel cells. On the other hand, the reliability over mechanical strength in support with piezoelectric characteristics offered by $BaZrO_3$-based ceramic composites has captivated the attention of the research and development sector to imply its material-oriented advantages in defense. The optical sensors and actuators across the fins of the bullets play a vital role in changing the course mid-flight while simultaneously transmitting the data with the precise estimation of the distance between the target and the bullet during its approach. The speculated results of $BaZrO_3$ modified KNN-based lead-free ceramic composite establish high piezoelectric properties with low hysteresis, high bipolar strain and low electric field. The background of which is immensely associated with R-O-T and relaxor–ferroelectric phase transitions. Meanwhile, the recent activity towards H_2 production *via* photocatalysis of H_2O by $BaZrO_3$-based ceramic composites offers enhanced chemical kinetics due to the reducing characteristics of molecular H_2. However, reduced particle size distribution within the host composition would improve the transfer of photogenerated charge carriers along with charge recombination. From the structural and physiochemical landscape, the high mechanical stability and chemical inertness of standalone and $BaZrO_3$-based composites have widened the gateway for their utility as suitable radiation shields. The dense and homogeneous morphology of $BaZrO_3$ composites regulates the penetration depth of high-frequency photons (X-rays and gamma rays) to prevent extended exposure to hazardous radiations to human health. A decrease in the mass absorption coefficient (MAC) and mean free path with high atomic number constituents within the $BaZrO_3$-based ceramic composites reveals a profound decrease in the degree of penetration of high-energy photons. However, high-temperature synthesis of $BaZrO_3$ complex invites shortcomings such as stoichiometric deviations and loss of phase purity. An appropriate synthesis strategy needs to be optimized and adopted to forbid inconsistent response. Modern-day technological advancement is measured based on the ease of telecommunication and network traffic. $BaZrO_3$-based ceramic and ceramic composites are used as core components in the antennas for data transmission and communication. Wireless data transmission and secure communication require a signal with selective frequency from a combination of frequencies in a signal. Dense and compact microstructure, high dielectric constant, Q x f product with the low-temperature coefficient of resonant frequency (τ_f) and low dielectric loss are fundamental factors offered by $BaZrO_3$-based ceramic composite matrix. Although harsh material processing techniques in terms of high sintering temperatures and additional sintering additives to obtain single-phase uniform composition with dense microstructure obstructs large-scale production. Additionally, the temperature dependence of the dielectric constant invokes further misery with a decline in the dielectric constant at elevated temperatures. On the other hand, the altered resonant frequency with differing parameters piles major concerns over dielectric fluctuations. The following analysis grants primary ideas regarding the multipurpose $BaZrO_3$-based ceramic composites with prevailing challenges. This proves a need for further introspection to extend the vision for state-of-the-art future technologies.

ACKNOWLEDGMENT

I thank DST-INSPIRE for the financial aid to carry out the following research.

REFERENCES

[1] Pandey, R.K. *Fundamentals of Electroceramics: Materials, Devices, and Applications*; John Wiley & Sons, **2019**.
[http://dx.doi.org/10.1002/9781119057093]

[2] de Lima, A.G.B.; Delgado, J.M.P.Q.; Nascimento, L.P.C.; de Lima, E.S.; de Oliveira, V.A.B.; Silva, A.M.V.; Silva, J.V. Clay Ceramic Materials: From fundamentals and manufacturing to drying process predictions. *Advanced Structured Materials,* **2021**, *133*, 1-29.
[http://dx.doi.org/10.1007/978-3-030-47856-8_1]

[3] Liu, Y.; Zhang, W.; Wang, B.; Sun, L.; Li, F.; Xue, Z.; Zhou, G.; Liu, B.; Nian, H. Theoretical and experimental investigations on high temperature mechanical and thermal properties of $BaZrO_3$. *Ceram. Int.,* **2018**, *44*(14), 16475-16482.
[http://dx.doi.org/10.1016/j.ceramint.2018.06.064]

[4] Zhang, H.; Wei, T.; Zhang, Q.; Ma, W.; Fan, P.; Salamon, D.; Zhang, S.T.; Nan, B.; Tan, H.; Ye, Z.G. A review on the development of lead-free ferroelectric energy-storage ceramics and multilayer capacitors. *J. Mater. Chem. C Mater. Opt. Electron. Devices,* **2020**, *8*(47), 16648-16667.
[http://dx.doi.org/10.1039/D0TC04381H]

[5] Yang, Y.; Zheng, W.; Zhao, L. RETRACTED: $BaZrO_3$/Au and $BaZrO_3$/Au-Pd hetero-structures: The characteristics and mechanism for their photocatalytic performance. *Ceram. Int.,* **2019**, *45*(17), 23808-

23814.
[http://dx.doi.org/10.1016/j.ceramint.2019.07.319]

[6] Antončík, F.; Lojka, M.; Hlásek, T.; Sedmidubský, D.; Jankovský, O.; Bartůněk, V. The effective synthesis of large volumes of the ultrafine $BaZrO_3$ nanoparticles. *Mater. Chem. Phys.,* **2021**, *259*, 124047.
[http://dx.doi.org/10.1016/j.matchemphys.2020.124047]

[7] Ultrasonic, H.; Niu, G.; Member, S.; Zhao, J.; Zhuang, J.; Zhang, J.; Zheng, K.; Environment-friendly, A. *Lead-Free KNN-Based Textured Ceramics for.,* **2021**, *68*(5), 1979-1987.

[8] Loureiro, F.J.A.; Nasani, N.; Reddy, G.S.; Munirathnam, N.R.; Fagg, D.P. A review on sintering technology of proton conducting $BaCeO_3$-$BaZrO_3$ perovskite oxide materials for Protonic Ceramic Fuel Cells. *J. Power Sources,* **2019**, *438*(August), 226991.
[http://dx.doi.org/10.1016/j.jpowsour.2019.226991]

[9] Khirade, P.P.; Raut, A.V.; Alange, R.C.; Barde, W.S.; Chavan, A.R. Structural, electrical and dielectric investigations of cerium doped barium zirconate ($BaZrO_3$) nano-ceramics produced via green synthesis: Probable candidate for solid oxide fuel cells and microwave applications. *Physica B,* **2021**, *613*(March), 412948.
[http://dx.doi.org/10.1016/j.physb.2021.412948]

[10] Tapia-Ruiz, N.; Robert Armstrong, A.; Alptekin, H. Roadmap on Inorganic Perovskites for Energy Applications Opportunities and Challenges of Inorganic Perovskites in High-Performance Photodetectors. 2021.

[11] Bi, L.; Da'as, E.H.; Shafi, S.P. Proton-conducting solid oxide fuel cell (SOFC) with Y-doped $BaZrO_3$ electrolyte. *Electrochem. Commun.,* **2017**, *80*(80), 20-23.
[http://dx.doi.org/10.1016/j.elecom.2017.05.006]

[12] Satapathy, A.; Sinha, E. A comparative proton conductivity study on Yb-doped $BaZrO_3$ perovskite at intermediate temperatures under wet N2 environment. *J. Alloys Compd.,* **2019**, *772*, 675-682.
[http://dx.doi.org/10.1016/j.jallcom.2018.08.329]

[13] Draber, F. M.; Ader, C.; Arnold, J. P.; Eisele, S.; Grieshammer, S.; Yamaguchi, S.; Martin, M. Publisher Correction: Nanoscale Percolation in Doped $BaZrO_3$ for High Proton Mobility (Nature Materials, (2020), 19, 3, (338-346), 10.1038/s41563-019-0561-7). Nat. Mater. 2020, 19 (5), 577. https://doi.org/10.1038/s41563-020-0654-3.

[14] Vignesh, D.; Rout, E. Structural and electrophysical perturbations in Y^{3+} doped $BaCeO_3$ proton conducting electrolyte: A first principles approach. *Computational Condensed Matter,* **2022**, *33*(October), e00763.
[http://dx.doi.org/10.1016/j.cocom.2022.e00763]

[15] Vignesh, D.; Rout, E. Analysis of symmetry variation as a function of rare earth dopant concentration in proton conducting solid oxide fuel cells. *Mater. Today Proc.,* **2022**, *66*, 3416-3421.
[http://dx.doi.org/10.1016/j.matpr.2022.08.373]

[16] Fu, J.; Zuo, R.; Qi, H.; Zhang, C.; Li, J.; Li, L. Low electric-field driven ultrahigh electrostrains in Sb-substituted (Na,K)NbO_3 lead-free ferroelectric ceramics. *Appl. Phys. Lett.,* **2014**, *105*(24), 242903.
[http://dx.doi.org/10.1063/1.4904476]

[17] Zuo, R.; Fu, J. Rhombohedral-Tetragonal Phase Coexistence and Piezoelectric Properties of (NaK)(NbSb)O_3-$LiTaO_3$-$BaZrO_3$ Lead-Free Ceramics. *J. Am. Ceram. Soc.,* **2011**, *94*(5), 1467-1470.
[http://dx.doi.org/10.1111/j.1551-2916.2010.04256.x]

[18] Wang, R.; Bando, H.; Katsumata, T.; Inaguma, Y.; Taniguchi, H.; Itoh, M. Tuning the orthorhombic–rhombohedral phase transition temperature in sodium potassium niobate by incorporating barium zirconate. *Phys. Status Solidi Rapid Res. Lett.,* **2009**, *3*(5), 142-144.
[http://dx.doi.org/10.1002/pssr.200903090]

[19] Khetkorn, W.; Rastogi, R.P.; Incharoensakdi, A.; Lindblad, P.; Madamwar, D.; Pandey, A.; Larroche,

C. Microalgal hydrogen production – A review. *Bioresour. Technol.,* **2017**, *243*, 1194-1206. [http://dx.doi.org/10.1016/j.biortech.2017.07.085] [PMID: 28774676]

[20] El-Shafie, M.; Kambara, S.; Hayakawa, Y. Hydrogen Production Technologies Overview. **2019**. [http://dx.doi.org/10.4236/jpee.2019.71007]

[21] Tao, Q.; Lu, T.; Sheng, Y.; Li, L.; Lu, W.; Li, M. Machine learning aided design of perovskite oxide materials for photocatalytic water splitting. *Journal of Energy Chemistry,* **2021**, *60*, 351-359. [http://dx.doi.org/10.1016/j.jechem.2021.01.035]

[22] Kolodiazhnyi, T.; Pulphol, P.; Vittayakorn, W.; Vittayakorn, N. Giant suppression of dielectric loss in $BaZrO_3$. *J. Eur. Ceram. Soc.,* **2019**, *39*(14), 4144-4148. [http://dx.doi.org/10.1016/j.jeurceramsoc.2019.06.037]

[23] Liu, F.; Cheng, L.; Li, H.; Liu, S. Ordering-induced domains in sub-micron-sized $Ba(Zn_1/_3Ta_2/_3)O_3$–$BaZrO_3$ microwave ceramics. *J. Mater. Sci. Mater. Electron.,* **2021**, *32*(21), 26126-26136. [http://dx.doi.org/10.1007/s10854-021-06341-3]

[24] Stetson, H.; Schwartz, B. Dielectric Properties of Zirconates. *J. Am. Ceram. Soc.,* **1961**, *44*(8), 420-421. [http://dx.doi.org/10.1111/j.1151-2916.1961.tb15476.x]

[25] Yamaguchi, T.; Komatsu, Y.; Otobe, T.; Murakami, Y. Newly developed ternary (Ca, Sr, Ba) zirconate ceramic system for microwave resonators. *Ferroelectrics,* **1980**, *27*(1), 273-276. [http://dx.doi.org/10.1080/00150198008226116]

[26] Sivasubramanian, V.; Murthy, V. R. K.; Viswanathan, B. Microwave Dielectric Properties of Certain Simple Alkaline Earth Perovskite Compounds as a Function of Tolerance Factor. Japanese J. Appl. Physics, Part 1 Regul. Pap. Short Notes Rev. Pap. 1997, 36 (1 A), 194–197. https://doi.org/10.1143/JJAP.36.194.

[27] Borja Urby, R. On the Study of Photoluminescence and Photocatalytic Properties of Doped $BaZrO_3$. 2012, 62. http://cio.repositorioinstitucional.mx/jspui/handle/1002/553

[28] Yuan, Y.; Zhang, X.; Liu, L.; Jiang, X.; Lv, J.; Li, Z.; Zou, Z. Synthesis and photocatalytic characterization of a new photocatalyst $BaZrO_3$. *Int. J. Hydrogen Energy,* **2008**, *33*(21), 5941-5946. [http://dx.doi.org/10.1016/j.ijhydene.2008.07.052]

[29] Szkliniarz, K.; Jastrzębski, J.; Bilewicz, A.; Chajduk, E.; Choiński, J.; Jakubowski, A.; Janiszewska, Ł.; Leszczuk, E.; Łyczko, M.; Sitarz, M.; Stolarz, A.; Trzcińska, A.; Wąs, B.; Zipper, W. Medical radioisotopes produced using the alpha particle beam from the warsaw heavy ion cyclotron. *Acta Phys. Pol. A,* **2015**, *127*(5), 1471-1474. [http://dx.doi.org/10.12693/APhysPolA.127.1471]

[30] Vassen, R.; Cao, X.; Tietz, F.; Basu, D.; Stöver, D. Zirconates as new materials for thermal barrier coatings. *J. Am. Ceram. Soc.,* **2000**, *83*(8), 2023-2028. [http://dx.doi.org/10.1111/j.1151-2916.2000.tb01506.x]

[31] Vignesh, D.; Rout, E. Technological Challenges and Advancement in Proton Conductors: A Review. *Energy Fuels,* **2023**, *37*(5), 3428-3469. [http://dx.doi.org/10.1021/acs.energyfuels.2c03926]

[32] Stotz, S.; Wagner, C. Die L{ö}slichkeit von Wasserdampf Und Wasserstoff in Festen Oxiden. *Berichte der Bunsengesellschaft f{ü}r Phys. Chemie,* **1966**, *70*(8), 781-788. [http://dx.doi.org/10.1002/bbpc.19660700804]

[33] Gu, Y.; Luo, G.; Chen, Z.; Huo, Y.; Wu, F. Enhanced chemical stability and electrochemical performance of $BaCe_{0\cdot8}Y_0.1Ni_{0\cdot04}Sm_{0\cdot06}0O_3$- perovskite electrolytes as proton conductors. *Ceram. Int.,* **2022**, *48*(8), 10650-10658. [http://dx.doi.org/10.1016/j.ceramint.2021.12.279]

[34] Karlsson, M.; Engberg, D.; Björketun, M.E.; Matic, A.; Wahnström, G.; Sundell, P.G.; Berastegui, P.;

Ahmed, I.; Falus, P.; Farago, B.; Börjesson, L.; Eriksson, S. Using Neutron Spin–Echo To Investigate Proton Dynamics in Proton-Conducting Perovskites. *Chem. Mater.*, **2010**, *22*(3), 740-742. [http://dx.doi.org/10.1021/cm901624v]

[35] Bello, I.T.; Zhai, S.; Zhao, S.; Li, Z.; Yu, N.; Ni, M. Scientometric review of proton-conducting solid oxide fuel cells. *Int. J. Hydrogen Energy,* **2021**, *46*(75), 37406-37428. [http://dx.doi.org/10.1016/j.ijhydene.2021.09.061]

[36] Vignesh, D.; Sonu, B.K.; Rout, E. Factors Constituting Proton Trapping in $BaCeO_3$ and $BaZrO_3$ Perovskite Proton Conductors in Fuel Cell Technology: A Review. *Energy Fuels,* **2022**, *36*(14), 7219-7244. [http://dx.doi.org/10.1021/acs.energyfuels.2c00650]

[37] Ivanov, S. Multiferroic Complex Metal Oxides: Main Features of Preparation, Structure, and Properties. *Science and Technology of Atomic, Molecular, Condensed Matter & Biological Systems*; Elsevier, **2012**, 2, pp. 163-238. [http://dx.doi.org/10.1016/B978-0-44-453681-5.00007-8]

[38] Mazzei, L.; Rukser, D.; Biebl, F.; Grimm-Lebsanft, B.; Neuber, G.; Pergolesi, D.; Börjesson, L.; Rübhausen, M.A.; Andreasson, J.; Karlsson, M. Phonon spectra of pure and acceptor doped $BaZrO_3$ investigated with visible and UV Raman spectroscopy. *J. Phys. Condens. Matter,* **2020**, *32*(40), 405403. [http://dx.doi.org/10.1088/1361-648X/ab95d1] [PMID: 32628643]

[39] Sakai, T.; Kato, T.; Katsui, H.; Tanaka, Y.; Goto, T. Preparation of Y-doped $BaZrO_3$ thin film electrolyte by laser chemical vapor deposition. *Mater. Today Commun.,* **2020**, *24*, 101184. [http://dx.doi.org/10.1016/j.mtcomm.2020.101184]

[40] Teng, H.P.; Huang, T.J.; Lu, F.H. Synthesis of crystalline $BaZrO_3$ thin films over preferred-oriented ZrN by a hydrothermal-galvanic couple method for photoelectrochemical evaluations. *Mater. Today Commun.,* **2022**, *33*(October), 104761. [http://dx.doi.org/10.1016/j.mtcomm.2022.104761]

[41] Sunarso, J.; Baumann, S.; Serra, J.M.; Meulenberg, W.A.; Liu, S.; Lin, Y.S.; Diniz da Costa, J.C. Mixed ionic–electronic conducting (MIEC) ceramic-based membranes for oxygen separation. *J. Membr. Sci.,* **2008**, *320*(1-2), 13-41. [http://dx.doi.org/10.1016/j.memsci.2008.03.074]

[42] Husain, J. DC Conductivity Studies on Polyethylene Oxide/Nickel Ferrite Composites. **2016**, *3*(2), 223–226.

[43] Alberti, G.; Casciola, M. Solid state protonic conductors, present main applications and future prospects. *Solid State Ion.,* **2001**, *145*(1-4), 3-16. [http://dx.doi.org/10.1016/S0167-2738(01)00911-0]

[44] Zhou, Y.; Guan, X.; Zhou, H.; Ramadoss, K.; Adam, S.; Liu, H.; Lee, S.; Shi, J.; Tsuchiya, M.; Fong, D.D.; Ramanathan, S. Strongly correlated perovskite fuel cells. *Nature,* **2016**, *534*(7606), 231-234. [http://dx.doi.org/10.1038/nature17653] [PMID: 27279218]

[45] Su, H.; Hu, Y.H. Progress in low-temperature solid oxide fuel cells with hydrocarbon fuels. *Chem. Eng. J.,* **2020**, *402*(April), 126235. [http://dx.doi.org/10.1016/j.cej.2020.126235]

[46] Vignesh, D.; Sonu, B.K.; Rout, E. Factors Constituting Proton Trapping in $BaCeO_3$ and $BaZrO_3$ Perovskite Proton Conductors in Fuel Cell Technology: A Review. *Energy Fuels,* **2022**, *36*(14), 7219-7244. [http://dx.doi.org/10.1021/acs.energyfuels.2c00650]

[47] Subotić, V.; Baldinelli, A.; Barelli, L.; Scharler, R.; Pongratz, G.; Hochenauer, C.; Anca-Couce, A. Applicability of the SOFC technology for coupling with biomass-gasifier systems: Short- and long-term experimental study on SOFC performance and degradation behaviour. *Appl. Energy,* **2019**, *256*, 113904.

[http://dx.doi.org/10.1016/j.apenergy.2019.113904]

[48] Li, G.; Gou, Y.; Qiao, J.; Sun, W.; Wang, Z.; Sun, K. Recent progress of tubular solid oxide fuel cell: From materials to applications. *J. Power Sources,* **2020**, *477*(June), 228693. [http://dx.doi.org/10.1016/j.jpowsour.2020.228693]

[49] Eichhorn Colombo, K.W.; Kharton, V.V. Start-up of a solid oxide fuel cell system with a view to materials science-related aspects, control and thermo-mechanical stresses. *Crystals,* **2021**, *11*(7), 732. [http://dx.doi.org/10.3390/cryst11070732]

[50] Fabbri, E.; Pergolesi, D.; Traversa, E. Materials challenges toward proton-conducting oxide fuel cells: A critical review. *Chem. Soc. Rev.,* **2010**, *39*(11), 4355-4369. [http://dx.doi.org/10.1039/b902343g] [PMID: 20818453]

[51] Lim, D.W.; Kitagawa, H. Rational strategies for proton-conductive metal–organic frameworks. *Chem. Soc. Rev.,* **2021**, *50*(11), 6349-6368. [http://dx.doi.org/10.1039/D1CS00004G] [PMID: 33870975]

[52] Gupta, A.; Goswami, S.; Elahi, S.M.; Konar, S. Role of framework–carrier interactions in proton-conducting crystalline porous materials. *Cryst. Growth Des.,* **2021**, *21*(3), 1378-1388. [http://dx.doi.org/10.1021/acs.cgd.0c01394]

[53] Xu, X.; Xu, Y.; Ma, J.; Yin, Y.; Fronzi, M.; Wang, X.; Bi, L. Tailoring electronic structure of perovskite cathode for proton-conducting solid oxide fuel cells with high performance. *J. Power Sources,* **2021**, *489*(January), 229486. [http://dx.doi.org/10.1016/j.jpowsour.2021.229486]

[54] Jing, J.; Pang, J.; Chen, L.; Zhang, H.; Lei, Z.; Yang, Z. Structure, synthesis, properties and solid oxide electrolysis cells application of Ba(Ce, Zr)O_3 based proton conducting materials. *Chem. Eng. J.,* **2022**, *429*, 132314. [http://dx.doi.org/10.1016/j.cej.2021.132314]

[55] Xu, X.; Bi, L. Proton-conducting electrolyte materials. *Proton-Conducting Electrolyte Materials,* **2020**, *96*, 81-111. [http://dx.doi.org/10.1016/B978-0-12-817445-6.00003-X]

[56] Radenahmad, N.; Afif, A.; Abdalla, A.M.; Saqib, M.; Park, J.Y.; Zaini, J.; Irvine, J.; Kim, J.H.; Azad, A.K. A New High-Performance Proton-Conducting Electrolyte for Next-Generation Solid Oxide Fuel Cells. *Energy Technol.,* **2020**, *8*(9), 2000486. [http://dx.doi.org/10.1002/ente.202000486]

[57] Yu, S.; Wang, Z.; Yang, L.; Liu, J.; Guan, R.; Xiao, Y.; He, T. Enhancing the sinterability and electrical properties of $BaZr_{0.1}Ce_{0.7}Y_{0.2}O_{3\square\delta}$ proton-conducting ceramic electrolyte. *J. Am. Ceram. Soc.,* **2021**, *104*(1), 329-342. [http://dx.doi.org/10.1111/jace.17467]

[58] Regalado Vera, C.Y.; Ding, H.; Peterson, D.; Gibbons, W.T.; Zhou, M.; Ding, D. A mini-review on proton conduction of $BaZrO_3$ -based perovskite electrolytes. *Journal of Physics: Energy,* **2021**, *3*(3), 032019. [http://dx.doi.org/10.1088/2515-7655/ac12ab]

[59] Uthayakumar, A.; Kavithanjali, M.; Sandhya, K.; Ponpandian, N.; Babu, K.S. The rare earth dopant (La, Gd, Sm & Y) modulated grain boundary energy barrier suppression in $BaZrO_3$-$BaCeO_3$ solid solution. *J. Alloys Compd.,* **2021**, *864*, 158098. [http://dx.doi.org/10.1016/j.jallcom.2020.158098]

[60] Minh, N.Q. Ceramic Fuel Cells. *J. Am. Ceram. Soc.,* **1993**, *76*(3), 563-588. [http://dx.doi.org/10.1111/j.1151-2916.1993.tb03645.x]

[61] Kaur, P.; Singh, K. Review of perovskite-structure related cathode materials for solid oxide fuel cells. *Ceram. Int.,* **2020**, *46*(5), 5521-5535. [http://dx.doi.org/10.1016/j.ceramint.2019.11.066]

[62] Fischer, S.A.; Gunlycke, D. Analysis of correlated dynamics in the grotthuss mechanism of proton diffusion. *J. Phys. Chem. B,* **2019**, *123*(26), 5536-5544.
[http://dx.doi.org/10.1021/acs.jpcb.9b02610] [PMID: 31180658]

[63] Kolokolov, D.I.; Lim, D.W.; Kitagawa, H. Characterization of proton dynamics for the understanding of conduction mechanism in proton conductive metal-organic frameworks. *Chem. Rec.,* **2020**, *20*(11), 1297-1313.
[http://dx.doi.org/10.1002/tcr.202000072] [PMID: 32959508]

[64] Wang, N.; Tang, C.; Du, L.; Zhu, R.; Xing, L.; Song, Z.; Yuan, B.; Zhao, L.; Aoki, Y.; Ye, S. Advanced cathode materials for protonic ceramic fuel cells: Recent progress and future perspectives. *Adv. Energy Mater.,* **2022**, *12*(34), 2201882.
[http://dx.doi.org/10.1002/aenm.202201882]

[65] Han, D.; Shinoda, K.; Sato, S.; Majima, M.; Uda, T. *Correlation between electroconductive and structural properties of proton conductive acceptor-doped barium zirconate,* **2015**.
[http://dx.doi.org/10.1039/C4TA05701E]

[66] Gilardi, E.; Fabbri, E.; Bi, L.; Rupp, J.L.M.; Lippert, T.; Pergolesi, D.; Traversa, E. Effect of dopant–host ionic radii mismatch on acceptor-doped barium zirconate microstructure and proton conductivity. *J. Phys. Chem. C,* **2017**, *121*(18), 9739-9747.
[http://dx.doi.org/10.1021/acs.jpcc.7b02163]

[67] Hossain, M.K.; Yamamoto, T.; Hashizume, K. Isotopic effect of proton conductivity in barium zirconates for various hydrogen-containing atmospheres. *J. Alloys Compd.,* **2022**, *903*, 163957.
[http://dx.doi.org/10.1016/j.jallcom.2022.163957]

[68] Mielewczyk-Gryń, A.; Subramani, T.; Jaworski, D.; Lilova, K.; Skubida, W.; Navrotsky, A.; Gazda, M. Water uptake and energetics of the formation of barium zirconate based multicomponent oxides. *Phys. Chem. Chem. Phys.,* **2023**, *25*(13), 9208-9215.
[http://dx.doi.org/10.1039/D2CP05265B] [PMID: 36919378]

[69] Fujii, S.; Shimizu, Y.; Hyodo, J.; Kuwabara, A.; Yamazaki, Y. Discovery of Unconventional Proton-Conducting Inorganic Solids *via* Defect-Chemistry-Trained, Interpretable Machine Learning. **2023**.

[70] Saini, D.S.; Ghosh, A.; Tripathy, S.; Kumar, A.; Sharma, S.K.; Kumar, N.; Majumdar, S.; Bhattacharya, D. A promising proton conducting electrolyte $BaZr_{1-x}Ho_xO_{3-\delta}$ ($0.05 \leq x \leq 0.20$) Ceramics for Intermediate Temperature Solid Oxide Fuel Cells. *Sci. Rep.,* **2020**, *10*(1), 3461.
[http://dx.doi.org/10.1038/s41598-020-60174-4] [PMID: 31913322]

[71] Ahmed, I.; Eriksson, S.; Ahlberg, E.; Knee, C.; Götlind, H.; Johansson, L.; Karlsson, M.; Matic, A.; Börjesson, L. Structural study and proton conductivity in Yb-doped $BaZrO_3$. *Solid State Ion.,* **2007**, *178*(7-10), 515-520.
[http://dx.doi.org/10.1016/j.ssi.2006.11.011]

[72] Ahmed, I.; Eriksson, S.; Ahlberg, E.; Knee, C.; Berastegui, P.; Johansson, L.; Rundlöf, H.; Karlsson, M.; Matic, A.; Börjesson, L.; Engberg, D. Synthesis and structural characterization of perovskite type proton conducting BaZr1−xInxO3−δ (0.0≤x≤0.75). *Solid State Ion.,* **2006**, *177*(17-18), 1395-1403.
[http://dx.doi.org/10.1016/j.ssi.2006.07.009]

[73] Kang, S-J.L. *Sintering: Densification, Grain Growth and Microstructure*; Elsevier, **2004**.

[74] Han, D.; Hatada, N.; Uda, T. Microstructure, Proton Concentration and Proton Conductivity of Barium Zirconate Doped with Ho, Er, Tm and Yb. *J. Electrochem. Soc.,* **2016**, *163*(6), F470-F476.
[http://dx.doi.org/10.1149/2.0551606jes]

[75] Sharova, N.V.; Gorelov, V.P. Electroconduction and the Nature of Ionic Transport in $BaZr_{0.95}Nd_{0.05}O_{3-\delta}$. *Russ. J. Electrochem.,* **2005**, *41*(10), 1130-1134.
[http://dx.doi.org/10.1007/s11175-005-0192-y]

[76] Satapathy, A.; Sinha, E.; Rout, S.K. Structural and proton conductivity study of $BaZr_{1-x}RE_xO_{3-\delta}$(RE = Dy, Sm) ceramics for intermediate temperature solid oxide fuel cell electrolyte. *J. Solid State*

Electrochem., **2020**, *24*(7), 1463-1473.
[http://dx.doi.org/10.1007/s10008-019-04423-1]

[77] Vignesh, D.; Rout, E. Proton Conductors: Physics and Technological Advancements for PC-SOFC. *Energy Materials: Structure, Properties and Applications*; Swain, B.P., Ed.; Springer Nature Singapore: Singapore, **2023**, pp. 1-55.
[http://dx.doi.org/10.1007/978-981-99-3866-7_1]

[78] Hossain, M.K.; Biswas, M.C.; Chanda, R.K.; Rubel, M.H.K.; Khan, M.I.; Hashizume, K. A review on experimental and theoretical studies of perovskite barium zirconate proton conductors. *Emergent Materials,* **2021**, *4*(4), 999-1027.
[http://dx.doi.org/10.1007/s42247-021-00230-5]

[79] Campos Covarrubias, M.S.; Sriubas, M.; Bockute, K.; Winiarz, P.; Miruszewski, T.; Skubida, W.; Jaworski, D.; Bartmański, M.; Szkodo, M.; Gazda, M.; Laukaitis, G. Properties of barium cerate thin films formed using E-Beam Deposition. *Crystals,* **2020**, *10*(12), 1152.
[http://dx.doi.org/10.3390/cryst10121152]

[80] Pergolesi, D.; Fabbri, E.; D'Epifanio, A.; Di Bartolomeo, E.; Tebano, A.; Sanna, S.; Licoccia, S.; Balestrino, G.; Traversa, E. High proton conduction in grain-boundary-free yttrium-doped barium zirconate films grown by pulsed laser deposition. *Nat. Mater.,* **2010**, *9*(10), 846-852.
[http://dx.doi.org/10.1038/nmat2837] [PMID: 20852619]

[81] Niwa, E.; Kluczny, M.; Kim, H.Y.; Song, J.T.; Watanabe, M.; Takagaki, A.; Ishihara, T. Proton conductivity in Yb-doped $BaZrO_3$-based thin film prepared by pulsed laser deposition. *Solid State Ion.,* **2023**, *396*(396), 116240.
[http://dx.doi.org/10.1016/j.ssi.2023.116240]

[82] Kumar, N. Bulletproof Vest and Its Improvement – A Review. *Int. J. Sci. Dev. Res.,* **2016**, *1*(1), 34-39.

[83] Bhatnagar, A. *Lightweight Ballistic Composites: Military and Law-Enforcement Applications*; Woodhead Publishing, **2016**.

[84] Samir, N.S.; Radwan, M.A.; Sadek, M.A.; Elazab, H.A. Preparation and Characterization of Bullet-Proof Vests Based on Polyamide Fibers. *Int. J. Eng. Technol.,* **2018**, *7*(3), 1290-1294.
[http://dx.doi.org/10.14419/ijet.v7i3.13175]

[85] Tepeduzu, B.; Karakuzu, R. Ballistic performance of ceramic/composite structures. *Ceram. Int.,* **2019**, *45*(2), 1651-1660.
[http://dx.doi.org/10.1016/j.ceramint.2018.10.042]

[86] Nurazzi, N.M.; Asyraf, M.R.M.; Khalina, A.; Abdullah, N.; Aisyah, H.A.; Rafiqah, S.A.; Sabaruddin, F.A.; Kamarudin, S.H.; Norrrahim, M.N.F.; Ilyas, R.A.; Sapuan, S.M. A review on natural fiber reinforced polymer composite for bullet proof and ballistic applications. *Polymers,* **2021**, *13*(4), 646.
[http://dx.doi.org/10.3390/polym13040646] [PMID: 33671599]

[87] Reddy, P. R. S.; Savio, S. G.; Madhu, V. Ceramic composite armour for ballistic protection. **2020**.
[http://dx.doi.org/10.1007/978-3-030-16347-1_10]

[88] Maystrenko, A.L.; Kushch, V.I.; Pashchenko, E.A.; Kulich, V.G.; Neshpor, O.V.; Bisyk, S.P. Ceramic Armour for Armoured Vehicles Against Large-Calibre Bullets. *Problems of Mechatronics Armament Aviation Safety Engineering,* **2020**, *11*(1), 9-16.
[http://dx.doi.org/10.5604/01.3001.0014.0279]

[89] Hart, M. Vulnerabilities and Challenges of Integrating AI into Future Air Force Intelligence Systems. **2020**.

[90] Raubenheimer, T.O.; Ruth, R.D. A dispersion-free trajectory correction technique for linear colliders. *Nucl. Instrum. Methods Phys. Res. A,* **1991**, *302*(2), 191-208.
[http://dx.doi.org/10.1016/0168-9002(91)90403-D]

[91] Haris, A.; Goo, N. S.; Park, H. C.; Yoon, K. J. Modeling and Analysis for the Development of Lightweight Piezoceramic Composite Actuators (LIPCA). Comput. Mater. Sci. 2004, 30 (3–4 SPEC.

ISS.), 474–481. https://doi.org/10.1016/j.commatsci.2004.02.039.

[92] Sun, X.; Gu, L.; Gong, C. Dynamics of a deflectable-nose missile. *Sci. China Technol. Sci.,* **2012**, *55*(12), 3483-3494.
[http://dx.doi.org/10.1007/s11431-012-4984-5]

[93] Lei, Y.; Zhang, B. *A Study on Piezoelectric Ammunition Based on Improved Single Neuron Adaptive PID Algorithm*; No. Mmebc, **2016**, pp. 1031-1035.
[http://dx.doi.org/10.2991/mmebc-16.2016.215]

[94] Sekhar, B. C.; Dhanalakshmi, B.; Rao, B. S.; Ramesh, S.; Prasad, K. V.; Rao, P. S. V. S.; Rao, B. P. Piezoelectricity and Its Applications. Multifunct. Ferroelectr. Mater. 2021, 71.

[95] Amol, P.; Dhole, R.N.; Waghmare, L.M. Smart Bullet System. **2015**. https://www.researchgate.net/publication/282770165

[96] Zhang, J.; Liu, Z.; Zhang, T.; Liu, Y.; Lyu, Y. High strain response and low hysteresis in $BaZrO_3$-modified KNN-based lead-free relaxor ceramics. *J. Mater. Sci. Mater. Electron.,* **2021**, *32*(12), 16715-16725.
[http://dx.doi.org/10.1007/s10854-021-06229-2]

[97] Yin, Y.; Tang, Y.; Pan, W.; Song, A.; Yu, J.; Zhang, B. Relaxor behaviors enhance piezoelectricity in lead-free $BiFeO_3$-$BaTiO_3$ ceramics incorporated with a tiny amount of $Bi(Mg_{1/2}Ti_{1/2})O_3$ near the morphotropic phase boundary. *Ceram. Int.,* **2021**, *47*(7), 9486-9494.
[http://dx.doi.org/10.1016/j.ceramint.2020.12.082]

[98] Li, F.; Wu, S.; Li, T.; Wang, C.; Zhai, J. Normal-relaxor ferroelectric phase transition induced morphotropic phase boundary accompanied by enhanced piezoelectric and electrostrain properties in strontium modulated $Bi_{0.5}K_{0.5}TiO_3$ lead-free ceramics. *J. Eur. Ceram. Soc.,* **2020**, *40*(12), 3918-3927.
[http://dx.doi.org/10.1016/j.jeurceramsoc.2020.05.030]

[99] Wu, L.; Zheng, T.; Wu, J. Excellent fatigue resistance in Sb nonstoichiometric KNN-based ceramics by engineering relaxor multiphase state. *J. Eur. Ceram. Soc.,* **2022**, *42*(12), 4888-4897.
[http://dx.doi.org/10.1016/j.jeurceramsoc.2022.05.028]

[100] Wu, B.; Wu, H.; Wu, J.; Xiao, D.; Zhu, J.; Pennycook, S.J. Giant piezoelectricity and high curie temperature in nanostructured alkali niobate lead-free piezoceramics through phase coexistence. *J. Am. Chem. Soc.,* **2016**, *138*(47), 15459-15464.
[http://dx.doi.org/10.1021/jacs.6b09024] [PMID: 27933925]

[101] Fu, J.; Zuo, R.; Xu, Z. High piezoelectric activity in $(Na,K)NbO_3$ based lead-free piezoelectric ceramics: Contribution of nanodomains. *Appl. Phys. Lett.,* **2011**, *99*(6), 062901.
[http://dx.doi.org/10.1063/1.3624704]

[102] Huan, Y.; Wang, X.; Shen, Z.; Kim, J.; Zhou, H.; Li, L. Nanodomains in KNN-based lead-free piezoelectric ceramics: Origin of strong piezoelectric properties. *J. Am. Ceram. Soc.,* **2014**, *97*(3), 700-703.
[http://dx.doi.org/10.1111/jace.12850]

[103] Koruza, J.; Bell, A.J.; Frömling, T.; Webber, K.G.; Wang, K.; Rödel, J. Requirements for the transfer of lead-free piezoceramics into application. *Journal of Materiomics,* **2018**, *4*(1), 13-26.
[http://dx.doi.org/10.1016/j.jmat.2018.02.001]

[104] Zhou, C.; Zhang, J.; Yao, W.; Liu, D.; He, G. Remarkably strong piezoelectricity, rhombohedral-orthorhombic-tetragonal phase coexistence and domain structure of $(K,Na)(Nb,Sb)O_3$–$(Bi,Na)ZrO_3$–$BaZrO_3$ ceramics. *J. Alloys Compd.,* **2020**, *820*, 153411.
[http://dx.doi.org/10.1016/j.jallcom.2019.153411]

[105] Zuo, R.; Fu, J.; Lv, D.; Liu, Y. Antimony tuned rhombohedral-orthorhombic phase transition and enhanced piezoelectric properties in sodium potassium niobate. *J. Am. Ceram. Soc.,* **2010**, *93*(9), 2783-2787.
[http://dx.doi.org/10.1111/j.1551-2916.2010.03804.x]

[106] Lee, M.K.; Yang, S.A.; Park, J.J.; Lee, G.J. Proposal of a rhombohedral-tetragonal phase composition for maximizing piezoelectricity of (K,Na)NbO_3 ceramics. *Sci. Rep.*, **2019**, *9*(1), 4195. [http://dx.doi.org/10.1038/s41598-019-40943-6] [PMID: 30626917]

[107] Sirivallop, A.; Areerob, T.; Chiarakorn, S. Enhanced visible light photocatalytic activity of N and Ag doped and co-doped TiO_2 synthesized by using an *in-situ* solvothermal method for gas phase ammonia removal. *Catalysts,* **2020**, *10*(2), 251. [http://dx.doi.org/10.3390/catal10020251]

[108] Zohuri, B. *Hydrogen Energy: Challenges and Solutions for a Cleaner Future*; Springer, **2019**. [http://dx.doi.org/10.1007/978-3-319-93461-7]

[109] Vinoth Kanna, I.; Paturu, P. A study of hydrogen as an alternative fuel. *International Journal of Ambient Energy,* **2020**, *41*(12), 1433-1436. [http://dx.doi.org/10.1080/01430750.2018.1484803]

[110] Singla, M.K.; Nijhawan, P.; Oberoi, A.S. Hydrogen fuel and fuel cell technology for cleaner future: a review. *Environ. Sci. Pollut. Res. Int.,* **2021**, *28*(13), 15607-15626. [http://dx.doi.org/10.1007/s11356-020-12231-8] [PMID: 33538968]

[111] Mathur, D. Central Asian Journal of Theoretical Hydrogen Energy : An Alternative Fuel for the Future. 2022, No. c, 16–21. https://cajotas.centralasianstudies.org

[112] Liao, G.; Gong, Y.; Zhang, L.; Gao, H.; Yang, G.J.; Fang, B. Semiconductor polymeric graphitic carbon nitride photocatalysts: the "holy grail" for the photocatalytic hydrogen evolution reaction under visible light. *Energy Environ. Sci.,* **2019**, *12*(7), 2080-2147. [http://dx.doi.org/10.1039/C9EE00717B]

[113] Ajiboye, T.O.; Oyewo, O.A.; Onwudiwe, D.C. The performance of bismuth-based compounds in photocatalytic applications. *Surf. Interfaces,* **2021**, *23*(January), 100927. [http://dx.doi.org/10.1016/j.surfin.2021.100927]

[114] Han, Y.; Chen, Y.; Fan, R.; Li, Z.; Zou, Z. Promotion effect of metal phosphides towards electrocatalytic and photocatalytic water splitting. *EcoMat,* **2021**, *3*(3), e12097. [http://dx.doi.org/10.1002/eom2.12097]

[115] Kim, J.H.; Hansora, D.; Sharma, P.; Jang, J.W.; Lee, J.S. Toward practical solar hydrogen production an artificial photosynthetic leaf-to-farm challenge. *Chem. Soc. Rev.,* **2019**, *48*(7), 1908-1971. [http://dx.doi.org/10.1039/C8CS00699G] [PMID: 30855624]

[116] Zhu, K.; Kang, S.Z.; Qin, L.; Han, S.; Li, G.; Li, X. Novel and highly active potassium niobate-based photocatalyst for dramatically enhanced hydrogen production. *ACS Sustain. Chem.& Eng.,* **2018**, *6*(7), 8591-8598. [http://dx.doi.org/10.1021/acssuschemeng.8b00908]

[117] Zhang, Y.; Heo, Y.J.; Lee, J.W.; Lee, J.H.; BajGai, J.; Lee, K.J.; Park, S.J. Photocatalytic Hydrogen Evolution *via* Water Splitting: A Short Review. *Catalysts,* **2018**, *8*(12), 655. [http://dx.doi.org/10.3390/catal8120655]

[118] Fu, C.F.; Wu, X.; Yang, J. Material design for photocatalytic water splitting from a theoretical perspective. *Adv. Mater.,* **2018**, *30*(48), 1802106. [http://dx.doi.org/10.1002/adma.201802106] [PMID: 30328641]

[119] Fajrina, N.; Tahir, M. A critical review in strategies to improve photocatalytic water splitting towards hydrogen production. *Int. J. Hydrogen Energy,* **2019**, *44*(2), 540-577. [http://dx.doi.org/10.1016/j.ijhydene.2018.10.200]

[120] Lin, L.; Hisatomi, T.; Chen, S.; Takata, T.; Domen, K. Visible-light-driven photocatalytic water splitting: recent progress and challenges. *Trends Chem.,* **2020**, *2*(9), 813-824. [http://dx.doi.org/10.1016/j.trechm.2020.06.006]

[121] Cao, S.; Piao, L. Considerations for a more accurate evaluation method for photocatalytic water

splitting. *Angew. Chem. Int. Ed.,* **2020**, *59*(42), 18312-18320.
[http://dx.doi.org/10.1002/anie.202009633] [PMID: 32902089]

[122] Mohamed, R.M.; Kadi, M.W.; Alhothali, H.T. Polyaniline@$BaZrO_3$ nanocomposites for hydrogen production under visible light. *Nanosci. Nanotechnol. Lett.,* **2019**, *11*(11), 1513-1521.
[http://dx.doi.org/10.1166/nnl.2019.3037]

[123] Gulen, B.; Demircivi, P.; Simsek, E.B. UV-A light irradiated photocatalytic performance of hydrothermally obtained W doped $BaZrO_3$ catalyst against the degradation of levofloxacin and tetracycline antibiotic. *J. Photochem. Photobiol. Chem.,* **2021**, *404*(404), 112869.
[http://dx.doi.org/10.1016/j.jphotochem.2020.112869]

[124] Ullah, H.M.N.; Rizwan, M.; Ali, S.S.; Usman, Z.; Cao, C. A DFT study of optical, elastic, mechanical, and overall water-splitting photocatalytic properties of pristine and Cd substituted $BaZrO_3$: A lead free environment friendly material. *Mater. Sci. Eng. B,* **2022**, *286*, 116041.
[http://dx.doi.org/10.1016/j.mseb.2022.116041]

[125] Chen, T.; Meng, J.; Lin, Q.; Wei, X.; Li, J.; Zhang, Z. One-step synthesis of hollow $BaZrO_3$ nanocrystals with oxygen vacancies for photocatalytic hydrogen evolution from pure water. *J. Alloys Compd.,* **2019**, *780*, 498-503.
[http://dx.doi.org/10.1016/j.jallcom.2018.11.367]

[126] Meng, J.; Lan, Z.; Lin, Q.; Chen, T.; Chen, X.; Wei, X.; Lu, Y.; Li, J.; Zhang, Z. Cubic-like $BaZrO_3$ nanocrystals with exposed 001/011 facets and tuned electronic band structure for enhanced photocatalytic hydrogen production. *J. Mater. Sci.,* **2019**, *54*(3), 1967-1976.
[http://dx.doi.org/10.1007/s10853-018-2995-8]

[127] Chen, Y.C.; Huang, Y.S.; Huang, H.; Su, P.J.; Perng, T.P.; Chen, L.J. Photocatalytic enhancement of hydrogen production in water splitting under simulated solar light by band gap engineering and localized surface plasmon resonance of $Zn_xCd_{1-x}S$ nanowires decorated by Au nanoparticles. *Nano Energy,* **2020**, *67*, 104225.
[http://dx.doi.org/10.1016/j.nanoen.2019.104225]

[128] Zhang, Y.; Park, M.; Kim, H.Y.; Park, S.J. *In-situ* synthesis of graphene oxide/BiOCl heterostructured nanofibers for visible-light photocatalytic investigation. *J. Alloys Compd.,* **2016**, *686*, 106-114.
[http://dx.doi.org/10.1016/j.jallcom.2016.06.004]

[129] Huang, J.; Li, G.; Zhou, Z.; Jiang, Y.; Hu, Q.; Xue, C.; Guo, W. Efficient photocatalytic hydrogen production over Rh and Nb codoped TiO_2 nanorods. *Chem. Eng. J.,* **2018**, *337*, 282-289.
[http://dx.doi.org/10.1016/j.cej.2017.12.088]

[130] Kim, W.; Monllor-Satoca, D.; Chae, W.S.; Mahadik, M.A.; Jang, J.S. Enhanced photoelectrochemical and hydrogen production activity of aligned CdS nanowire with anisotropic transport properties. *Appl. Surf. Sci.,* **2019**, *463*, 339-347.
[http://dx.doi.org/10.1016/j.apsusc.2018.08.127]

[131] Dai, F.; Wang, Y.; Zhao, R.; Zhou, X.; Han, J.; Wang, L. $ZnIn_2S_4$ modified $CaTiO_3$ nanocubes with enhanced photocatalytic hydrogen performance. *Int. J. Hydrogen Energy,* **2020**, *45*(53), 28783-28791.
[http://dx.doi.org/10.1016/j.ijhydene.2020.07.228]

[132] Khan, Z.; Qureshi, M. Tantalum doped $BaZrO_3$ for efficient photocatalytic hydrogen generation by water splitting. *Catal. Commun.,* **2012**, *28*, 82-85.
[http://dx.doi.org/10.1016/j.catcom.2012.08.002]

[133] Patra, A.S.; Chauhan, M.S.; Keene, S.; Gogoi, G.; Reddy, K.A.; Ardo, S.; Prasad, D.L.V.K.; Qureshi, M. Combined experimental and theoretical insights into the synergistic effect of cerium doping and oxygen vacancies in bazro $_{3-\delta}$ hollow nanospheres for efficient photocatalytic hydrogen production. *J. Phys. Chem. C,* **2019**, *123*(1), 233-249.
[http://dx.doi.org/10.1021/acs.jpcc.8b10626]

[134] Khan, Z.; Khannam, M.; Vinothkumar, N.; De, M.; Qureshi, M. Hierarchical 3D NiO–CdS heteroarchitecture for efficient visible light photocatalytic hydrogen generation. *J. Mater. Chem.,*

2012, *22*(24), 12090-12095.
[http://dx.doi.org/10.1039/c2jm31148h]

[135] Khan, Z.; Chetia, T.R.; Qureshi, M. Rational design of hyperbranched 3D heteroarrays of SrS/CdS: synthesis, characterization and evaluation of photocatalytic properties for efficient hydrogen generation and organic dye degradation. *Nanoscale,* **2012**, *4*(11), 3543-3550.
[http://dx.doi.org/10.1039/c2nr30666b] [PMID: 22562405]

[136] Barpuzary, D.; Khan, Z.; Vinothkumar, N.; De, M.; Qureshi, M. Hierarchically grown urchinlike CdS@ZnO and CdS@Al_2 O_3 heteroarrays for efficient visible-light-driven photocatalytic hydrogen generation. *J. Phys. Chem. C,* **2012**, *116*(1), 150-156.
[http://dx.doi.org/10.1021/jp207452c]

[137] Gundeboina, R.; Perala, V.; Muga, V. Perovskite material-based photocatalysts. *Revolution of Perovskite*; Springer, **2020**, pp. 251-287.
[http://dx.doi.org/10.1007/978-981-15-1267-4_9]

[138] Luo, M.; Lu, P.; Yao, W.; Huang, C.; Xu, Q.; Wu, Q.; Kuwahara, Y.; Yamashita, H. Shape and composition effects on photocatalytic hydrogen production for Pt–Pd alloy cocatalysts. *ACS Appl. Mater. Interfaces,* **2016**, *8*(32), 20667-20674.
[http://dx.doi.org/10.1021/acsami.6b04388] [PMID: 27439590]

[139] Qin, J.; Zeng, H. Photocatalysts fabricated by depositing plasmonic Ag nanoparticles on carbon quantum dots/graphitic carbon nitride for broad spectrum photocatalytic hydrogen generation. *Appl. Catal. B,* **2017**, *209*, 161-173.
[http://dx.doi.org/10.1016/j.apcatb.2017.03.005]

[140] Rather, R.A.; Singh, S.; Pal, B. A C_3N_4 surface passivated highly photoactive Au-TiO_2 tubular nanostructure for the efficient H_2 production from water under sunlight irradiation. *Appl. Catal. B,* **2017**, *213*, 9-17.
[http://dx.doi.org/10.1016/j.apcatb.2017.05.002]

[141] Akhtar, S.; Alay-e-Abbas, S.M.; Batool, J.; Zulfiqar, W.; Laref, A.; Abbas, G.; Amin, N. Investigation of structural, electronic and optical properties of (V+P)-doped $BaZrO_3$ for photocatalytic applications using density functional theory. *J. Phys. Chem. Solids,* **2020**, *147*, 109662.
[http://dx.doi.org/10.1016/j.jpcs.2020.109662]

[142] Saeed, S.W.; Norby, T.; Bjørheim, T.S. Charge-carrier enrichment at $BaZrO_3$/$SrTiO_3$ interfaces. *J. Phys. Chem. C,* **2019**, *123*(34), 20808-20816.
[http://dx.doi.org/10.1021/acs.jpcc.9b06296]

[143] Maeda, K.; Domen, K. Water oxidation using a particulate $BaZrO_3$-$BaTaO_2N$ solid-solution photocatalyst that operates under a wide range of visible light. *Angew. Chem. Int. Ed.,* **2012**, *51*(39), 9865-9869.
[http://dx.doi.org/10.1002/anie.201204635] [PMID: 22951897]

[144] Tasleem, S.; Tahir, M. Recent progress in structural development and band engineering of perovskites materials for photocatalytic solar hydrogen production: A review. *Int. J. Hydrogen Energy,* **2020**, *45*(38), 19078-19111.
[http://dx.doi.org/10.1016/j.ijhydene.2020.05.090]

[145] Maksoud, M.I.A.A.; Kassem, S.M.; Bekhit, M.; Fahim, R.A.; Ashour, A.H.; Awed, A.S. Gamma radiation shielding properties of poly(vinyl butyral)/Bi_2O_3@$BaZrO_3$ nanocomposites. *Mater. Chem. Phys.,* **2021**, *268*, 124728.
[http://dx.doi.org/10.1016/j.matchemphys.2021.124728]

[146] Abdel Maksoud, M.I.A.; Kassem, S.M.; Sallam, O.I. Structural, optical, and radiation shielding features of newly developed $BaZrO_3$/Na_2O–B_2O_3 glass. *Ceram. Int.,* **2022**, *48*(20), 30938-30950.
[http://dx.doi.org/10.1016/j.ceramint.2022.07.051]

[147] Verma, S.; Sarma, B.; Chaturvedi, K.; Malvi, D.; Srivastava, A.K. Emerging graphene and carbon nanotube-based carbon composites as radiations shielding materials for X-rays and gamma rays: a

review. *Compos. Interfaces,* **2023**, *30*(2), 223-251.
[http://dx.doi.org/10.1080/09276440.2022.2094571]

[148] Sakhil, M.D. Study of the effect of radiology x-ray and its risks on human. *Int. J. Sci. Trends,* **2022**, *1*(1), 1-11.

[149] Abulyazied, D.E.; Saudi, H.A.; Zakaly, H.M.H.; Issa, S.A.M.; Henaish, A.M.A. Novel nanocomposites based on polyvinyl alcohol and molybdenum nanoparticles for Gamma irradiation shielding. *Opt. Laser Technol.,* **2022**, *156*, 108560.
[http://dx.doi.org/10.1016/j.optlastec.2022.108560]

[150] Khozemy, E.E.; Salem, E.F.; El-Hag Ali, A. Radiation shielding and enhanced thermal characteristics of high-density polyethylene reinforced with Al $(OH)_3/Pb_2O_3$ for radioactive waste management. *Radiat. Phys. Chem.,* **2022**, *193*, 109976.
[http://dx.doi.org/10.1016/j.radphyschem.2022.109976]

[151] Abbasova, N.; Yüksel, Z.; Abbasov, E.; Gülbiçim, H.; Tufan, M.Ç. Investigation of gamma-ray attenuation parameters of some materials used in dental applications. *Results Phys.,* **2019**, *12*, 2202-2205.
[http://dx.doi.org/10.1016/j.rinp.2019.02.068]

[152] Sayyed, M.I.; Abdalsalam, A.H.; Taki, M.M.; Mhareb, M.H.A.; Alim, B.; Baltakesmez, A.; Şakar, E. MoO3 reinforced ultra high molecular weight PE for neutrons shielding applications. *Radiat. Phys. Chem.,* **2020**, *172*, 108852.
[http://dx.doi.org/10.1016/j.radphyschem.2020.108852]

[153] Daw, H.T.; Jaeger, R.G. The international atomic energy agency and radiology. *Can. Med. Assoc. J.,* **1962**, *87*(8), 389-397.
[PMID: 13884012]

[154] Sievert, R.M. *Report on Decisions at the 1959 Meeting of the International Commission on Radiological Protection (ICRP). Radiology*; The Radiological Society of North America, **1960**, pp. 116-119.
[http://dx.doi.org/10.1148/74.1.116]

[155] Abdel-Gawwad, H.A.; Mohammed, M.S.; Zakey, S.E. Preparation, performance, and stability of alkali-activated-concrete waste-lead- bearing sludge composites. *J. Clean. Prod.,* **2020**, *259*, 120924.
[http://dx.doi.org/10.1016/j.jclepro.2020.120924]

[156] Abdel-Haseiba, A.; Ahmeda, Z.; Hassanb, M.M. Investigation of the gamma rays attenuation coefficients by experimental and mcnp simulation for polyamide 6/acrylonitrile-butadiene--styrene blends. *J Nucl Radiat Phys,* **2018**, *13*(1), 81-89.https://www.researchgate.net/publication/328164349

[157] Kacal, M.R.; Akman, F.; Sayyed, M.I. Investigation of radiation shielding properties for some ceramics. *Radiochim. Acta,* **2019**, *107*(2), 179-191.
[http://dx.doi.org/10.1515/ract-2018-3030]

[158] Akman, F.; Khattari, Z.Y.; Kaçal, M.R.; Sayyed, M.I.; Afaneh, F. The radiation shielding features for some silicide, boride and oxide types ceramics. *Radiat. Phys. Chem.,* **2019**, *160*, 9-14.
[http://dx.doi.org/10.1016/j.radphyschem.2019.03.001]

[159] Jawad, A.A.; Demirkol, N.; Gunoğlu, K.; Akkurt, I. Radiation shielding properties of some ceramic wasted samples. *Int. J. Environ. Sci. Technol.,* **2019**, *16*(9), 5039-5042.
[http://dx.doi.org/10.1007/s13762-019-02240-7]

[160] Usta, M.; Tozar, A. The effect of the ceramic amount on the radiation shielding properties of metal-matrix composite coatings. *Radiat. Phys. Chem.,* **2020**, *177*, 109086.
[http://dx.doi.org/10.1016/j.radphyschem.2020.109086]

[161] Malekzadeh, R.; Zali, V.S.; Jahanbakhsh, O.; Okutan, M.; Mesbahi, A. The preparation and characterization of silicon-based composites doped with $BaSO_4$, WO_3, and PbO nanoparticles for shielding applications in positron emission tomography (PET) and nuclear medicine facilities.

Nanomed. J., **2020**, *7*(4), 324-334.
[http://dx.doi.org/10.22038/nmj.2020.07.00009]

[162] Ghrib, T.; Mhareb, M.H.A.; Sayyed, M.I.; Alajerami, Y.S.M.; Dwaikat, N.; Ben Ali, A.; Gondal, M.A. Structural, optical and radiation shielding properties of Zirconium–Titanium–Thallium Ternary Oxide (0.5ZrO2-(0.5-x)TiO2-xTl2O3). *Ceram. Int.,* **2021**, *47*(15), 21837-21847.
[http://dx.doi.org/10.1016/j.ceramint.2021.04.201]

[163] Marí, B.; Cembrero-Coca, P.; Singh, K.C.; Kaushik, R.D.; Om, H. Preparation and luminescence properties of MZrO3:Eu3+, A (M=Ca2+, Ba2+; A=Li+, Na+, K+) phosphors with perovskite structure. *Wuli Huaxue Xuebao,* **2013**, *29*(6), 1357-1362.
[http://dx.doi.org/10.3866/PKU.WHXB201304032]

[164] Ozel, F.; Akman, F.; Kaçal, M.R.; Ozen, A.; Arslan, H.; Polat, H.; Yurtcan, S.; Agar, O. Production of microstructured $BaZrO_3$ and $Ba_2P_2O_7$-based polymer shields for protection against ionizing photons. *J. Phys. Chem. Solids,* **2021**, *158*(January), 110238.
[http://dx.doi.org/10.1016/j.jpcs.2021.110238]

[165] More, C.V.; Alsayed, Z.; Badawi, M.S.; Thabet, A.A.; Pawar, P.P. Polymeric composite materials for radiation shielding: a review. *Environ. Chem. Lett.,* **2021**, *19*(3), 2057-2090.
[http://dx.doi.org/10.1007/s10311-021-01189-9] [PMID: 33558806]

[166] Maksoud, M.I.A.A.; Kassem, S.M.; Bekhit, M.; Fahim, R.A.; Ashour, A.H.; Awed, A.S. Gamma radiation shielding properties of poly(vinyl butyral)/Bi_2O_3@$BaZrO_3$ nanocomposites. *Mater. Chem. Phys.,* **2021**, *268*(March), 124728.
[http://dx.doi.org/10.1016/j.matchemphys.2021.124728]

[167] McKeen, L.W. *Introduction to the Properties of Plastic and Elastomer Films*; Film Prop. Plast. Elastomers, **2012**, pp. 19-55.
[http://dx.doi.org/10.1016/B978-1-4557-2551-9.00002-5]

[168] Rizwan, M.; Aleena, S.; Shakil, M.; Mahmood, T.; Zafar, A.A.; Hussain, T.; Farooq, M.H. A computational insight of electronic and optical properties of Cd-doped $BaZrO_3$. *Zhongguo Wuli Xuekan,* **2020**, *66*, 318-326.
[http://dx.doi.org/10.1016/j.cjph.2020.04.022]

[169] Azar, S.A.; Al-Zoubi, I.; Mousa, A.A.; Masharfe, R.S.; Jaradat, E.K. Investigation of electronic, optical and thermoelectric properties of perovskite $BaTMO_3$ (TM=Zr, Hf): First principles calculations. *J. Alloys Compd.,* **2021**, *887*, 161361.
[http://dx.doi.org/10.1016/j.jallcom.2021.161361]

[170] Kajfez, D.; Kishk, A.A. Dielectric resonator antenna - possible candidate for adaptive antenna arrays Darko Kajfez and Ahmed A. Kishk. *Proceedings VITEL 2002, International Symposium on Telecommunications, Next Generation Networks and Beyond,* **2002**, pp. 13-14.

[171] Zhang, Y.; Liu, M.; Wang, J.; Shimada, T.; Kitamura, T. Strain tunable ferroelectric and dielectric properties of $BaZrO_3$. *J. Appl. Phys.,* **2014**, *115*(22), 224107.
[http://dx.doi.org/10.1063/1.4883298]

[172] Khirade, P.P.; Birajdar, S.D.; Humbe, A.V.; Jadhav, K.M. Structural, electrical and dielectrical property investigations of Fe-Doped $BaZrO_3$ nanoceramics. *J. Electron. Mater.,* **2016**, *45*(6), 3227-3235.
[http://dx.doi.org/10.1007/s11664-016-4472-y]

[173] Charoonsuk, T.; Kolodiazhnyi, T.; Vittayakorn, N. The modification of surface, size and shape of barium zirconate powder *via* salt flux. *J. Am. Ceram. Soc.,* **2019**, *102*(10), 5772-5785.
[http://dx.doi.org/10.1111/jace.16495]

[174] Parida, S.; Rout, S.K.; Cavalcante, L.S.; Sinha, E.; Li, M.S.; Subramanian, V.; Gupta, N.; Gupta, V.R.; Varela, J.A.; Longo, E. Structural refinement, optical and microwave dielectric properties of $BaZrO_3$. *Ceram. Int.,* **2012**, *38*(3), 2129-2138.
[http://dx.doi.org/10.1016/j.ceramint.2011.10.054]

[175] Srikanth, C.; Madhu, G.M. Synthesis, characterization and properties evaluation of ZrO_2 and its composites A review. *SSRN,* **2018**.
[http://dx.doi.org/10.2139/ssrn.3101419]

[176] Suhaib, S.A.; Zhu, A.K.; Sun, J.; Guo, Y.M.; Wang, S.T.; Wang, C.C. High-temperature dielectric relaxations in (Ce, Y) codoped $BaZrO_3$ ceramics. *Mater. Res. Express,* **2019**, *6*(7), 075010.
[http://dx.doi.org/10.1088/2053-1591/ab10be]

CHAPTER 3

A High-capacity Anode Material for Lithium-ion Batteries is Sili-graphene Type SiC_3

M. Houmad[1,*], M. Khuili[2], A. El Kenz[1] and **A. Benyoussef[1]**

[1] *Laboratory of Condensed Matter and Interdisciplinary Sciences (LaMScI), Faculty of Science, Mohammed V University of Rabat, Rabat, Morocco*

[2] *Superior School of Technology (EST-Khenifra), University of Sultan Moulay Slimane, Khenifra, 54000, Morocco*

Abstract: In this study, we have proposed an anode material based on Silicon doped graphene (Siligraphene) for developing the Li-ion batteries (LIBs). We have predicted that Siligraphene can be an anode material for lithium batteries. In particular, we have found that the Siligraphene sheet can adsorb lithium atoms in different sites in a hexagonal structure. Also, we have found that Lithium atoms can be diffused along the plane of siligraphene. The energy of diffusion of siligraphene (SiC_3) is about 0.095eV, and for Li on top of silicon atoms is about 0.223eV, indicating rapid charging/discharging processes. During charging and discharging, the electrode Li_xSiC_3 exhibits small variations in voltage, making them a potential candidate for Li-ion batteries.

Keywords: Capacity, Charging/discharging processes, Diffusion barrier, DFT, Lithium batteries, Voltage.

INTRODUCTION

A key challenge for rechargeable Li batteries is to propose an appropriate theory to realize the experimental results for a novel application as an anode material with high efficiency and good performance [1 - 3]. The research in this area is devoted to developing a new rechargeable Libattery to overcome the energy crisis and provide sustainable energy resources [4, 5]. The anode of Li-battery was developed in 1979 by Yazami *et al.* [6] and successfully launched in the market during the early 1990s by Sony Corporation [7]. After that, great efforts have been ade to improve the stability and efficiency [8, 9]. Also, it has paved the way in a

* **Corresponding author M. Houmad:** Laboratory of Condensed Matter and Interdisciplinary Sciences (LaMScI), Faculty of Science, Mohammed V University of Rabat, Rabat, Morocco; Tel: +212678660742;
E-mail: houmadmohamed@gmail.com

great deal for the development and investigation of new materials that can possibly be integrated into Li batteries. Intensive studies were done after the discovery due to the following reasons; their good life [10], higher reversible capacity [11, 12], and higher energy density [13, 14]. Graphite is used in commercial batteries; the Li-ion has a storage capacity between two layers of graphite because it has a charge capacity of ~372 mA h/g with acceptable reversibility as well [15, 16]. In addition, the demand for its application is increasing in different areas, such as portable electronic devices and electric vehicles [17, 18]. The growing demand for market valuation and rigorous research conducted in parallel may improve the capacity of lithium-ion batteries [19].

The recent development for fabricating the Li-battery used 2D materials, such as the transition metal dichalcogenides (TMDs) [20, 21], which are used thanks to their high charge capacity, dimensionality, and structural flexibility [22]. Graphene battery is similar to traditional lead-batteries in that it is composed of an electrolyte solution between the two electrodes to facilitate the transfer of ions in solution between the electrodes. Graphene-based batteries are quickly becoming more favorable than their graphite predecessors. A variety of graphene-based batteries has been developing in their nanocomposites, such as Li-ion batteries [23 - 26], Na-ion batteries [27 - 29], Zn-batteries [30, 31], K-ion battery [32] and Li-S battery [33 - 35]. The batteries are composed of three parts: the anodes, cathodes, and interface, to improve the performance of batteries. It is necessary to improve all three parts for better performance. The mechanical instability and the structural phase transition due to Li insertion in 2D materials have outclassed the performance of all batteries. Moreover, the improvement of life durability and reversible capacity is still a focal point of core research.

The presence of remarkable chemical and physical properties makes this material a promising component in different technological devices such as conversion systems and advanced energy storage [36]. The graphene anodes have already been used for Lithium batteries, and several research groups are involved in the development of their storage capacity [37, 38]. These successes in the application of graphene have encouraged the scientific community to invent a new class of graphene-like 2D materials [39 - 41]. At the same time, the discovery of graphene has brought a great change in the field of electronics. For example, the construction of the novel nano-transistors is based on graphene [42, 43]. However, graphene is a gapless material, and many efforts are made to open the band gap. Avetisyan *et al.* and Mak *et al.* apply the electric field to open the band gap of a graphene [44]. Denis *et al.* and Dai *et al.* studied the variation of physical properties of heteroatom-doped graphene material [45, 46]. Houmad *et al.* have doped silicon on graphene to tune the electronic properties [47] by opening the band gaps from zero to 2.51 eV as a function of silicon concentration. Similar facts have already been confirmed by Z. Shi *et al.* [48, 49]. Wu *et al.* experimentally verified the results of the opening of band gap (0.13-0.25eV) on silicon doped with a concentration level of 2.7-4.7% by using electrical measurements of field effect transistor. The presence of a band gap in the range of ultraviolet has been confirmed by the photoelectron spectroscopy measurement. This result gives a new hope for its possible application in solar cell devices. In addition, it is possible to control the band gap with silicon doping graphene. Also, they found that silicon has increased the optical conductivity in the visible light region. Denis *et al.* have successfully demonstrated the opening of the band gap in graphene (monolayer or bilayer) by doping with aluminum, silicon, phosphorus or sulfur [50]. Other groups used the chemical functionalization [51].

Here, we have proposed sili-graphene as a novel anode material for Li-ion batteries (LIBs). Based on the first principles calculation, we will study the diffusion of Li-ion on the siligraphene as an anode material. The energy of diffusion of siligraphene (SiC_3) is rapid in charging/discharging processes. Also, the electrode of the siligraphene has a small variation of voltage in Li_xSiC_3 siligraphene during charging and discharging. This best advantage makes this 2D-material modified (g-SiC_3) a good candidate for Li-ion batteries.

COMPUTATIONAL DETAILS

The structural and electronic properties are investigate using the plane-wave projector-augmented wave method as implemented in the QUANTUM-ESPRESSO package [52]. Electron–ion interactions are described using ultrasoft pseudo potentials with a kinetic energy cut-off of 38 Ry, and the convergence criteria for energy are assumed to be 10^{-8} Ry. We perform Brillouin-zone integrations using Monkhorst-Pack grids of special points with 10x10x2 meshes for the calculations of structural and electronic properties. For structural relaxation, lattice constants and atomic coordinates were optimized based on the PBE Perdew Burke Ernzerhof [53, 54]. The band structures obtained from the first principles are key factors in determining all the carrier transport properties.

RESULTS AND DISCUSSION

Electronic Properties of Li Adsorption on the Siligraphene

After relaxing the structure of Li adsorption on the Siligraphene we have got the following information: the first observation on the surface of siligraphene has become a buckled structure. The buckling distance between the carbon and silicon atoms along a Z-axis is given by $\Delta l = +0.36A°$, and the distance between Lithium atoms and the surface of siligraphene is $\Delta l = +1.75A°$ (Fig. **1**).

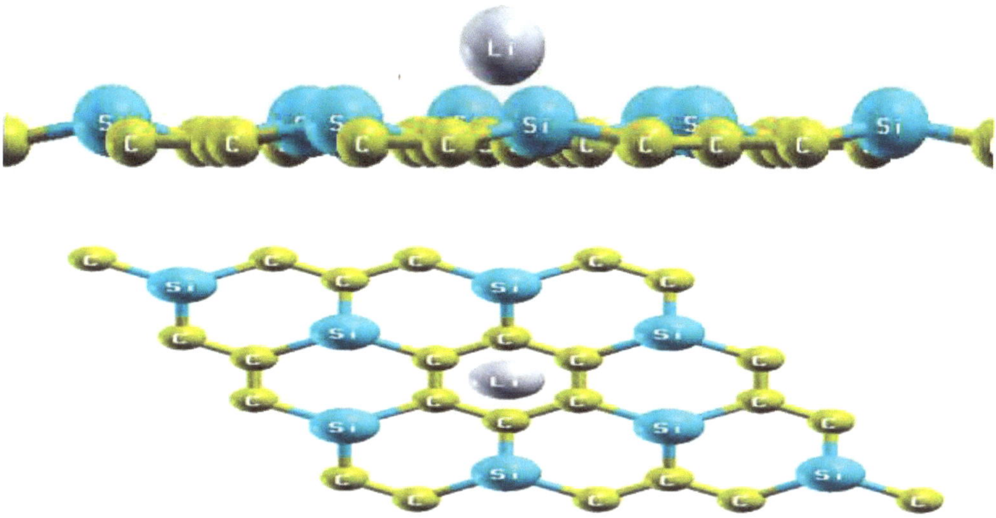

Fig. (1). Schematics of optimized free siligraphene SiC_3**(a)** Top and **(b)** Side views, respectively. H1 hollow sites contain the atoms Carbon (C) And Silicon (Si). H_2 hollow sites contain the Carbon (C) atoms.

To estimate the strength of the interaction between Li and Sili-graphene type SiC_3, we have calculated the adsorption energy (E_{ads}) [55] by using Eq. (1).

$$nE_{ads} = E_{system} - E_{SiC_3} - nE_x \quad (1)$$

where E_{SiC_3} is the energy of siligraphene SiC_3, E_{Li} is the energy of isolated atoms of Li calculated in a cubic lattice cell, "n" a number of Li atoms adsorb on the surface of SiC_3 siligraphene, and E_{system} is the total energy of the Li_xSiC_3 system.

Our study shows that the binding energies for Li_xSiC_3, ($0<x<1$) are presented by the blue line in Fig. (**2**). The Li adsorption on the structures of siligraphene (Li_xSiC_3) has a range of binding energies between −1.80 to −0.2eV. Analyzing the calculated results, we have concluded that 2D SiC_3siligraphene is a favorable material to adsorb Li-atoms on their surface. We have seen that the Li atoms adsorbed on siligraphene SiC_3 are energetically stable and favourable for anode materials.

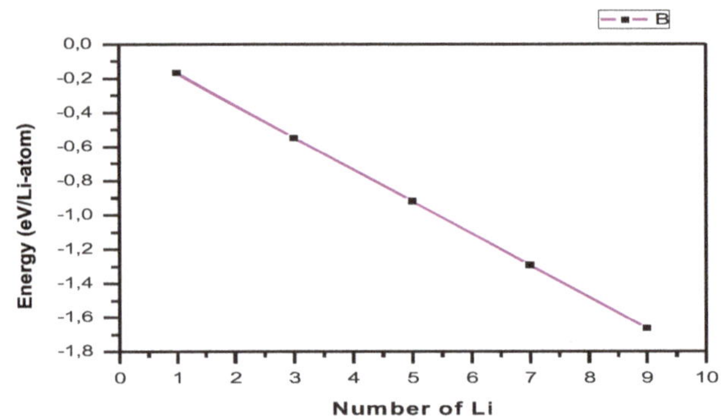

Fig. (2). Calculated adsorption energy for Li on the surface of siligraphene (SiC_3).

When comparing the total density of states of Li adsorption on the siligraphene SiC_3. Fig. (**3b**) with the total density of states of pure siligraphene SiC_3. Fig. (**3a**); we have observed a change in the Fermi level for Li-SiC_3 exhibiting metallic behavior.

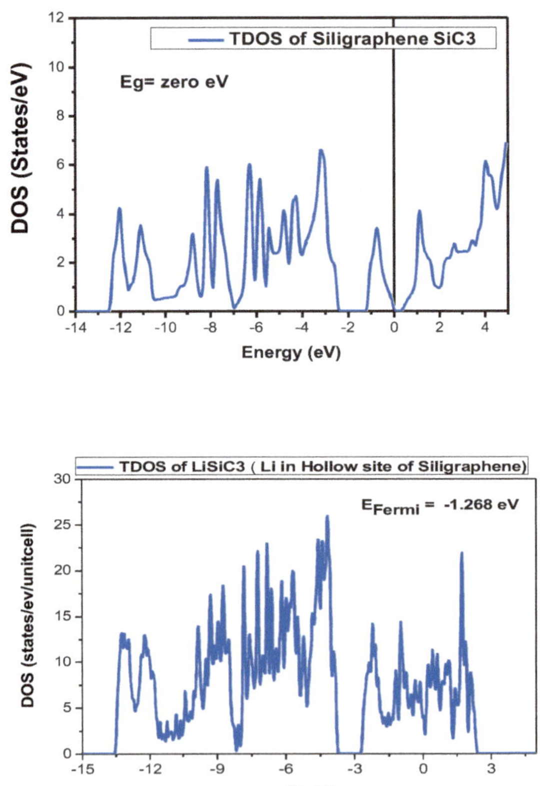

Fig. (3). **(a)** Total density of states of siligraphene SiC_3, **(b)** Total density of Li adsorption on the siligraphene SiC_3.

Li Adsorption and Diffusion on the Siligraphene

In this section, we have plotted the different positions of the Li-atom in the Siligraphene [56 - 59] structure (thickness of a single layer). In order to examine the binding energy of the siligraphene surface with Li, we have placed a single ad-atom at five different adsorption sites, as shown in Fig. (**4**). We have reported the most stable configurations as: (i) vertical distances between the Li ad-atom and Carbon (C) Top d_{Li-C} = 1.7115A°, (ii) between Li and Si Top d_{Li-Si} = 1.3459A°, (iii) silicon atom H2 d_{Li-Si} = 2.3374A°, (iv) carbon atom H2 d_{Li-C} = 2.4604A°, (v) carbon Atom H1 d_{Li-C} = 2.2018A°. Vertical height (h) at all sites and distances (dP – A) from the siligraphene to the Li ad-atom position at the different sites (T1: on top of Carbon atoms; T2: on top of Silicon atoms; H1: hollow of Carbon hexagonal on siligraphene; H2: hollow of Silicon-carbon hexagonal on siligraphene) are all displayed in Table **1**.

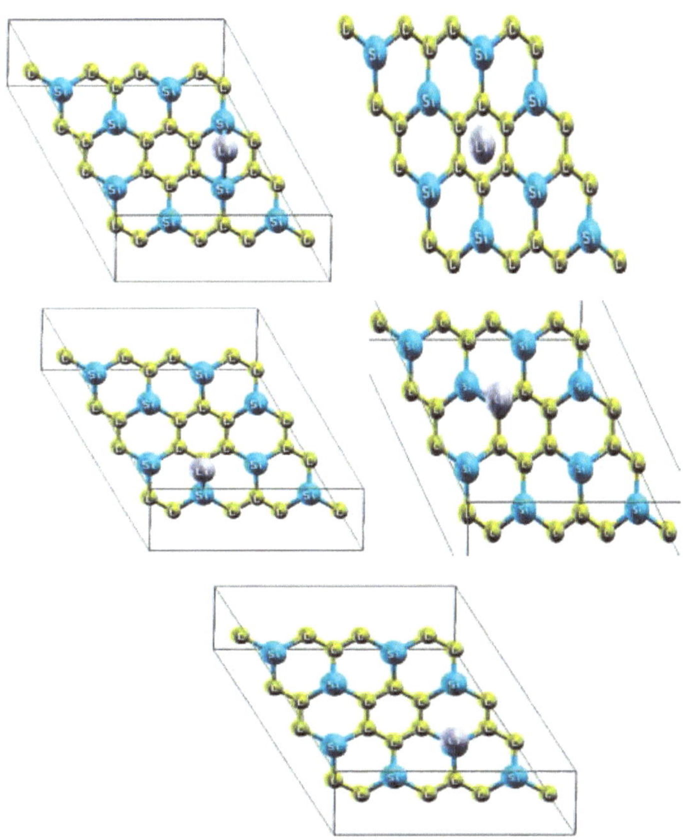

Fig. (4). Schematics of optimized all positions of lithium onsiligraphene (g-SiC_3).

Table 1. Absolute binding energies *E*b at different adsorption sites, vertical distance (*h*) at the H-site.

-	System (Ry)	Substrate (Ry)	Eb (Ry)	Eb (eV)
H1	-335,95747831	-335,22843455	-0,17262468	-2,3486792748
T1 (C)	-335,88921143	-335,22843455	-0,1043578	-1,4198607176
B2 (C-Si)	-335,82106147	-335,22843455	-0,03620784	-0,4926329387
T2 (Si)	-335,33632645	-335,22843455	0,44852718	6,1025253855
H2	-335,84680505	-335,22843455	-0,06195142	-0,8428923153

Li Diffusion Process

The actual mechanism of a Lithium battery is based on the mobility of lithium-ion in Siligraphene and the formation of the energy barrier. The possible diffusion path of Li on siligraphene is crucial. Before calculating the diffusion of lithium ion on siligraphene, it's necessary to deduce the most stable site of Li on siligraphene. We have four principal sites: two bridges, between atoms C-C and between Si-C and two-hole in hexagonal constructed with Carbone atoms called H_1 and the second hole, called $H_{2,}$ is constituted of a hexagonal with silicon and carbon atoms. For calculating the diffusion between two adjacent sites; there should exist a difference in the potential energy between the two sites. The diffusion barrier is very low when Li is placed at the top of the carbon atom ~0.095eV, but it is much higher for Li on the top of the silicon atom ~0.223 eV.

The diffusion barriers along H-H, b -H and H-B-H direction are 0.09 eV and 0.223eV, respectively, which is consistent with the previous study [56]. It has been observed that Li can diffuse faster along the groove than between two adjacent grooves, like the water flow inside the groove (Fig. **6**).

Voltage Profiles as a Function of Capacity for Siligraphene $LixSiC_3$

While comparing with other 2D materials, siligraphene (SiC_3) shows a fast charge-discharge capability for Lithium. We have also calculated the voltage profile of lithium adsorption on siligraphene (Li_xSiC_3) using the following equation (2) [60, 61]:

$$V = [-E(Li_{x2}SiC_3) + E(Li_{x1}SiC_3) + (x_2-x_1)E(M)]/(x_2-x_1) \quad \textbf{(2)}$$

$E_{(LixSiC3)}$, $E_{(Lix0SiC3)}$ and $E_{(Li)}$ are the total energies of modified materials $Li_{x2}SiC_3$, $Li_{x1}SiC_3$ respectively. Where x1 and x2 are the number of Li-ions adsorbed on the on the surface of SiC_3 unit cell. We calculated the specific capacities for $LixSiC_3$ using the following equation [62].

$$C = (nN_a e)/(3.6*M(Li_xSiC_3)) \quad (3)$$

In which $M_{LixSiC3}$ represents the molar mass of Li_xSiC_3 and, N_a is the Avogadro constant, n is the adsorbed ion Li, finally, the elementary charge (e) in coulomb, our results are plotted in Fig. (7).

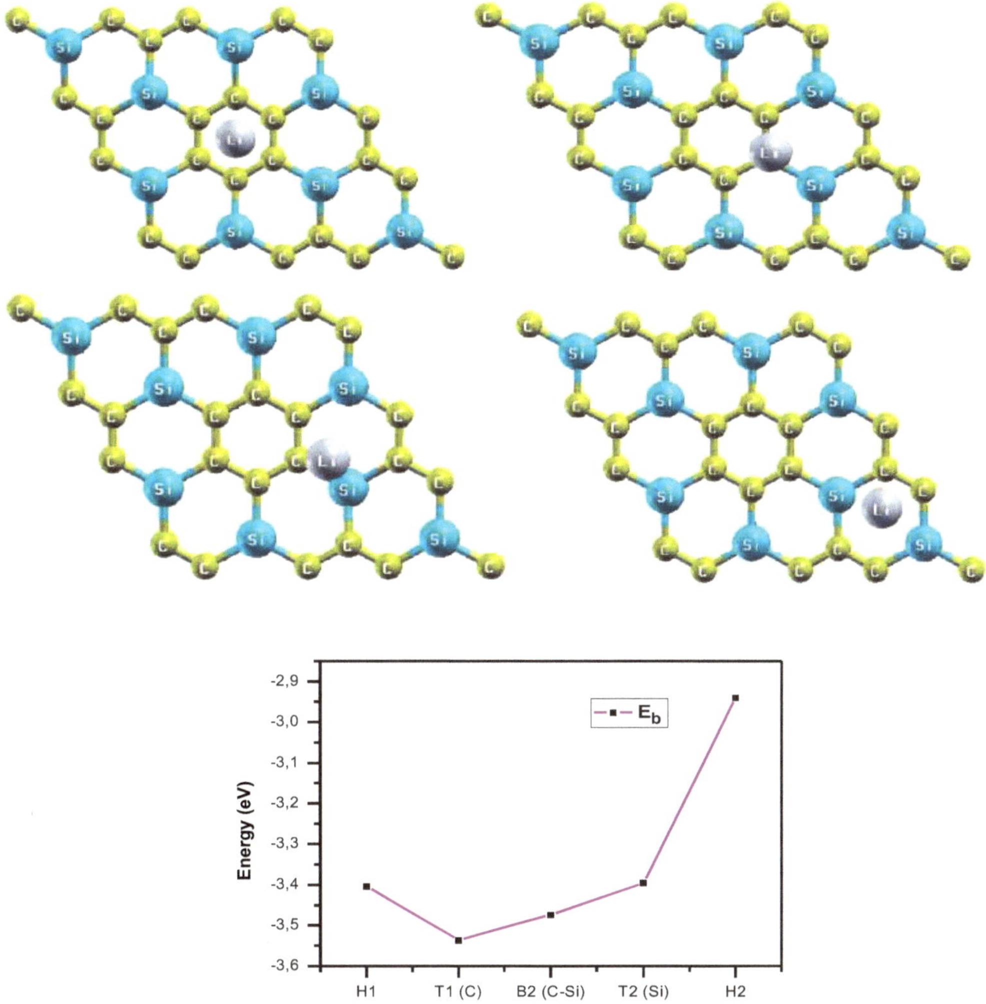

Fig. (5). The total binding energies of Li adatoms at all sites on siligraphene.

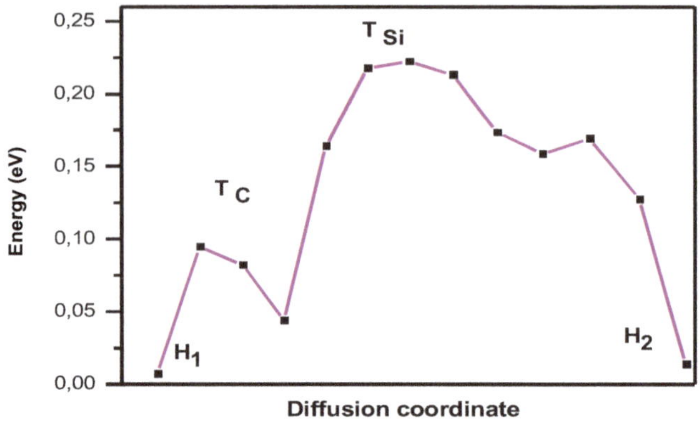

Fig. (6). The diffusion barriers.

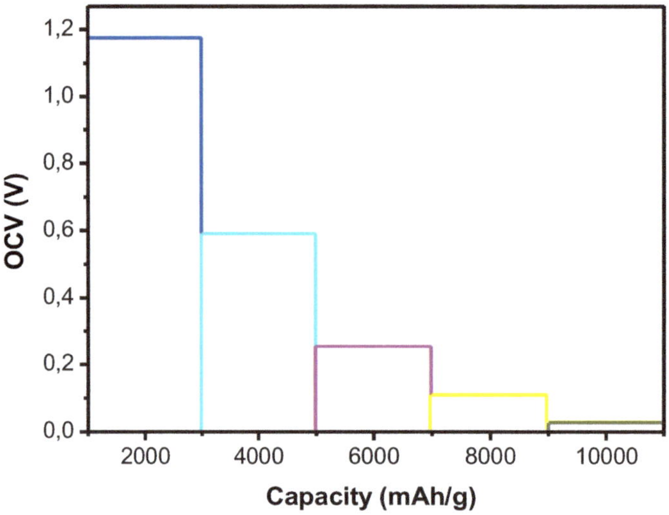

Fig. (7). Voltage profiles as a function of capacity for siligraphene Li_xSiC_3.

To fulfill this aim, we have created a series of configurations of siligraphene with the chemical stoichiometry of Li_xSiC_3 ($0 < x < 1$) by insertion of different numbers of metal atoms on the surface of siligraphene SiC_3 (Fig. **8**). The calculated voltage profiles of a siligraphene LixSiC_3 are all positive (the value of voltage decrease in the range -1,66242V and -0,16488V) which confirms its suitability as an anode material. The voltage profile of siligraphene type SiC_3 electrode has an average potential in the range of -0,54633V and -1,66242V.

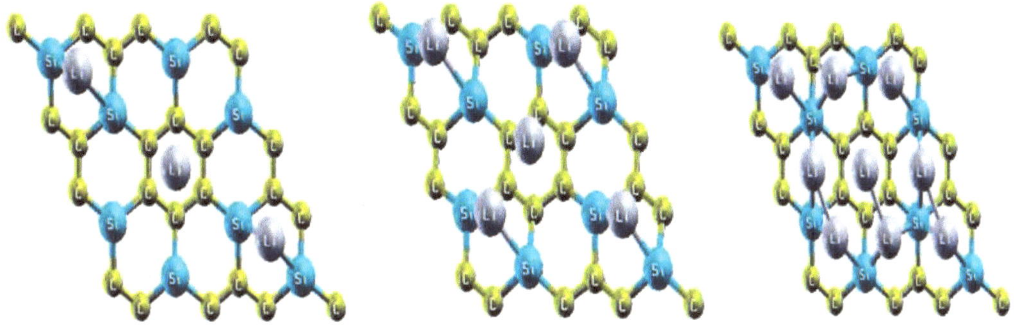

Fig. (8). The optimized atomic structures for (a) $Li_{0.33}SiC_3$, (b) $Li_{0.55}SiC_3$, (c) $LiSiC_3$.

The capacity of siligraphene $LiSiC_3$ is 1904,499 mAh/g. These values are very large as compared to the conventional electrode capacities, for example, the theoretical capacities for the Li_2C_6 are about 744 mAh/g [63]. Thus, we have predicted that siligraphene SiC_3 can be a promising 2D material as anode for Li-ion batteries.

CONCLUSION

In summary, we have investigated the Li-ion battery from a first-principles calculation based on density functional theory. We have examined the potential of Siligraphene as a new anode material for battery applications. The adsorption and diffusion of the Li atom on a siligraphene monolayer are studied. We report that the structural anisotropy in Siligraphene highly affects the diffusion of Lithium ions in all positions via exerting attractive forces. Our results of stability show that Li atoms prefer to occupy the hexagonal centers with minimum energy. The energy of diffusion of siligraphene (SiC_3) is about 0.095eV, while that of Li atoms on top of silicon atoms is about 0.223eV, indicating rapid charging/discharging processes. During the charging and discharging process, the electrode shows a small variation in voltage, which makes Li-SiC_3 a good candidate for Li-ion batteries. The presence of metallic behavior on the adsorption of Li-atoms on the surface makes SiC_3 a prospective electrode in the development of next-generation Li-batteries.

REFERENCES

[1] Goodenough, J.B.; Kim, Y. Challenges for rechargeable li batteries. *Chem. Mater.*, **2010**, *22*(3), 587-603.
[http://dx.doi.org/10.1021/cm901452z]

[2] ARMAND, ; Michel, ; John , B.; PADHI, ; Akshaya, K. Cathode materials for secondary (rechargeable) lithium batteries. U.S. Patent No 6,514,640, 4 févr. 2003.

[3] Kaboli, S.; Demers, H.; Paolella, A.; Darwiche, A.; Dontigny, M.; Clément, D.; Guerfi, A.; Trudeau,

M.L.; Goodenough, J.B.; Zaghib, K. Behavior of solid electrolyte in li-polymer battery with nmc cathode *via in-situ* scanning electron microscopy. *Nano Lett.,* **2020**, *20*(3), 1607-1613. [http://dx.doi.org/10.1021/acs.nanolett.9b04452] [PMID: 32017575]

[4] Kong, X.; Plett, G.L.; Trimboli, M.S.; Zhang, Z.; Zheng, Y. An exact closed-form impedance model for porous-electrode lithium-ion cells. *J. Electrochem. Soc.,* **2020**, *167*(1), 013539. [http://dx.doi.org/10.1149/1945-7111/ab67c7]

[5] Chu, Z.; Plett, G.L.; Trimboli, M.S.; Ouyang, M. A control-oriented electrochemical model for lithium-ion battery, Part I: Lumped-parameter reduced-order model with constant phase element. *J. Energy Storage,* **2019**, *25*, 100828. [http://dx.doi.org/10.1016/j.est.2019.100828]

[6] Yazami, R.; Touzain, P. A reversible graphite-lithium negative electrode for electrochemical generators. *J. Power Sources,* **1983**, *9*(3), 365-371. [http://dx.doi.org/10.1016/0378-7753(83)87040-2]

[7] Yoshio, NISHI; Hideto, AZUMA; Atsuo, et OMARU Non-aqueous electrolyte cell. U.S. Patent No 4,959,281 1990.

[8] Hua, ZHANG; Shuwu, LIU; Xiaofang, YU Improving rate capacity and cycling stability of Si-anode lithium-ion battery by using copper nanowire as conductive additive. *J. Alloys Compd.,* **2020**, 153664.

[9] Nayoung, KANG; Hyeon-Woo, YANG; Woo Seung, KANG Improvement of Reversibility and Cyclic Stability: A Monolithic Solid Electrolyte Interphase in SiO x-Based Anode for Lithium-Ion Batteries. *J. Phys. Chem. C,* **2020**.

[10] Ji, L.; Lin, Z.; Alcoutlabi, M.; Zhang, X. Recent developments in nanostructured anode materials for rechargeable lithium-ion batteries. *Energy Environ. Sci.,* **2011**, *4*(8), 2682-2699. [http://dx.doi.org/10.1039/c0ee00699h]

[11] Xie, F.; Zhang, L.; Chen, B.; Chao, D.; Gu, Q.; Johannessen, B.; Jaroniec, M.; Qiao, S-Z. Revealing the origin of the improved reversible capacity of dual-shell bismuth boxes anode for potassium-ion batteries. *Matter,* **2019**, *1*(6), 1681-1693. [http://dx.doi.org/10.1016/j.matt.2019.07.006]

[12] Aguilar-Eseiza, N.; Ramos-Sánchez, G.; González, F.; González, I. High voltage – Improved reversible capacity in Ni+2/+3 modified copper-based cathodes for lithium ion batteries. *Electrochem. Commun.,* **2018**, *96*, 32-36. [http://dx.doi.org/10.1016/j.elecom.2018.09.002]

[13] Shen, X.; Liu, H.; Cheng, X-B.; Yan, C.; Huang, J-Q. Beyond lithium ion batteries: Higher energy density battery systems based on lithium metal anodes. *Energy Storage Mater.,* **2018**, *12*, 161-175. [http://dx.doi.org/10.1016/j.ensm.2017.12.002]

[14] Shiqiang, L.I.U.; Dapeng, X.U.; Tianyi, M.A. Thermal safety studies of high energy density lithium-ion batteries under different states of charge. *Int. J. Energy Res.,* **2020**.

[15] Zheng, LI; Chengzhi, ZHANG; Fei, HAN Towards high-volumetric performance of Na/Li-ion batteries: A better anode material with molybdenum pentachloride–graphite intercalation compounds (MoCl5-GICs). *J. Mater. Chem. A Mater. Energy Sustain.,* **2020**.

[16] Liu, M.; Xing, L.; Xu, K.; Zhou, H.; Lan, J.; Wang, C.; Li, W. Deciphering the paradox between the Co-intercalation of sodium-solvent into graphite and its irreversible capacity. *Energy Storage Mater.,* **2020**, *26*, 32-39. [http://dx.doi.org/10.1016/j.ensm.2019.12.026]

[17] Tyler, OR.; GOURLEY, ; WD, Storm ; Karthikeyan, KALIYAPPAN Recycling of mixed cathode lithium-ion batteries for electric vehicles: Current status and future outlook. *Carbon Energy,* **2020**.

[18] Lu, M.; Zhang, X.; Ji, J.; Xu, X.; Zhang, Y. Research progress on power battery cooling technology for electric vehicles. *J. Energy Storage,* **2020**, *27*, 101155. [http://dx.doi.org/10.1016/j.est.2019.101155]

[19] Mircea, RUBA; Raul, NEMEŞ; Sorina, CIONEI Parameter identification, modeling and testing of li-ion batteries used in electric vehicles. In: *Applied Electromechanical Devices and Machines for Electric Mobility Solutions*; IntechOpen, **2020**.

[20] Pam, M.E.; Huang, S.; Fan, S.; Geng, D.; Kong, D.; Chen, S.; Ding, M.; Guo, L.; Ang, L.K.; Yang, H.Y. Interface engineering by atomically thin layer tungsten disulfide catalyst for high performance Li–S battery. *Mater. Today Energy,* **2020**, *16*, 100380. [http://dx.doi.org/10.1016/j.mtener.2019.100380]

[21] Khamlich, S.; Nuru, Z.Y.; Bello, A.; Fabiane, M.; Dangbegnon, J.K.; Manyala, N.; Maaza, M. Pulsed laser deposited Cr_2O_3 nanostructured thin film on graphene as anode material for lithium-ion batteries. *J. Alloys Compd.,* **2015**, *637*, 219-225. [http://dx.doi.org/10.1016/j.jallcom.2015.02.155]

[22] Zhang, Z.; Yang, M.; Zhao, N.; Wang, L.; Li, Y. Two-dimensional transition metal dichalcogenides as promising anodes for potassium ion batteries from first-principles prediction. *Phys. Chem. Chem. Phys.,* **2019**, *21*(42), 23441-23446. [http://dx.doi.org/10.1039/C9CP03948A] [PMID: 31616881]

[23] Loveridge, M.J.; Lain, M.J.; Johnson, I.D.; Roberts, A.; Beattie, S.D.; Dashwood, R.; Darr, J.A.; Bhagat, R. Towards high-capacity Li-ion batteries based on silicon-graphene composite anodes and sub-micron V-doped $LiFePO_4$ cathodes. *Sci. Rep.,* **2016**, *6*(1), 37787. [http://dx.doi.org/10.1038/srep37787] [PMID: 27898104]

[24] Li, Y.; Zhao, Y.; Ma, C.; Shi, J.; Zhao, Y. Highly monodispersed graphene/SnO_2 hybrid nanosheets as bifunctional anode materials of Li-ion and Na-ion batteries. *J. Alloys Compd.,* **2020**, *821*, 153506. [http://dx.doi.org/10.1016/j.jallcom.2019.153506]

[25] Fu, J.; Kang, W.; Guo, X.; Wen, H.; Zeng, T.; Yuan, R.; Zhang, C. 3D hierarchically porous NiO/Graphene hybrid paper anode for long-life and high rate cycling flexible Li-ion batteries. *J. Energy Chem.,* **2020**, *47*, 172-179. [http://dx.doi.org/10.1016/j.jechem.2019.11.022]

[26] Belasfar, K.; Houmad, M.; Boujnah, M.; Benyoussef, A.; Kenz, A.E.L. First-principles study of BC3 monolayer as anodes for lithium-ion and sodium-ion batteries applications. *J. Phys. Chem. Solids,* **2020**, *139*, 109319. [http://dx.doi.org/10.1016/j.jpcs.2019.109319]

[27] Jiajie et SCHWINGENSCHLÖGL, ZHU; Udo, Silicene for Na-ion battery applications. *2D Materials,* **2016**, *3*(3), 035012.

[28] Broux, T.; Fauth, F.; Hall, N.; Chatillon, Y.; Bianchini, M.; Bamine, T.; Leriche, J-B.; Suard, E.; Carlier, D.; Reynier, Y.; Simonin, L.; Masquelier, C.; Croguennec, L. High rate performance for carbon-coated $Na_3\ V_2\ (PO_4)_2\ F_3$ in Na-ion batteries. *Small Methods,* **2019**, *3*(4), 1800215. [http://dx.doi.org/10.1002/smtd.201800215]

[29] Rybarczyk, M.K.; Li, Y.; Qiao, M.; Hu, Y-S.; Titirici, M-M.; Lieder, M. Hard carbon derived from rice husk as low cost negative electrodes in Na-ion batteries. *J. Energy Chem.,* **2019**, *29*, 17-22. [http://dx.doi.org/10.1016/j.jechem.2018.01.025]

[30] Tan, P.; Chen, B.; Xu, H.; Cai, W.; He, W.; Liu, M.; Shao, Z.; Ni, M. Co_3O_4 nanosheets as an active material for hybrid Zn batteries. *Small,* **2018**, *14*(21), 1800225. [http://dx.doi.org/10.1002/smll.201800225]

[31] Zhang, R.; Pan, C.; Nuzzo, R.G.; Gewirth, A.A. $CoS\ _2$ as a sulfur redox-active cathode material for high-capacity nonaqueous Zn batteries. *J. Phys. Chem. C,* **2019**, *123*(14), 8740-8745. [http://dx.doi.org/10.1021/acs.jpcc.9b02142]

[32] Umer, YOUNIS; Imran, MUHAMMAD; Yoshiyuki, KAWAZOE *Design of Tetracene-Based Metallic 2D Carbon Materials for Na-and K-Ion Batteries*; Applied Surface Science, **2020**, p. 146456.

[33] Ding, N.; Li, X.; Chien, S.W.; Liu, Z.; Zong, Y. *In situ* monitoring the viscosity change of an

electrolyte in a Li–S battery. *Chem. Commun.,* **2017**, *53*(73), 10152-10155.
[http://dx.doi.org/10.1039/C7CC04841F] [PMID: 28848940]

[34] Haipeng, LI; Liancheng, SUN; Yongguang, ZHANG Enhanced cycle performance of Li/S battery with the reduced graphene oxide/activated carbon functional interlayer. *J. Energy Chem.,* **2017**, *26*(6), 1276-1281.

[35] Muhammad, I.; Younis, U.; Xie, H.; Kawazoe, Y.; Sun, Q. Graphdiyne-based monolayers as promising anchoring materials for lithium–sulfur batteries: A theoretical study. *Adv. Theory Simul.,* **2020**, *3*(3), 1900236.
[http://dx.doi.org/10.1002/adts.201900236]

[36] Khuili, M.; Fazouan, N.; Abou El Makarim, H.; Atmani, E.H.; Houmad, M. Improvement of optical properties of Mg doped ZnO by nanostructuring for applications in optoelectronics. *Mater. Res. Express,* **2020**, *7*(2), 025043.
[http://dx.doi.org/10.1088/2053-1591/ab748b]

[37] Yu-Jen, TSAI; Chin-Lung, et KUO Effect of structural disorders on the li storage capacity of graphene nanomaterials: A first-principles study. *ACS Appl. Mater. Interfaces,* **2020**.

[38] Lavin-Lopez, M.P.; Valverde, J.L.; de Lucas-Consuegra, A.; Calcerrada, A.B.; Paton-Carrero, A.; Romero, A. Influence of the synthesis method on electrical storage capacity of graphene-related materials. *Mater. Sci. Technol.,* **2019**, *35*(3), 361-367.
[http://dx.doi.org/10.1080/02670836.2018.1560106]

[39] Rai, D.P.; Vu, T.V.; Laref, A.; Hossain, M.A.; Haque, E.; Ahmad, S.; Khenata, R.; Thapa, R.K. Electronic properties and low lattice thermal conductivity (κ_l) of mono-layer (ML) MoS_2 : FP-LAPW incorporated with spin–orbit coupling (SOC). *RSC Advances,* **2020**, *10*(32), 18830-18840.
[http://dx.doi.org/10.1039/D0RA02585B] [PMID: 35518316]

[40] Houmad, M.; Mohammed, M.H.; Masrour, R.; El Kenz, A.; Benyoussef, A. Electronic and electrical properties of siligraphene (g-SiC_3) in the presence of several strains. *J. Phys. Chem. Solids,* **2019**, *127*, 231-237.
[http://dx.doi.org/10.1016/j.jpcs.2018.12.016]

[41] RAI, D. P.; Tuan V, VU.; Amel, LAREF *Promising optoelectronic response of 2D monolayer MoS2: A first principles study*; Chemical Physics, **2020**, p. 110824.

[42] Wang, X.; Ouyang, Y.; Li, X.; Wang, H.; Guo, J.; Dai, H. Room-temperature all-semiconducting sub-10-nm graphene nanoribbon field-effect transistors. *Phys. Rev. Lett.,* **2008**, *100*(20), 206803.
[http://dx.doi.org/10.1103/PhysRevLett.100.206803] [PMID: 18518566]

[43] Mak, K.F.; Lui, C.H.; Shan, J.; Heinz, T.F. Observation of an electric-field-induced band gap in bilayer graphene by infrared spectroscopy. *Phys. Rev. Lett.,* **2009**, *102*(25), 256405.
[http://dx.doi.org/10.1103/PhysRevLett.102.256405] [PMID: 19659105]

[44] Mak, K.F.; Lui, C.H.; Shan, J.; Heinz, T.F. Observation of an electric-field-induced band gap in bilayer graphene by infrared spectroscopy. *Phys. Rev. Lett.,* **2009**, *102*(25), 256405.
[http://dx.doi.org/10.1103/PhysRevLett.102.256405] [PMID: 19659105]

[45] Avetisyan, A.A.; Partoens, B.; Peeters, F.M. Electric field tuning of the band gap in graphene multilayers. *Phys. Rev. B Condens. Matter Mater. Phys.,* **2009**, *79*(3), 035421.
[http://dx.doi.org/10.1103/PhysRevB.79.035421]

[46] Dai, J.; Yuan, J.; Giannozzi, P. Gas adsorption on graphene doped with B, N, Al, and S: A theoretical study. *Appl. Phys. Lett.,* **2009**, *95*(23), 232105.
[http://dx.doi.org/10.1063/1.3272008]

[47] Denis, P.A.; Faccio, R.; Mombru, A.W. Is it possible to dope single-walled carbon nanotubes and graphene with sulfur? *ChemPhysChem,* **2009**, *10*(4), 715-722.
[http://dx.doi.org/10.1002/cphc.200800592] [PMID: 19189365]

[48] Houmad, M.; Zaari, H.; Benyoussef, A.; El Kenz, A.; Ez-Zahraouy, H. Optical conductivity

enhancement and band gap opening with silicon doped graphene. *Carbon,* **2015**, *94*, 1021-1027. [http://dx.doi.org/10.1016/j.carbon.2015.07.033]

[49] Shi, Z.; Zhang, Z.; Kutana, A.; Yakobson, B.I. Predicting two-dimensional silicon carbide monolayers. *ACS Nano,* **2015**, *9*(10), 9802-9809. [http://dx.doi.org/10.1021/acsnano.5b02753] [PMID: 26394207]

[50] Denis, P.A. Band gap opening of monolayer and bilayer graphene doped with aluminium, silicon, phosphorus, and sulfur. *Chem. Phys. Lett.,* **2010**, *492*(4-6), 251-257. [http://dx.doi.org/10.1016/j.cplett.2010.04.038]

[51] Zhang, Y-Y.; Pei, Q-X.; Cheng, Y.; Zhang, Y-W.; Zhang, X. Thermal conductivity of penta-graphene: The role of chemical functionalization. *Comput. Mater. Sci.,* **2017**, *137*, 195-200. [http://dx.doi.org/10.1016/j.commatsci.2017.05.042]

[52] Zanella, I.; Guerini, S.; Fagan, S.B.; Mendes Filho, J.; Souza Filho, A.G. Chemical doping-induced gap opening and spin polarization in graphene. *Phys. Rev. B,* **2008**, *77*, 073404.

[53] Wang, X.; Li, X.; Yoon, Y.; Weber, P.K.; Wang, H.; Guo, J.; Dai, H. N-doping of graphene through electrothermal reactions with ammonia. *Science,* **2009**, *324*(5928), 768-771.

[54] Sofo, J.O.; Chaudhari, A.S.; Barber, G.D. Graphane: A two-dimensional hydrocarbon. *Phys. Rev. B Condens. Matter Mater. Phys.,* **2007**, *75*(15), 153401. [http://dx.doi.org/10.1103/PhysRevB.75.153401]

[55] Chan, K.T.; Neaton, J.B.; Cohen, M.L. First-principles study of metal adatom adsorption on graphene. *Phys. Rev. B Condens. Matter Mater. Phys.,* **2008**, *77*(23), 235430. [http://dx.doi.org/10.1103/PhysRevB.77.235430]

[56] Hussain, T.; Farokh Niaei, A.H.; Searles, D.J.; Hankel, M. Three-dimensional silicon carbide from siligraphene as a high capacity lithium ion battery anode material. *J. Phys. Chem. C,* **2019**, *123*(45), 27295-27304. [http://dx.doi.org/10.1021/acs.jpcc.9b06151]

[57] Pereyra Huelmo, C.; Denis, P.A. Silicon carbide induced doping of graphene: A new potential synthetic route for SiC_3 siligraphene. *J. Phys. Chem. C,* **2019**, *123*(50), 30341-30350. [http://dx.doi.org/10.1021/acs.jpcc.9b07978]

[58] Wang, H.; Wu, M.; Lei, X.; Tian, Z.; Xu, B.; Huang, K.; Ouyang, C. Siligraphene as a promising anode material for lithium-ion batteries predicted from first-principles calculations. *Nano Energy,* **2018**, *49*, 67-76. [http://dx.doi.org/10.1016/j.nanoen.2018.04.038]

[59] Dong, H.; Ji, Y.; Hou, T.; Li, Y. Two-dimensional siligraphenes as cathode catalysts for nonaqueous lithium-oxygen batteries. *Carbon,* **2018**, *126*, 580-587. [http://dx.doi.org/10.1016/j.carbon.2017.10.077]

[60] Ling, C.; Mizuno, F. Chen et MIZUNO, Fuminori. Capture lithium in αMnO_2: Insights from first principles. *Chem. Mater.,* **2012**, *24*(20), 3943-3951. [http://dx.doi.org/10.1021/cm302347j]

[61] Chan, K.T.; Neaton, J.B.; Cohen, M.L. First-principles study of metal adatom adsorption on graphene. *Phys. Rev. B Condens. Matter Mater. Phys.,* **2008**, *77*(23), 235430. [http://dx.doi.org/10.1103/PhysRevB.77.235430]

[62] Lv, X.; Wei, W.; Sun, Q.; Yu, L.; Huang, B.; Dai, Y. Sc_2C as a promising anode material with high mobility and capacity: A first-principles study. *ChemPhysChem,* **2017**, *18*(12), 1627-1634. [http://dx.doi.org/10.1002/cphc.201700181] [PMID: 28383808]

[63] Liu, Y.; Xue, J.S.; Zheng, T.; Dahn, J.R. Mechanism of lithium insertion in hard carbons prepared by pyrolysis of epoxy resins. *Carbon,* **1996**, *34*(2), 193-200. [http://dx.doi.org/10.1016/0008-6223(96)00177-7]

CHAPTER 4

An Introduction to the Fabrication of White Light-emitting Diodes

Naorem Premjit Singh[1,*] and **Yengkhom Rangeela Devi**[2]

[1] *A. R. S. D. College, University of Delhi, Dhaula Kuan, New Delhi, India*

[2] *Department of Physics, Pachhunga University College (PUC), Aizawl, Mizoram, India*

Abstract: Light-emitting diodes, especially white light-emitting diodes are very attractive and fascinating lighting sources at this present time because they have the potential for high energy saving and environmental friendliness as compared to conventional lighting sources such as incandescent and fluorescent lamps and also have wide applications in a variety of fields including in lighting, architectural and medical *etc*. Among the various applications, the lighting sector is one of the most important fields because it consumes a large amount of electricity. About 15-22% of total electricity production in the world is consumed in the lighting sector. Therefore, understanding how to fabricate a white light-emitting diode is very necessary in order to improve its practical application further. Basically, there are two methods of fabrication for white light emitting diode, mixing of multiple LEDs and phosphor converted white light emitting diode (pc-WLED). The luminous efficiency and rendering index is influenced by the type of fabrication. In this chapter, the general introduction of light emitting diode (LED), its working principle, characteristics of light including CIE, color temperature and rendering index, the different modes of fabrication for white light-emitting diodes, and their advantages and disadvantages have been discussed.

Keywords: LED, Fabrication methods, pc-WLED.

INTRODUCTION

Electrical energy is one of the very essential parts of modern life, which is required in every household for various applications. The lighting sector is one of the fields in which a substantial quantity of electricity is being consumed. Around (15-22) % of the total electricity production is being used in the lighting sector [1 - 3]. At this point, light emitting diode (LED), especially white light emitting diode (WLED), is highly fascinating because they have the potential for low

* **Corresponding Author Naorem Premjit Singh:** A. R. S. D. College, University of Delhi, Dhaula Kuan, New Delhi, India; Tel: +919366139124; E-mail: premjit.naorem@yahoo.in

Dibya Prakash Rai, Kingsley O. Obodo, & Jitendra Pal Singh (Eds.)

energy consumption, high efficiency, long lifetime, no heat production, compactness and environmental friendliness [4]. Therefore LED, LED-based lighting sources are considered as next-generation lighting sources and nowadays, conventional lighting sources such as incandescent light and halogen lamps have been replaced gradually by WLED in many sectors, including street lamps.

LED is a semiconductor-based lighting source. The light emission in LED is due to the recombination of electrons and holes inside the diode. LED consists of at least two semiconductors, n-type and p-type and a diode junction is formed between them. When electric current is passed in forward biased condition, the negative terminal of the battery is joined to the n-type and the positive terminal is joined to the p-type semiconductor. The electron from the n-type combines with the hole of the p-type; which results generation of light in different electromagnetic waves depending on the energy gap between the electron and the hole to be combined [5]. The working of an LED is diagrammatically shown in Fig. (**1**).

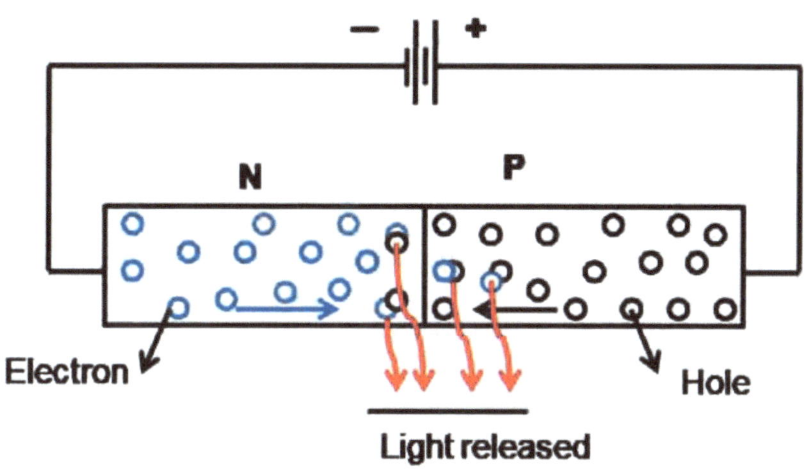

Fig. (1). The schematic diagram of an LED showing how it generates light.

The amount of energy released from an LED material can be calculated by using the formula given below:

$$\Delta E = hv = \frac{hc}{\lambda} \quad \textbf{(1)}$$

Here, ΔE represents the emission energy, h is Planck's constant, c is the velocity of light, which is equal to the velocity of emitted electromagnetic light, and λ is the wavelength of the emitted light.

TYPES OF LEDS AND THEIR APPLICATIONS

Based on the amount of energy released, LEDs can be classified into different types such as UV- LEDs (λ_{Emi} < 400 nm), Visible-LEDs (λ_{Emi} = 400-800 nm) and Infra-Red-LEDs (λ_{Emi} > 800 nm). They have applications in different fields. For instance, Infrared-LEDs are used in remote control circuits such as in Televisions, DVD players and other domestic appliances *etc*., UV-LEDs are used in industries for curing and hardening such as printing and manufacturing, and as disinfectants or sterilizers in the purification of air and water *etc*. [5]. whereas visible light emitting LEDs are found applications in several fields. Some of the applications of visible light-emitting LEDs are shown in Fig. (**2**).

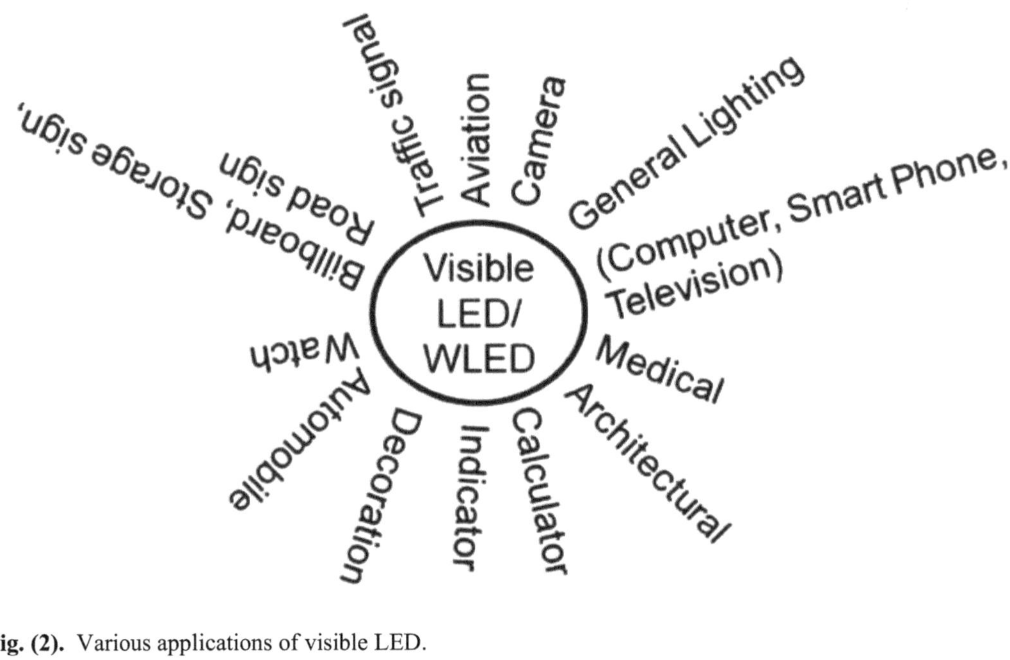

Fig. (2). Various applications of visible LED.

It is known that the visible light electromagnetic spectrum is extended in the wavelength (λ) range between 400 to 800 nm. The color of emitted light from a visible LED depends on the magnitude of the energy gap between the electron and hole to be combined. In Table **1**, the variation of color with wavelength of visible light is shown [6].

Among the various applications, the lighting sector is one of the most interesting and attractive fields. The first practical visible light-emitting LED (red) was invented by Nick Holonyak Jr. in 1962 while he was working at General Electric

Company [7]. The light was observed from the material gallium arsenide phosphide (GaAsP), and he is known as the "Father of light emitting diode" [7]. However, the efficiency of the emitted light was very low and, therefore, not suitable for several applications; the brightness was enough for application only as an indicator, seven segment numeric displays [5 - 10]. Nevertheless, growth and advances in visible LED technology have been observed since Shuji Nakamura developed a highly efficient and environment--friendly blue-emitting LED from InGaN material in 1996. For this invention, he was awarded the Novel Prize in Physics along with two other scientists, Hiroshi Amano and Isamu Akasaki, in 2014 [5, 10]. After the development of this highly efficient blue light-emitting LED, the highly efficient and high power-saving WLED was developed very soon. Nowadays, such LED based lighting sources are considered as an important and essential lighting source all over the world.

Table 1. Variation color with wavelength of light.

Color	Wavelength (nm)
Blue	435-480 nm
Green	500-560 nm
yellow	565-590 nm
Orange	595-610 nm
Red	610-700 nm

CHARACTERISTICS OF A LIGHTING SOURCE

There are some basic parameters to characterize or describe the quality of a light source. Some of the characteristics are color rendering index (CRI), CIE coordinates, Co-related color temperature (CCT) and quantum efficiency. A brief discussion of these characteristics is given here.

Color Rendering Index (CRI)

The color rendering index of a lighting source indicates how correctly the color of an object is produced as compared to the illumination of the object by natural lighting sources such as sunlight or black body radiation, and its value is indicated in the scale between 0-100. If a Light source extends its emission spectral power distribution to the whole of visible light wavelengths, then the light source will have a high CRI value. Natural lighting sources are rated their CRI value 100 and other lighting sources are measured in comparison to it. Incandescent and halogen lights are found to have higher CRI values as compared to LED lighting sources to date. Most of the LED light sources have CRI values in the range between 80 –

90. If a light source produces the color of an object similar to that produced by natural lighting sources, then the light source will be said to have a 100 CRI value.

CIE Color Co-ordinates

Any color can be described by three variables defined by the CIE XYZ 1931 scheme called the color matching functions. x(λ), y(λ), and z(λ). It can be thought of as the spectral sensitivity curve of three linear light detectors that can yield the tristimulus value X, Y, Z [6, 11]. The spectral sensitivity distribution curve of color-matching function is given in Fig. (**3**).

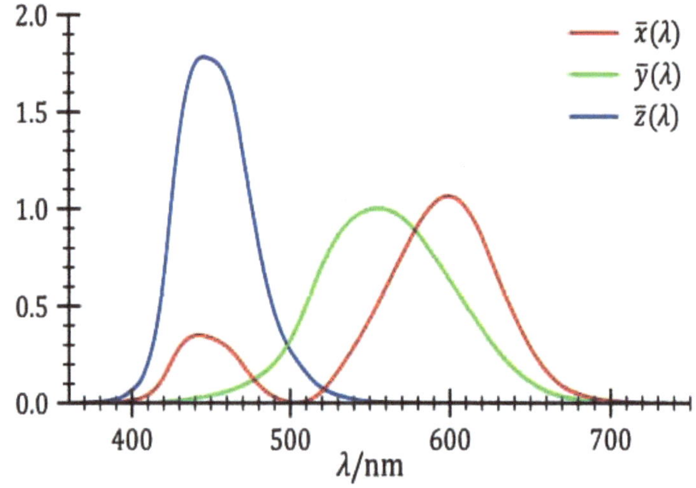

Fig. (3). Spectral sensitivity distribution curve of color matching function [11].

For a given spectral distribution P (λ), the tristimulus values X, Y, and Z can be related mathematically to the color-matching function as given below [12].

$$\left.\begin{aligned} X &= \int \bar{x}(\lambda)P(\lambda)d(\lambda) \\ Y &= \int \bar{y}(\lambda)P(\lambda)d(\lambda) \\ Z &= \int \bar{z}(\lambda)P(\lambda)d(\lambda) \end{aligned}\right\} \quad (2)$$

Further, a color can be divided into its luminescence (brightness) and chromaticity. For example, one can say that white and grey colors have the same chromaticity, but luminescence is different. The chromaticity can be specified by two derived parameters, say *x* and *y* which is called the chromaticity coordinates and are functions of all the three tristimulus values X, Y and Z by the following relation.

$$x = \frac{X}{X+Y+Z}, \quad y = \frac{Y}{X+Y+Z}, \tag{3}$$

By taking this *x* and *y* chromaticity coordinates, the color of any spectral distribution can be identified in a horseshoe shape chromaticity diagram called the gamut. The 1931 *x y* chromaticity diagram is shown in Fig. (**4**).

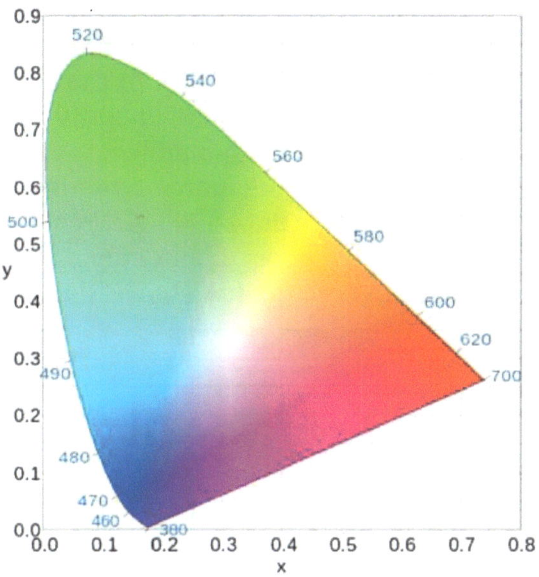

Fig. (4). The 1931 *x y* chromaticity diagram.

The purity of color is very significant when considering the performance of a device. It can be estimated by using the equation given below [13 - 15].

$$\text{Color purity} = \frac{\sqrt{(x - x_i)^2 + (y - y_i)^2}}{\sqrt{(x_d - x_i)^2 + (y_d - y_i)^2}} \tag{4}$$

Where *x* and *y* represent the CIE color coordinates, x_i and y_i indicate the color coordinates of the standard white source (0.310, 0.316) [15], and x_d and y_d indicate the co-ordinates of the dominant wavelength, which can be obtained by extending the straight line between the color co-ordinates (x, y) and the standard white source (x_i, y_i) to the other site [13 - 16].

Color Co-related Temperature (CCT)

This is another characteristic of a light source and is defined as the temperature of an ideal black body radiator that radiates light of comparable color to that of a light source. Simply, we can say that a particular temperature (T_c) is shown in a specific color. It is a characteristic of visible light and has significant applications in various fields, such as lighting, photography, videography and publishing *etc* [6]. Most artificial lighting sources do not produce the color of an object the same as the black body radiation at the same temperature. This is because at the same temperature the radiant energy emitted from any artificial lighting sources is always less than that of the black body radiation. Therefore, the term color co-related (CCT) temperature is used, which is defined as the temperature of black body radiation whose color is closest to the color of the test lighting source. CCT is expressed in the unit of Kelvin (K) and is generally calculated by using the McCamy equation [17, 18] given in equation number 5. Light color with CCT less than 5000 K is called warm light, which is more of yellow light and is suitable for outdoor applications, whereas greater than 5000 K is called cool color, which is more blue light and is suitable for indoor applications and lighting source with CCT = 6500 K is called daylight. Fig. (**5**) shows the variation of color with temperature.

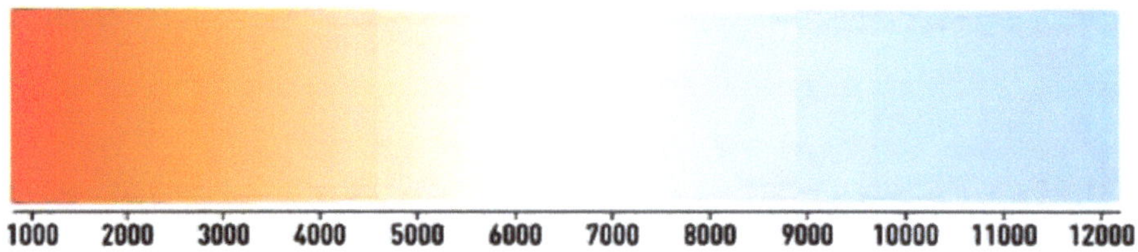

Fig. (5). Variation of color appearance with temperature in Kelvin (Wikipedia) [19].

$$T_c = -449n^3 + 3525\,n^2 - 6823.3\,n + 5520.33 \quad \textbf{(5)}$$

Where n=(x- x_e)/(y-y_e), x_e=0.3320; y_e=0.1858, x and y are CIE color co-ordinates.

Luminous Efficacy (K)

It indicates how well a light source produces light in the visible region using a given amount of power. It is defined as the ratio of total luminous flux (lumen) to the radiant flux (watt), which means the total amount of light output in all directions (in the visible region) from the device to the total amount of electricity input and measured in the unit of lumen per watt. Mathematically, it can be formulated as:

$$K = \frac{\Phi_v}{\Phi_e} \quad \textbf{(6)}$$

Where K = Luminous efficacy, Φv = luminous flux and Φe = radiant flux

The luminous efficacy is always in contradiction with the colour rendering index because the high CRI is achieved by the broadband spectra distributed throughout the entire visible region, whereas high luminous efficacy is obtained by a narrow spectrum of monochromatic light.

FABRICATIONS OF WHITE LIGHT-EMITTING DIODE

White light is a polychromatic light and it can be generated by using different fabrication methods. Basically, there are two primary fabrication methods for the generation of a white light-emitting diode.

(1) Mixing of multiple LEDs emitting different monochromatic light.

(2) Combination of blue or UV/NUV emitting LED chip with phosphor.

Mixing of Multiple LEDs Emitting Different Monochromatic Color

The device can be constructed by combining different LED chips emitting different colors of monochromatic light within a package. When electricity is passed in forward biased condition, the LED chips emit light of different colors depending on the energy gap of the electron and hole to be combined. When these emitted lights are mixed in an appropriate ratio, white light can be generated. This type of white light emission can be observed by the different types of mixing given below.

Dichromatic (Mixing of blue and yellow emitting LED chips)

Trichromatic (Mixing of blue, green and red light-emitting LED chips)

Tetrachromatic (Mixing of blue, cyan, green and red light-emitting LEDs)

Amongst them, the dichromatic type of mixing generally produces light of high efficiency and low CRI (approx. 40) as compared to the other two types of mixing, whereas the tetrachromatic type produces light of low efficiency and high CRI (> 95) [5]. On the other hand, the trichromatic type of mixing, also called

RGB WLED, produces light efficiency in the range between the other two types of mixing (dichromatic and tetrachromatic) and CRI value greater than 80 [5]. From this observation, it is seen that there is some contradiction between CRI and light efficiency. For some applications like general indoor lighting such as at home, office *etc*, CRI value is very significant and in such cases, the lighting sources having a CRI value greater than 80 is highly preferred. Therefore the trichromatic type of mixing is considered as the most suitable mixing type for such applications. As an example, Nakamura *et al.* have shown that white light emission can be obtained by the combination of blue emitting InGaN SQW [single quantum well] LED chip, green-emitting InGaN SQW LED chip and red emitting GaAlAs LED chip [20].

So far, several LED materials that emit light in the visible, infrared or UV region have been developed. Some of the conventional LED materials and their characteristic emission light/color are listed in Table **2** [5, 10].

Table 2. Some commonly used LED materials and their characteristic emission color.

LED Materials	Emitting Light/color
GaAs	Infra red (IR)
AlGaAs	Red and Infrared
GaAsP	Red, Orange, Orange-Red and Yellow, Infrared
AlGaP	Green
AlGaInP	High brightness, orange-red, green and yellow
AlGaAsP	Red, orange-red, orange, yellow
GaInP	Green
GaP	Red, orange, Yellow and Green
GaN	Green, Emerald green, Blue
GaInN	Blue, Green, violet, Near ultraviolet (NUV)
SiC	Blue
ZnSe	Blue
AlGaN	Ultraviolet (UV)
BN	Ultraviolet
AlN	Ultraviolet
AlGaInN	Near to far ultraviolet

Advantages and Disadvantages

The most attractive advantage of this type of fabrication is its ability to produce light of high efficiency. However, it has suffered from several limitations, including difficulty in manufacturing and high expenditure. It is known that different drive voltage is required for light emission in different wavelengths. Therefore different independent circuits are required to apply for each of the LEDs in the package. Furthermore, it is also very complicated to control the blending and diffusion of light of different colors for the generation of white light [21, 22]. In order to overcome these issues, the device needs to be manufactured with a sophisticated design. Moreover, the rate of degradation of different colors varies with time and temperature. Consequently, the lumen output and emission wavelength are changed with temperature and time, which leads to inconsistency in the color point [22]. Because of all these drawbacks, even though highly efficient white light emission can be produced by this technique, it is hardly used in industries for mass-scale production.

Mixing of an LED Chip with a Phosphor

This type of fabrication for the generation of white light is called phosphor-converted white light emitting diode (pc-WLED) because the phosphor material absorbs the light emitted from the LED chip and then converts it into white light. In this case, the efficiency or quality of the generated light depends on the attached phosphor material and the LED. Therefore, the attached phosphor plays a very important role in the performance of the device. The efficiency of a phosphor generally depends on several factors including crystallinity, morphology, and particle size *etc* of the phosphor. The device consists of an LED chip and phosphor materials, which are generally encapsulated by a transparent material. The commonly used encapsulant is epoxy resin made from polymer material [23, 24]. The function of this resin is to provide protection of the LED material from the mechanical and outside environment and also to provide high light extraction to improve efficiency [23, 24]. The typical schematic diagram of phosphor converted white light-emitting diode is shown in Fig. (**6**).

The device can be constructed by using the following modes of fabrication.

(a) Combination of blue emitting LED chip with a phosphor emitting yellow light.

(b) Combination of UV/NUV emitting LED chip with multiple phosphors.

(c) Combination of UV/NUV emitting LED with a single phosphor emitting blue and yellow lights or blue, green and red lights.

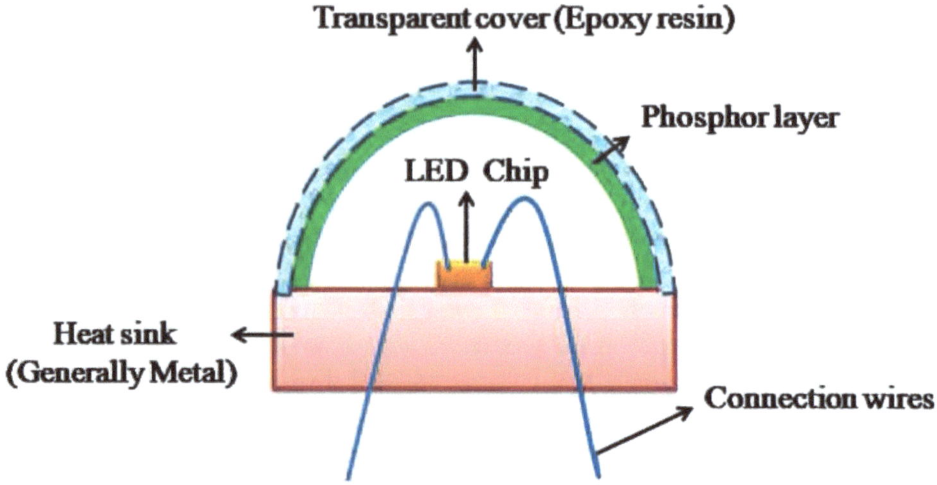

Fig. (6). A typical fabrication of a phosphor-converted white light emitting diode.

In the phosphor-converted white light emitting diode, the amount of energy released from the LED material has to match the energy required for excitation of the attached phosphor material. Therefore selection of LED and phosphor materials for combination is very important. There are some important terms related to the pc-WLED, such as luminescence, phosphor and energy transfer and concentration and thermal quenching of luminescence. Brief discussions of these terms are given below.

Luminescence

Luminescence is defined as the subsequent emission of light from the excited atom or molecule after excitation by absorption of energy. It is called cold radiation because it neither requires nor produces high temperatures. Based on the source of excitation energy, luminescence can be classified into different types. Photoluminescence is a type of luminescence which involves light radiation by exciting the atom/molecule with a photon. Photoluminescence can be seen in two ways: Fluorescence and Phosphorescence. Fluorescence is the radiation of photons from the excited singlet state to the ground state. It has a short lifetime, generally in the range of nanoseconds [25, 26]. The fluorescence radiation stops immediately after the excitation source is cut off. On the other hand, phosphorescence involves the radiation of photons from the excited triplet state to the ground state. It has a longer lifetime, and the phosphorescence radiation continues even when the excitation source is cut off [26, 27]. Further, the luminescence can be divided into down-conversion (DC) and up-conversion (UC) based on the excitation and emission energies. Down-conversion luminescence involves the conversion of high energy, say UV light, to lower energy, say visible light, whereas up-conversion luminescence involves the conversion of low energy, say infrared (IR) light into a high energy, say UV or Visible light. For the up-conversion luminescence, two or more photons of lower energy have to be supplied in order to excite the phosphor material.

Phosphor

Phosphors are solid materials that exhibit light emission (luminescence) when excited by radiation. This word comes from the Greek word "Phosphoros" meaning light bearer. Italian Alchemist Vincentinus Casciarolo first used this word in 1603 to describe a solid substance that glows red light in the dark after exposure to daylight. In fact, the solid substance was $BaSO_4$ doped with metal ions such as Zn^{2+}, Sn^{2+} and Cd^{2+} [27, 28]. Phosphor usually consists of a host lattice and intentionally doped impurity ions. Insulator or Semiconductor material generally works as host lattice, and rare earth ions or transition metal ions work as activator ions. The emission of light takes place from the activator ion. So far, several phosphor materials emitting light in different color regions of visible lights have been developed by using different host materials. Some of the commonly used host materials are tungstate, molybdate, vanadate, silicate, borate, fluorite, nitride *etc*. The absorption of energy for excitation may take place either by host lattice (indirect excitation) or doping impurity ion (direct excitation). Sometimes sensitizer can also be added in order to enhance the absorption co-efficient.

DIFFERENT SYNTHETIC METHODS OF PHOSPHOR

The synthetic method is very important because it gives strong influences on the properties and performance of the phosphor, such as in microstructure, luminescence property, quantum efficiency *etc*. [22]. There are various synthetic methods for the preparation of phosphor. Some of the commonly used methods are briefly given here.

Co-Precipitation Method

The co-precipitation method is one of the most commonly used synthetic methods. It is very simple and cost-effective, moreover, particles of highly uniform morphology can be prepared, which is a very important characteristic of a phosphor, especially for application in display devices [28]. In this technique, two or more salts are dissolved in water/solvent to get a homogeneous solution, and then an insoluble product is formed. Capping agents or surfactants are usually added to control the particle size and morphology. In some cases, phosphors were enabled to prepare at very low temperatures, as low as room temperature [17, 29], and in some other cases, annealing of the product formed is required in order to improve the crystallinity [30]. Parameters including concentrations of the starting reactants, pH of the solution, and temperature of heating/annealing give strong influences on the size and morphology of the product formed [31]. By this method, most of the phosphor types, such as oxide, sulphide, and fluorides, can be prepared.

Hydrothermal/Solvothermal Method

This method involves the occurrence of a chemical reaction of a substance in a sealed vessel called an autoclave above the ambient temperature and pressure. If the solvent used is water only, then it is called hydrothermal and if a solvent other than water is used, it is called solvothermal. It is a very attractive and promising method for the synthesis of phosphor since the morphology and size of the prepared samples can be properly controlled, and the reaction can also be achieved at lower temperatures and shorter time [22]. A large number of phosphors, including oxides and fluorides [22, 32 - 37], have been prepared by using this method and shown that the properties of the products, shape and size are significantly dependent on the composition of the reactant, reaction temperature and time, pH of the solution [22, 31, 37]. The main drawback of this method is its limitation in the large-scale production which limits its application in industries. Some of the phosphors can include $ZnWO_4:Dy^{3+}$, Eu^{3+} [38], and $CdWO_4:Eu^{3+}$ [33].

Sol-gel Method

This method is also one of the commonly used wet methods for the preparation of phosphor. It involves a series of chemical processes, including hydrolysis, condensation, gelatin, gel and heat treatment [22, 28]. This method allows the formation of high homogeneity of the chemical composition, and also the microstructure can be controlled. It can be used for large-scale production, and usually, the product possesses high luminescence intensity [39]. This method is widely applied for the synthesis of glass, ceramics *etc.* [28].

Combustion Method

It is another commonly used synthetic method for the preparation of phosphor. In this method, the oxidizer (usually metal nitride) and reducer (urea, glycine or citric acid as fuel) undergo self-sustainable combustion reaction upon heating, which is an exothermic reaction and hence produces heat [22, 39]. The high-temperature heat produced from this exothermic reaction heats up the precursor at high temperatures, leading to the crystallization of the phosphor. This process can be performed at short processing times and at low temperatures, with less expenditure and also can achieve a high yield of the product with high purity. The obtained product generally possesses more homogeneity, less impurity and large surface area as compared to other conventional methods [22]. It has been applied for the synthesis of many of the phosphors, including oxides, borates, aluminates and silicates [22, 40 - 44].

Sonochemical Route

This method employs a powerful tool that provides high-intensity ultrasound waves to induce the chemical reaction of the precursors. When it is irradiated to a solution of precursors, due to the chemical and physical effect of the ultrasound, the formation of the nanophosphor is facilitated in room temperature liquid without being subjected to high temperature and pressure treatment [45]. In the process, the solution of a precursor is set at a specific pH and temperature in a water bath, and a precipitating agent is added slowly, drop by drop [45]. Some of the phosphors synthesized by using this method can include $SrSO_4$:Sm^{3+} [46] and Bi_2O_3:Tb^{3+}/Eu^{3+} [47].

Microwave-assisted Synthesis

This method has now emerged as one of the attractive methods for substitution of conventional methods because it can produce strong heat and a rapid rate of reaction. So, the chemical reaction of reactants can take place at lower temperatures and shorter time. In this technique, the heat for inducing the chemical reaction is provided by microwave. The microwave is able to produce high temperatures at ambient conditions. When microwave is absorbed by the polar solvent molecule, they are oriented in the direction of an external field, which causes the emission of strong heat and results in strong agitation and strong internal heating [45]. But the required condition is that the solvent for dissolving the precursor has to be a polar liquid like H_2O since the microwave is active only in polar molecules, and the precursor has to be dissolved properly to the solvent. The amount of heat produced is strongly dependent on the strength of the polarity of the solvent [45]. Many of the phosphors have been synthesized by this method. Some of them are Gd_2O_2S: Tb^{3+} [48], Sr_3SiO_5:Eu^{2+} [49], and LaF_3:CeTb [50].

Solid State Method

This method is one of the most widely used methods for the preparation of phosphor, especially for oxide, nitride, and fluorides *etc.*, because of its simplicity in operation and feasibility for mass-scale production [12, 22, 51]. In this process, the solid powder reactants are mixed together and heated for an extended period of time at very high temperatures, normally above 1000°C, to induce interphase atomic diffusion [22]. Since the reaction takes place at the solid state level, the rate of the reaction depends on several parameters, including reaction conditions (Temperature and Pressure), structural property of the reactants, surface area, surface contact *etc.* [39, 52]. As the atomic diffusion controls the reaction, repeated grinding and heating are required in the process and, normally, results in a thermodynamically stable product with poor homogeneity and large particle size [22]. Some of the phosphors can include $BaNb_2O_6:Dy^{3+}$ [12], $ZnWO_4:Eu^{3+}$ [53], $Sr_2V_2O_7:Eu^{3+}$ [54], $Sr_2La_{1-x-y}GaO_5$:xDy, ySm^{3+} [55], $Sr_2LaTaO_6:Mn^{4+}$ [13].

Energy Transfer

It is the process involving the transfer of energy from a certain species partially or completely to another species. This type of energy transfer may take place between similar or dissimilar species. The species in which energy is donated is called donor (D), and the species in which energy is accepted is called acceptor (A). Such energy transfer can be observed only if the resonance condition between the donor and acceptor species is satisfied means that if the energy difference between the ground and excited states of the donor species is equal to the acceptor species or if the distance between the donor and acceptor species is short enough for interaction [9, 56, 57].

$$D^* + A \rightarrow A^* + D$$

Energy transfer may take place in between host and activator ion or in between activator ions. Concentration quenching is due to the increase of non-radiative energy transfer between the activator ions.

Quenching of Luminescence

What is concentration quenching? Concentration quenching is the reduction of emission intensity with the increase of activator ion concentration. Generally, the luminescence intensity increases with the increase of activator ion concentrations as a result of an increase in emission sites. However, the intensity decreases after reaching a particular concentration called the optimum concentration; this is called the concentration quenching effect. As said above, concentration quenching is due to the increase of non-radiative energy transfer among the activator ions. This energy transfer may take place either through exchange interaction or electric or magnetic multi–polar interaction [13, 14, 16]. The exchange interaction can take place only if the donor and acceptor species are very close; the critical distance (R_c) is within 5 Å, and if there exists an overlapping of emission of the donor and the excitation of the acceptor [56 - 58]. The critical distance can be calculated from the equation given by Blasse [13, 14, 16].

$$R_c = 2\left(\frac{3V}{4\pi x_c Z}\right)^{1/3} \tag{7}$$

Where V is the volume of unit cell, x_c is the activator ion concentration, and Z is the number of cations per unit cell.

On the other hand, no overlapping of emission and excitation spectrum is required in multi-polar interactions. This energy transfer can be due to the dipole - dipole, dipole - quadrupole or quadrupole – quadrupole [13, 55]. The type of interaction involved can be identified by using the formula shown below given by Dexter [13, 55].

$$log\left(\frac{I}{x}\right) = -\left(\frac{Q}{3}\right)\log x + A \tag{8}$$

Where I is the luminescence intensity, A is a constant, its value is the same for a host material under the same conditions of excitation, Q represents the interaction characteristics, its value provides the type of interaction and can be determined from the slope of the curve of $log\left(\frac{I}{x}\right)$ $vs \log x$. The Q value 6, 8 and 10 corresponds to the dipole – dipole, dipole – quadrupole and quadrupole – quadrupole interaction respectively [13, 55].

Thermal Quenching

It is another important type of quenching in luminescence. It refers reduction of luminescence intensity with the increase in temperature. Because of the thermal quenching, the brightness and life span of a phosphor decrease with time, which affects the performance of the phosphor. So, an efficient phosphor should have high thermal stability apart from its high intensity. The thermal quenching of a phosphor can be understood by calculating the activation energy from the equation given by Arrhenius [13, 14].

$$I_T = \frac{I_0}{1+CExp^{\left(-\frac{E_a}{KT}\right)}} \tag{9}$$

I_T and I_0 represent the emission intensity measured at temperatures T and 300K, respectively. E_a is the activation energy; K is the Boltzmann constant and C is a constant, depending on the host material. The above equation can be modified as:

$$\ln\left(\frac{I_0}{I_T} - 1\right) = \ln C - \frac{E_a}{KT} \quad \textbf{(10)}$$

So, from the plot of $\ln\left(\frac{I_0}{I_T} - 1\right)$ vs $\frac{1}{KT}$, the value of activation energy can be determined. The larger the value of activation energy, the more the thermal stability of the phosphor.

COMBINATION OF BLUE EMITTING LED WITH A PHOSPHOR EMITTING YELLOW LIGHT

In this technique, the LED material emits blue light, which is partially absorbed by the coated phosphor and converted into yellow light while the remaining part of blue light is transmitted. When these radiated yellow light and transmitted blue light from the phosphor are mixed at appropriate proportions, white light is generated. This concept of white light generation was first demonstrated by Shimizu *et al.* of Nichia company, Anan, of Japan, in the year 1996 [59]. The device was constructed by combining efficient blue light-emitting InGaN LED chip with down-converted yellow light emitting cerium doped yttrium aluminum garnet phosphor ($Y_3Al_5O_{12}$:Ce) (YAG). The typical spectral distribution of white light emission by this combination is shown in Fig. (7).

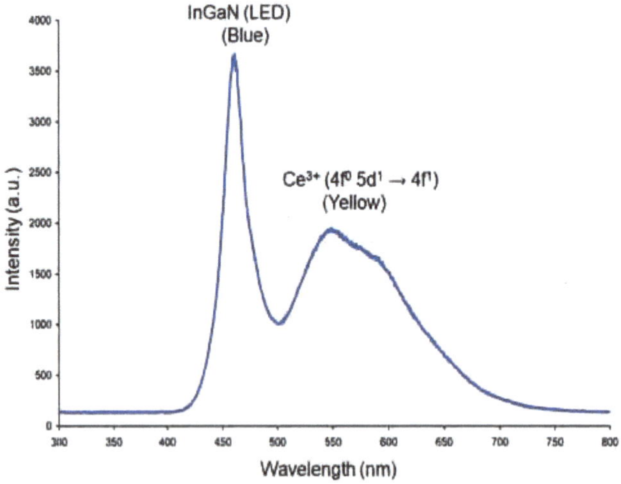

Fig. (7). Typical spectral distribution of white light emission formed by the combination of blue emitting InGaN LED chip with yellow emitting $Y_3Al_5O_{12}$:Ce phosphor (Modified from Wikipedia [5]).

The spectrum consists of a strong emission band with a maximum at around 460 nm in the blue region originating from the InGaN LED chip and a broadband covering a wide range of wavelengths with the maximum at around 560 nm (yellow) originating from the $Y_3Al_5O_{12}$:Ce phosphor. It is known that those rare earth ions that love to oxidize showed 4f-5d electronic transition. Therefore, the trivalent rare earth ions, such as Ce^{3+}, Pr^{3+} and Tb^{3+} and divalent rare earth ions, such as Sm^{2+}, Eu^{2+} and Yb^{2+}, show 4f-5d electronic transition. Ce^{3+}, Pr^{3+} and Tb^{3+} show absorption bands in ultraviolet electromagnetic waves because of 4f-5d electronic transition [56]. Among these rare earth ions, Ce^{3+} and Eu^{2+} ions generally showed parity allowed d-f electronic transition because their $4f^65d^1$ excited state is in lower energy than the lowest energy of excited 4f orbital. As the d-orbital is strongly affected by the outside chemical environment and the f-orbital is hardly affected by the outside chemical environment, the excited d-orbital of Ce^{3+}/Eu^{2+} will be split differently in different host material depending on the crystal field around the ion, which will give a strong effect on the emission light and color. Hence the emission band originated from the 5d → 4f transition can be exhibited into a wide range of wavelengths depending on the surrounding environment. Ce^{3+} ion in YAG: Ce^{3+} phosphor existed in a distorted cubic structure, and the observed yellow emission originated from the 5d → 4f electronic transition of Ce^{3+} ion existing in a distorted cubic structure [60]. The schematic diagram of a d-splitting pattern in a distorted cubic environment modified from J. L. Wu *et al.* [60] is, shown in Fig. (**8**).

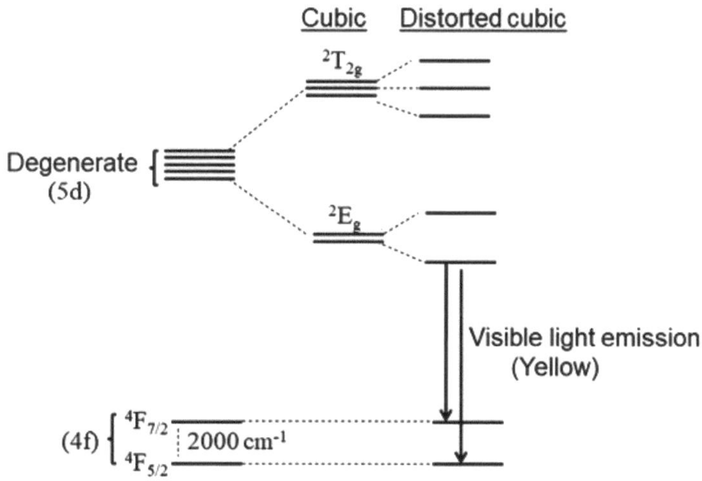

Fig. (8). The d-splitting of Ce^{3+} ion under distorted cubic structure and d-f electronic transition (Modified from J. L. Wu *et al.* [60]).

Advantages and Disadvantages

There are many advantages of this type of fabrication. Some of the important advantages can include high efficiency, simplicity in manufacturing, low cost, high chemical stability, capability to withstand the high radiation intensity of blue LED chip and requirement of only one LED chip.

Although this type of fabrication has several advantages, like other types of fabrications, it also suffers from several drawbacks, such as low color rendering index (Ra < 80) due to the absence of sufficient red component, high correlated color temperature (CCT > 6000 K) due to the instability of blue light released from the LED material, less optical stability and change of color spot with input power. Nevertheless, this technique is still the most commonly used fabrication method in industries for large-scale production because of the above aforementioned advantages.

COMBINATION OF UV/NUV EMITTING LED WITH MULTIPLE PHOSPHORS EMITTING RED, GREEN AND BLUE COLOR

In this fabrication, an LED material emitting UV/NUV light is coated by phosphor materials emitting red, green and blue light. The UV/NUV light released from the LED material is absorbed by the coated phosphor materials and then converted it into red, green and blue light. As UV/NUV light is invisible light, the white light emission in this technique is due to the combination of red, green and blue lights radiated from the attached phosphors only. One example of such type of white light emission is the combination of 400 nm NUV light emitting GaN LED chip with blue emitting $CaPO_4Cl$: Eu^{2+}, green-emitting $(Ba, Sr)_2SiO_4:Eu^{2+}$ and red emitting $CaAlN_3:Eu^{2+}$ phosphors [61]. All these attached phosphors are excited to a higher energy state by absorbing the 400 nm UV light released from the GaN LED material. Fig. (**9**) shows the typical spectral distribution of white light emission observed from this type of fabrication.

In the above example, all the attached phosphors contain the Eu^{2+} ion as an activator ion, therefore, the required blue, green and red lights for white light emission originate from the electronic transition (d-f) within the Eu^{2+} ion of attached phosphor materials. This can be accomplished since the d-f electronic transition of Eu^{2+} is strongly influenced by the crystal structure, composition, bond length, bond energy, covalency and the crystal field environment around Eu^{2+} ion [61]. The Eu^{2+} ion in blue-emitting Ca_2PO_4Cl: Eu^{2+} phosphor exists in two different crystallographic sites, C_2 and C_s symmetry surrounded by 8 neighboring ions, 6 oxide and 2 chloride ions with different bond lengths of Metal-oxide and Metal- Chloride [61, 62]. In the case of green-emitting $(Ba, Sr)_2$ SiO_4: Eu^{2+} phosphor, the Eu^{2+} ion existed in two different distorted crystallographic sites surrounded by 10 and 9 oxide ions [63 - 65]. On the other hand, the Eu^{2+} ion in red-emitting $CaAlSiN_3:Eu^{2+}$ phosphor exists in distorted tetrahedral geometry surrounded by five N^{3-} anions with different bond lengths of Eu^{2+}-N bond [66, 67].

So, because of the presence of different environments around the Eu^{2+} ion in these phosphors, the pattern and amount of d-splitting of the Eu^{2+} ion in each of these phosphors are different, which leads to the emission of light in different color regions.

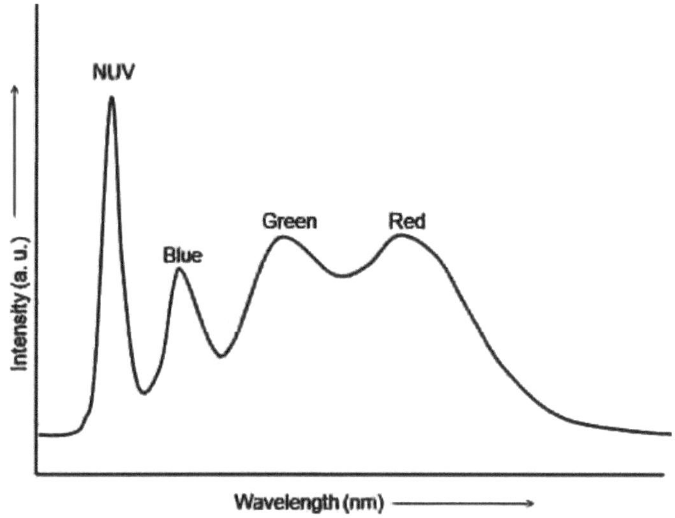

Fig. (9). A typical spectral distribution of pc-WLED constructed by a combination of NUV emitting LED chips with red, green and blue emitting phosphors (Modified from Yi-Chen Chiu *et al.* [61]).

Other combinations are also reported in the literature. Some of the examples are a combination of $LiCaPO_4$: Eu^{2+} (Blue), (Ba, $Sr)_2SiO_4$:Eu^{2+} (Green), and (Sr, Ca)$AlSiN_3$:Eu^{2+} (Red) with 395 nm NUV light emitted LED chip [68], $Sr_5(PO_4)_3Cl$: Eu^{2+} (blue) (Ba, $Sr)_2SiO_4$:Eu^{2+} (Green) and $CaAlSiN_3$:Eu^{2+} (Red) phosphor with 395 nm NUV light emitted LED Chip [69], KMg_4PO_4Cl: Eu^{2+} (Blue), (Ba, $Sr)_2SiO_4$:Eu^{2+} (Green), $CaAlSiN_3$:Eu^{2+} (Red) with 385 nm NUV light emitting chip [70].

Advantages and Disadvantages

The main advantages of this type of fabrication are the generation of high color rendering index light due to the presence of sufficient red components and low color temperature. However, there also existed several drawbacks, including lower efficiency due to the absorption of high energy blue light by the green and red-emitting phosphors and high manufacturing cost since the device requires multiple phosphors of different phases, and those phosphors need to be prepared separately. Moreover, since it is the conversion of high energy into lower energy, there exists a large stock shift which leads to loss of energy as heat.

COMBINATION OF UV/NUV EMITTING LED WITH A SINGLE PHOSPHOR EMITTING BLUE AND YELLOW LIGHTS OR BLUE, GREEN AND RED LIGHTS

This type of white light emission is called single-phase white light emission because the white light is generated from a single phosphor only without any phase segregation. In this fabrication, an LED material emitting UV/NUV light is coated with phosphor emitting multiple colors. The phosphor material absorbed the light released from the LED material and then excited it to a higher energy level. On returning to the ground state, the phosphor material released lights of multiple colors, say blue and yellow or blue, green and red. By combining those released multiple lights from the attached phosphor materials at appropriate proportions, white light is generated. It is a very interesting and attractive fabrication type because the light of highly efficient and high color rendering index could be observed at low costs. In this case, energy transfer from the host material/sensitizer to the activator ion plays an important role in achieving white light emission. Sensitizer is the substance that enhances/improves the absorption coefficient. They are generally added to those phosphors that show low absorption coefficient. Therefore, host material and activator ions are very significant in this fabrication. By this mode of fabrication, white light can be observed by exciting the phosphor material directly or indirectly. Direct excitation involves the excitation of the phosphor through the activator ion, whereas the indirect method involves the excitation of the phosphor material through the host material/sensitizer, followed by control energy transfer to the activator ion. Control of energy transfer can be achieved usually by altering the host composition and concentration of doping ions.

Host Composition Change: With the change of host composition, there will be a change in the crystal field, bond length between M-L, electronegativity, covalency, bond energy *etc*, around the activator ion, which leads to the change of luminescence property and emitting color [9].

Concentration Change: With the change of dopant activator ion concentrations, there will be a change in the energy transfer efficiency from host/sensitizer to activator due to different amounts of acceptor ions present, which will influence the luminescence property and eventually the emitting color. This effect is much less if the activator ion replaces the cation of the same valency. However, it is very significant if the activator ion replaces the cation of different valency because it will create a vacancy in the lattice, which affects the crystal field around the activator ion.

Besides these above factors, the energy transfer can also be influenced by parameters like crystallinity, morphology, particle size *etc* [33, 53]. By this fabrication mode, white light emission can be obtained through the following different modes of mixing [9].

i. Doping of single activator ion into the suitable host material.

ii. Doping of multiple activator ions into the suitable host material.

iii. Controlling the concentrations of defect and reaction conditions of the defect-related luminescent material.

Doping of a Single Activator Ion into the Host Material

It is already known that a phosphor or luminescence material consists of a host material and an activator ion. Usually, rare earth ions such as Dy^{3+}, Eu^{3+}, Sm^{3+} and Eu^{2+} are used as activator ions in this method because they have abundant energy levels, consequently, multiple colors can be produced.

Dy^{3+} ion with $4f^9$ electronic configuration has abundant energy levels and generally exhibits two primary emission bands at blue (470 500 nm) and yellow (570-600 nm) regions due to f-f electronic transition. The yellow emission is originated from ${}^4F_{9/2}$-${}^6H_{13/2}$ transition, which is called the electric dipole transition ($\Delta l = 2$, $\Delta J = 2$) and is hypersensitive to the local environment [9, 17]. Therefore its intensity depends on its local environment. On the other hand, the blue emission originates from ${}^4F_{9/2}$-${}^6H_{15/2}$ transition, which is called magnetic dipole transition ($\Delta J = 3$, magnetic dipole transition) and is insensitive to its local environment [9]. So, the white light can be obtained by a combination of these emitted blue and yellow colors at appropriate proportions. This appropriate combination for producing white light emission can be achieved by controlling the intensity ratio of yellow to blue (Y/B), and it can be done by altering the host composition, synthetic method, covalency, site symmetry, contents of Dy^{3+} ion and excitation wavelength [9]. Some of the examples are $Ca_2Ga_2GeO_7{:}Dy^{3+}$ [71], $ZnWO_4{:}Dy^{3+}$ [72], $CdMoO_4{:}Dy^{3+}$ [17], *etc*. Fig. (**10**) shows the spectral power distribution of white light emission from $CdMoO_4{:}Dy^{3+}$ phosphor synthesized by co-precipitation method at room temperature followed by annealing at 300 °C for 2h and its CIE diagram [17].

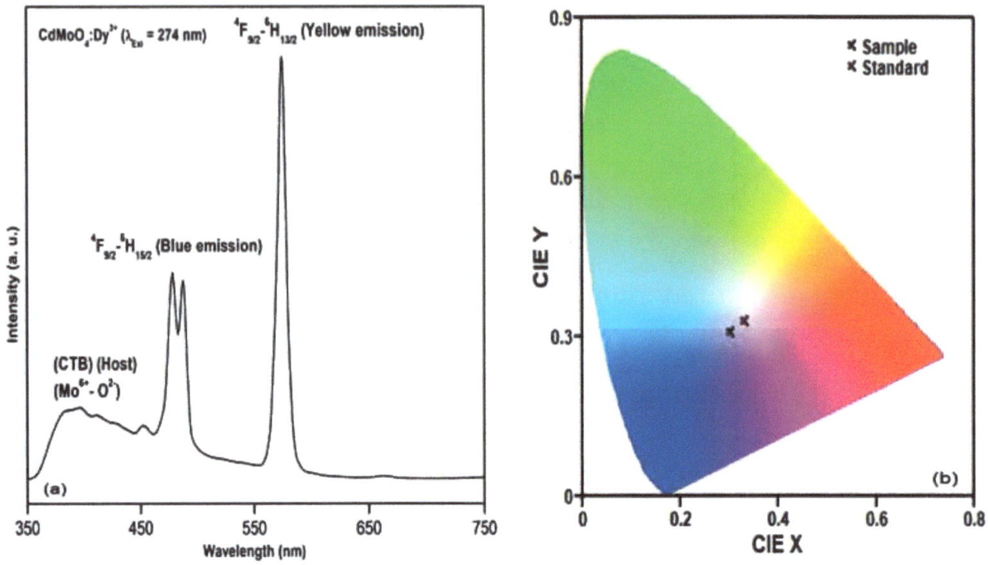

Fig. (10). (a) Emission spectra of white light emitting $CdMoO_4:Dy^{3+}$ phosphor and **(b)** Corresponding CIE chromaticity diagram along with standard chromaticity.

The spectra consist of two sharp peaks at blue and yellow regions originating from the electronic transition within the 4f of Dy^{3+} ion and a broad band in the blue region originated from the charge transfer transition Mo^{6+} — O^{2-} of the host lattice upon excitation at 274 nm through indirect method. The emission from Dy^{3+} ion is observed due to the transfer of energy from host material to Dy^{3+} ion. The CIE color coordinate of the sample is found to be (0.292, 0.312), which is very close to the chromaticity coordinates of the National standard television committee system (0.33, 0.33). The probable mechanism for electronic transition, including excitation, emission, non-radiative transition, radiative transition and energy transfer, is diagrammatically shown in Fig. (**11**).

White light emission can also be obtained by solely doping of Eu^{3+} ion to the suitable host lattice. Eu^{3+} ions having the $4f^6$ electronic configuration contain abundant energy levels in the ground state. Consequently, the emission of light from Eu^{3+} ions can be extended to a whole range of visible light. But, generally, it shows emission in the red region due to the $^5D_0$–^7F_J (J = 0, 1, 2, 3, 4) electronic transitions. The $^5D_0$-$^7F_2$ electronic transition is an electric dipole transition; therefore is hypersensitive to the local environment and emits light in the red region [9]. On the other hand, the transition due to $^5D_0$-$^7F_1$ is a magnetic dipole transition that is hardly affected by the local environment and falls in the orange region [9]. So, one can find out the site symmetry around the Eu^{3+} ion (inversion site or non-inversion) from the intensity ratio of red to orange. In addition to these transitions, Eu^{3+} ions can also exhibit emission in blue-green regions due to electronic transition from the higher ^5D levels such as $^5D_1$ (Blue), $^5D_2$ (Green), $^5D_3$ (Blue) [9]. Thus, white light emission can be obtained by mixing of emitted blue, green color and red color at appropriate proportion. However the appearance of the emission band in blue blue-green region is usually controlled by the crystal structure of the host material, phonon energy and concentration of the doping activator Eu^{3+} ion [9]. With the increase of phonon energy of the host material, this emission is quenched by the multi-phonon relaxation process, and it is quenched by cross relaxation process with the increase of Eu^{3+} ion concentration [9]. Therefore, such type of white light emission is usually observed in the host materials having low phonon energy, such as fluoride, oxyfluoride *etc*. Examples $NaYF_4:Eu^{3+}$ [73], $LaOF:Eu^{3+}$ [32].

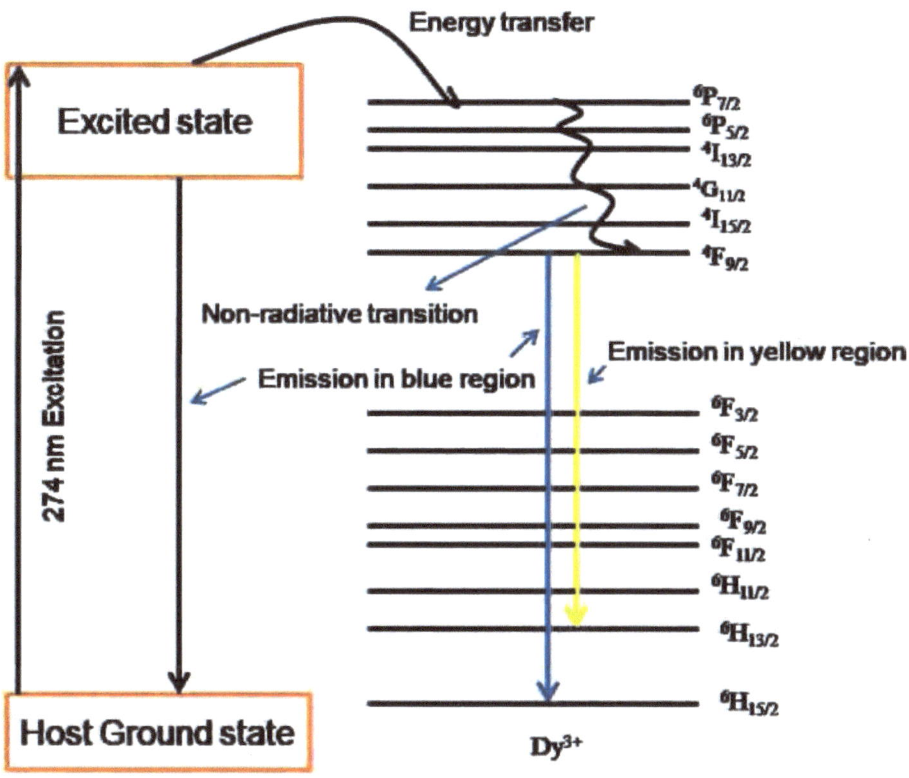

Fig. (11). Schematic diagram showing excitation, emission, radiative and non-radiative transition and energy transfer of $CdMoO_4:Dy^{3+}$ under excitation at 274 nm.

Further, white light emission can also be reported by the appropriate combination of emission in the blue green region from the host material and red emission from Eu^{3+} ion. Such type of white light emission is generally observed from the host materials having intrinsic luminescence in blue-green regions such as tungstate, vanadate, molybdate *etc*. In such cases, the electron is excited through host material and the red emission from Eu^{3+} is obtained by partial transfer of energy from the host material to the Eu^{3+} ion. This energy transfer can be controlled by altering the host composition, the contents of the activator Eu^{3+} ion, particle size, morphology, pH of the synthetic solution, the synthetic method and synthetic temperature *etc*. [33, 53, 54]. For instance, when UV/NUV light from LED material is absorbed by a tungstate host lattice, say $ZnWO_4$, the electron from filled 2p orbital of O^{2-} is excited to empty 5d orbital of W^{6+} ion. This excited energy is partially transferred to the dopant Eu^{3+} ion and then emission is observed from both the host lattice in blue blue-green region and Eu^{3+} ion in the red region [53]. Some examples of such types of white light emission include $CdWO_4:Eu^{3+}$ [33], $ZnWO_4:Eu^{3+}$ [53], $Sr_2V_2O_7:Eu^{3+}$ [54] *etc.*

Similarly, white light emission is also reported from solely doping of Sm^{3+} ion into the appropriate host lattice. Sm^{3+} ion with $4f^5$ electronic configuration usually emits light in the orange region due to electronic transition from excited state $^4G_{5/2}$ to ground states $^6H_{5/2}$, $^6H_{7/2}$ and $^6H_{9/2}$ within the 4f orbital. When this emitted orange color is combined with blue-green color radiated from host material in appropriate proportion, white light can be emitted. Example $Y_2WO_6:Sm^{3+}$ [34].

It is already known that Eu^{2+} ion extends its emission band in a wide range of wavelengths, say from UV to visible light to IR depending on the crystal lattice or crystal field of the surrounding. Therefore, white light can be obtained from Eu^{2+} ions solely doping into a suitable host material. This is generally observed in the host material containing two or more cationic sites for replacement by Eu^{2+} ion, which is normally found in the materials like phosphate, aluminates and silicate *etc* [9], and it creates different crystal field environments around Eu^{2+} ion in the phosphor, consequently, the emission can be extended to whole of visible light. An example of such type of white light emitting phosphor is BaSrMg $(PO_4)_2:Eu^{2+}$ phosphor [35]. In this phosphor, Ba^{2+} and Sr^{2+} ions exist in different crystal field environments, and Eu^{2+} ion is supposed to substitute both these cations due to the closeness of their size [35]. So, the emission from Eu^{2+} ion because of the $4f^65d^1 \rightarrow 4f^7$ electronic transition will fall in different color regions. It is reported that the Eu^{2+} ion substituted for Sr^{2+} ion gives an emission band in yellow color, whereas the Eu^{3+} ion substituted for Ba^{2+} ion gives blue color [35]. Thus, by the appropriate combination of yellow light and blue light emitted from the Eu^{2+} ion existing in different crystal field environments, white light is generated.

Doping of Multiple Activator Ions into the Host Material

In this mode of fabrication, two or more activator ions emitting light at different color regions are co-doped simultaneously into a suitable host material. Generally rare earth ions such as Eu^{3+}, Eu^{2+}, Tb^{3+}, Sm^{3+}, Dy^{3+}, Ce^{3+}, Tm^{3+}, Ho^{3+}, Yb^{3+}, Er^{3+}, Pr^{3+} are considered as the suitable activator ions for this mode of fabrication because they have abundant emitting colors [9]. In this process, the activator ions can be excited simultaneously by absorption of energy released from the LED material to their respective excited states or through energy transfer from the host/sensitizer ion. The doping activator ions will produce their characteristic emission colors on returning to the ground state, and on combining the emitted colors at appropriate proportions, white light will be produced. The control of light intensity for combining to produce white light can be adjusted by altering the concentration of the activator ion. Since white light emission can be obtained by the combination of red, green and blue light or blue and yellow light at appropriate proportions, co-doping of activator ions emitting light at blue, green and red color regions can produce white light. One such example is the co-doping of Tb^{3+}, Dy^{3+}, Eu^{3+} ions into $CaWO_4$ [36]. Generally, Tb^{3+} ion emits light in blue and green regions due to the electronic transition within the 4f orbital, Eu^{3+} ion emits in red region and Dy^{3+} ion in blue and yellow regions. It is reported that upon excitation at 378 nm, both Tb^{3+} and Dy^{3+} ions are excited simultaneously, and on returning to the ground state characteristics, emissions are observed while the red emission from the Eu^{3+} ion is observed through energy transfer from Tb^{3+} ion [36]. Thus, by the combination of emitted blue (Tb^{3+} and Dy^{3+}), green (Tb^{3+}), and red (Eu^{3+}) colors, white light is generated.

White light emission can also be obtained by an energy transfer process from sensitizer or host materials to activator ion. In this process, the sensitizer ion absorbs the energy radiated from the LED material and then partially transfers to the activator ion which makes the excitation of the activator ion. White light is obtained by combining the radiated light at an appropriate proportion on returning to the ground state. For example, $Sr_3B_2O_6$:Ce^{3+}, Eu^{2+} phosphor [74]. It is reported that the Eu^{2+} ion is excited through energy transfer from the Ce^{3+} ion when the phosphor is excited by 351 nm UV light *via* electronic transition within Ce^{3+} ion. The reported emission band consists of the broadband with a maximum in blue and yellow regions originating from the Ce^{3+} and Eu^{2+} ions, respectively. Thus, by controlling the ratio of yellow and blue ratio through altering the concentration ratio of Ce^{3+} and Eu^{2+}, white light emission could be obtained.

Such type of white light emission *via* multiple activator ion doping has been extensively investigated in several host materials. Examples includes $ZnWO_4$:Dy^{3+}, Eu^{3+} [38], $Ca_9Gd(PO_4)_7$:Eu^{2+}, Mn^{2+} [75] *etc.*

CONTROLLING THE CONCENTRATIONS OF DEFECT AND REACTION CONDITIONS OF THE DEFECT-RELATED LUMINESCENT MATERIAL

In this type of phosphor, the luminescence is generated by the presence of defects, mainly vacancies, impurities, radicals *etc*, within specific sites in a lattice without the presence of doping activator ion. Due to the presence of these defects, electronic states are created in between the valence band and conduction band. As a result, several excitations and emissions are possible. The extent of light emissions, such as peak position and intensity, depends on the types of defects and concentrations present in the material, which in turn is induced by various factors, including reaction conditions, such as temperature, pressure, synthetic solvent, reactant composition, synthetic method, morphology *etc.* [9]. The optical emission due to defects gives a broadband and can be extended in a wide range of wavelengths depending on the types of defects and concentrations present. Fig. (**12**) shows typical conduction and valence bands along with energy state induced by defects and different types of possible electronic transition (Excitation and emission). The remarkable growth in defect-related luminescence came after W. H. Green *et al.* developed white light-emitting phosphor from silicate-carboxyllate sol-gel precursor in 1997 [76, 77]. This fabrication is very interesting and attractive because white light can be generated without using expensive activator ions such as rare earth ions or transition metal ions.

Other examples of defect-induced white light emitting phosphors can include Ba_3BPO_7 synthesized by solid-state reaction method [78], Tetragonal phase ZrO_2 synthesized by pechini sol-gel method [79], Al_2O_3 synthesized by pechini sol-gel method with annealing at 500 and 600 ° C [80].

ADVANTAGES AND DISADVANTAGES

The main advantages of this mode of fabrication, as compared to other modes of fabrication include its ability to generate lights of high efficiency and high color rendering index. As only one phosphor is used, the problems of color re-absorption and color mixing ratio can be avoided and will also be cost-effective. However, still, more research is still needed for an effective method, and thermal stability is still a challenging area.

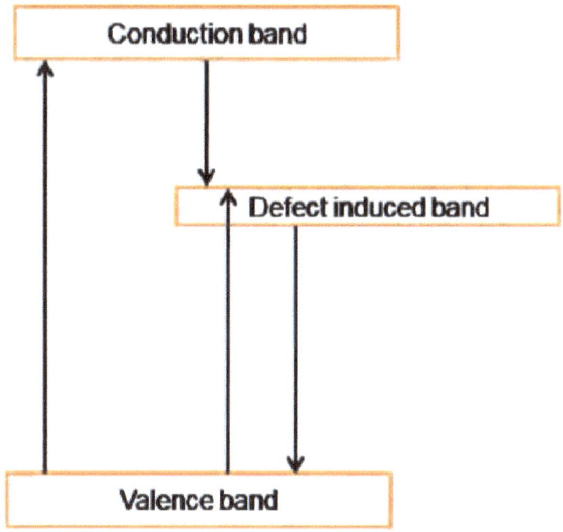

Fig. (12). Typical energy bands of conduction, valence and defect-induced band and possible electronic transition (Excitation and emission).

CURRENT LIMITATIONS AND CHALLENGING

It is seen that phosphor-converted white light emitting diode gains significant advantages in terms of fabrication, cost effectiveness *etc.*, as compared to a combination of monochromatic light emitting LEDs. Many of the phosphors with high photoluminescence quantum yield also have been synthesized and reported. However, the phosphors still do not meet the requirement of commercial standards. The price of the WLED in the market is still high, so it is difficult to afford by the public. In addition, the issue of thermal instability and synthesis of phosphor for high-resolution applications are yet being highly challenging areas. Hence, more research work is still needed.

CONCLUSION

The chapter summarizes the different methods of fabrication of white light emitting diode (WLED). Basically, there are two approaches for the generation of white light: A combination of monochromatic light emitting LEDs and phosphor-converted white light emitting diode. It is revealed that phosphor-converted white light emitting diode has gained advantages of easy fabrication, cost-effective *etc*, and therefore widely used in industries for large scale production. Further, the different methods of fabrication for phosphor-converted white light emitting diode, such as by combining blue emitting LED with yellow emitting phosphor, UV/NUV emitting LED with multiple phosphors and UV/NUV emitting LED with a phosphor emitting multiple colors, are summarized. Single-phase white light emitting diode fabricated by a combination of UV/NUV emitting LED with a phosphor emitting multiple colors gains lots of advantages, including high efficiency, high color rendering index and cost-effectiveness.

REFERENCES

[1] Bergh, A.; Craford, G.; Duggal, A.; Haitz, R. The promise and challenge of solid-state lighting. *Phys. Today,* **2001**, *54*(12), 42-47.
[http://dx.doi.org/10.1063/1.1445547]

[2] Feezell, D.; Nakamura, S. Invention, development, and status of the blue light-emitting diode, the enabler of solid-state lighting. *C. R. Phys.,* **2018**, *19*(3), 113-133.
[http://dx.doi.org/10.1016/j.crhy.2017.12.001]

[3] Joanna, M.; Lauren, E.S.R. Review: Down conversion materials for solid-state lighting. *J. Am. Ceram. Soc.,* **2014**, *97*, 1-26.

[4] Lei, F.; Yin, L.; Shi, Y.; Xie, J.; Zhang, L.; Fan, L. Hydrothermal synthesis of uniform $CdWO_4:Eu^{3+}$ microrods as single component for UV-based WLEDs. *J. Rare Earths,* **2016**, *34*(12), 1179-1187.
[http://dx.doi.org/10.1016/S1002-0721(16)60151-5]

[5] Available from: https://en.wikipedia.org/wiki/Light-emitting_diode

[6] Asim, K.R.C. *Principles of color appearance and measurement,* 1st ed; Elsevier Science B. V: Amsterdam, **2014**.

[7] Holonyak, N., Jr; Bevacqua, S.F. Coherent (VISIBLE) light emission from Ga(As1− *x* P *x*) junctions. *Appl. Phys. Lett.,* **1962**, *1*(4), 82-83.
[http://dx.doi.org/10.1063/1.1753706]

[8] Fret Schubert, E. *Light emitting diode,* 2nd ed; Cambridge University Press: UK, **2006**.
[http://dx.doi.org/10.1017/CBO9780511790546]

[9] Shang, M.; Li, C.; Lin, J. How to produce white light in a single-phase host? *Chem. Soc. Rev.,* **2014**, *43*(5), 1372-1386.
[http://dx.doi.org/10.1039/C3CS60314H] [PMID: 24346246]

[10] Subash, C.S. *Handbook of Light Emitting and Schottky Diode Research,* 1st ed; N P Chen, **2009**, pp. 1-37.

[11] Available from: https://en.wikipedia.org/wiki/CIE_1931_color_space.

[12] Vishwakarma, A.K.; Jha, K.; Jayasimhadri, M.; Sivaiah, B.; Gahtori, B.; Haranath, D. Emerging cool white light emission from Dy^{3+} doped single phase alkaline earth niobate phosphors for indoor lighting applications. *Dalton Trans.,* **2015**, *44*(39), 17166-17174.
[http://dx.doi.org/10.1039/C5DT02436F] [PMID: 26374377]

[13] Hua, Y.; Li, Z. Synthesis and photoluminescence $Sr_2LaTaO_6:Mn^{4+}$ phosphors with ultrahigh color purity for good CRI warm-WLED devices. *Optik,* **2021**, *242*, 167348.
[http://dx.doi.org/10.1016/j.ijleo.2021.167348]

[14] Hua, Y.; Ran, W.; Yu, J.S. Strong red emission with excellent thermal stability in double-perovskite type $Ba_2GdSbO_6:Eu^{3+}$ phosphors for potential field-emission displays. *J. Alloys Compd.,* **2020**, *835*, 155389.
[http://dx.doi.org/10.1016/j.jallcom.2020.155389]

[15] Hua, Y.; Qiu, X.; Sonne, C.; Brown, R.J.C.; Kim, K.H. Construction of novel luminescent thermometers based on dual-emission centers of rare-earth and bismuth ions. *Chemosphere,* **2022**, *303*(Pt 2), 135150.
[http://dx.doi.org/10.1016/j.chemosphere.2022.135150] [PMID: 35660390]

[16] Wang, X.; Zhao, Z.; Wu, Q.; Wang, C.; Wang, Q.; Yanyan, L.; Wang, Y. Structure, photoluminescence and abnormal thermal quenching behavior of Eu^{2+} -doped $Na_3Sc_2(PO_4)_3$: A

novel blue-emitting phosphor for n-UV LEDs. *J. Mater. Chem. C Mater. Opt. Electron. Devices,* **2016**, *4*(37), 8795-8801.
[http://dx.doi.org/10.1039/C6TC01049K]

[17] Singh, N.P.; Singh, N.R.; Devi, Y.R.; Singh Sh, B.; Singh, T.D.; Singh, N.R.; Singh, N.M. Effects of annealing temperature on structural and luminescence properties of $CdMoO_4$:Dy^{3+} phosphor synthesized at room temperature by co-precipitation method. *Solid State Sci.,* **2020**, *102*, 106172.
[http://dx.doi.org/10.1016/j.solidstatesciences.2020.106172]

[18] McCamy, C.S. Correlated color temperature as an explicit function of chromaticity coordinates. *Color Res. Appl.,* **1992**, *17*(2), 142-144.
[http://dx.doi.org/10.1002/col.5080170211]

[19] Available from: https://en.wikipedia.org/wiki/Color_temperature

[20] Nakamura, S.; Senoh, M.; Iwasa, N.; Nagahama, S.; Yamada, T.; Mukai, T. Superbright green InGaN single-quantum-well structure light emitting diodes. *Jpn. J. Appl. Phys.,* **1995**, *34*(10B), L1332-L1335.
[http://dx.doi.org/10.1143/JJAP.34.L1332]

[21] Zhanchao, W.; Zhiguo, X. *Nitride semiconductor light emitting diode; Jianjanh, H.; Hao-Chung, K.; Shyh-Ching, S,* 2nd ed; Elsevier Science B. V: Amsterdam, **2018**, pp. 123-208.

[22] Ye, S.; Xiao, F.; Pan, Y. X.; Ma, Y. Y.; Zh, Q. Y. Phosphors in phosphor converted white light emitting diodes: Recent advances in materials, techniques and properties. *Mater. Sci. Eng. R Rep,* **2010**, *R 71*, 1-34.

[23] Morita, Y.; Tajima, S.; Suzuki, H.; Sugino, H. Cationic copolymerization of epoxy siloxane monomer with liquid poly-butadiene and its light emitting diode encapsulation. *J. Appl. Polym. Sci.,* **2008**, *109*(3), 1808-1813.
[http://dx.doi.org/10.1002/app.28212]

[24] Masui, H.; Nakamura, N. *Encyclopedia of Materials*; Science and Technology, Masui H., **2010**, pp. 1-6.

[25] Kartik, N.Sh.; Dhoble, S.J.; Swart, H.C.; Kyeongsoon, P. *Phosphate phosphors for solid-state lighting*; Springer, **2012**.

[26] Valeur, B.; Berberan-Santos, M.N. A brief history of fluorescence and phosphorescence before the emergence of quantum theory. *J. Chem. Educ.,* **2011**, *88*(6), 731-738.
[http://dx.doi.org/10.1021/ed100182h]

[27] Leverenz, H.W. *An Introduction to Luminescence of Solids*; Wiley and Sons: New York, **1950**.
[http://dx.doi.org/10.1063/1.3067011]

[28] Shigeo, Sh.; William, M.Y.; Hajime, Y. *Phophor Handbook*; CRC Press, Taylor and Francis, **2006**.

[29] Singh, N.P.; Singh, N.R.; Singh, N.M. Synthesis of CdMoO4:Sm3+ phosphors at room temperature and investigation on photoluminescence properties. *Optik,* **2018**, *156*, 365-373.
[http://dx.doi.org/10.1016/j.ijleo.2017.11.045]

[30] He, H.Y. Preparation and luminescence property of Sm-doped $ZnWO_4$ powders and films with wet chemical methods. *Phys. Status Solidi, B Basic Res.,* **2009**, *246*(1), 177-182.
[http://dx.doi.org/10.1002/pssb.200844218]

[31] Ahmad, A.; Ahmad, I.; Ramzan, Sh.; Kiyani, M.Z.; Dubal, D.; Mubarak, N.M. *Nanomedicine manufacturing and application; Francis Verproot, Awais Ahmad, Ching Yern Chee, Ikram Ahmad, Anish Khan*; Elsevier, **2021**, pp. 73-85.

[32] Shang, M.; Li, G.; Kang, X.; Yang, D.; Geng, D.; Peng, C.; Cheng, Z.; Lian, H.; Lin, J. LaOF : Eu^{3+} nanocrystals: Hydrothermal synthesis, white and color-tuning emission properties. *Dalton Trans.,* **2012**, *41*(18), 5571-5580.
[http://dx.doi.org/10.1039/c2dt30082f] [PMID: 22415433]

[33] You, M.; Xu, J.; Zhang, Z.; Zhou, Y. Eu3+:CdWO4 single-phased white-light-emitting phosphor: pH

controlled luminescence properties. *Ceram. Int.,* **2015**, *41*(6), 7698-7703. [http://dx.doi.org/10.1016/j.ceramint.2015.02.099]

[34] Qian, H.; Zhang, J.; Yin, L. Crystal structure and optical properties of white light-emitting Y_2WO_6:Sm^{3+} phosphor with excellent color rendering. *RSC Advances,* **2013**, *3*(23), 9029-9034. [http://dx.doi.org/10.1039/c3ra40742j]

[35] Wu, Z-C.; Liu, J.; Hou, W-G.; Xu, J.; Gong, M-L. A new single-host white-light-emitting $BaSrMg(PO_4)_2$: Eu^{2+} phosphor for white-light-emitting diodes. *J. Alloys Compd.,* **2010**, *498*(2), 139-142. [http://dx.doi.org/10.1016/j.jallcom.2010.03.136]

[36] Barbosa, H.P.; Silva, I.G.N.; Felinto, M.C.F.C.; Teotonio, E.E.S.; Malta, O.L.; Brito, H.F. Photoluminescence of single-phased white light emission materials based on simultaneous Tb3+, Eu3+ and Dy3+ doping in CaWO4 matrix. *J. Alloys Compd.,* **2017**, *696*, 820-827. [http://dx.doi.org/10.1016/j.jallcom.2016.11.378]

[37] You, F.; Huang, S.; Liu, S.; Tao, Y. Synthesis, structure and VUV luminescent properties of rubidium rare-earth fluorides. *J. Solid State Chem.,* **2004**, *177*(8), 2777-2782. [http://dx.doi.org/10.1016/j.jssc.2004.04.040]

[38] Zhai, Y.; Wang, M.; Zhao, Q.; Yu, J.; Li, X. Fabrication and Luminescent properties of ZnWO 4 :Eu 3+, Dy 3+ white light-emitting phosphors. *J. Lumin.,* **2016**, *172*, 161-167. [http://dx.doi.org/10.1016/j.jlumin.2015.11.037]

[39] Kumar, V.; Tiwari, S.P.; Ntwaeaborwa, O.M.; Swart, H.C. *Sectroscopy of lanthanide doped oxide materials; S. J. Dhoble, Vijay B. Pawade, Hendrik C. Swart, Vibha Chopra*; Elsevier, **2020**, pp. 345-364. [http://dx.doi.org/10.1016/B978-0-08-102935-0.00010-1]

[40] Yang, C.H.; Pan, Y.X.; Zhang, Q.Y.; Jiang, Z.H. Cooperative energy transfer and frequency upconversion in Yb^{3+}-Tb^{3+} and Nd^{3+}-Yb^{3+}-Tb^{3+} codoped $GdAl_3(BO_3)_4$ phosphors. *J. Fluoresc.,* **2007**, *17*(5), 500-504. [http://dx.doi.org/10.1007/s10895-007-0200-6] [PMID: 17609865]

[41] Pan, Y.X.; Zhang, Q.Y. White upconverted luminescence of rare earth ions codoped $Gd_2(MoO_4)_3$ nanocrystals. *Mater. Sci. Eng. B,* **2007**, *138*(1), 90-94. [http://dx.doi.org/10.1016/j.mseb.2006.12.015]

[42] Bhatkar, V.B.; Omanwar, S.K.; Moharil, S.V. Combustion synthesis of silicate phosphors. *Opt. Mater.,* **2007**, *29*(8), 1066-1070. [http://dx.doi.org/10.1016/j.optmat.2006.04.006]

[43] Peng, T.; Yang, H.; Pu, X.; Hu, B.; Jiang, Z.; Yan, C. Combustion synthesis and photoluminescence of $SrAl_2O_4$:Eu,Dy phosphor nanoparticles. *Mater. Lett.,* **2004**, *58*(3-4), 352-356. [http://dx.doi.org/10.1016/S0167-577X(03)00499-3]

[44] Wako, A.H.; Dejene, F.B.; Swart, H.C. Combustion synthesis, characterization and luminescence properties of barium aluminate phosphor. *J. Rare Earths,* **2014**, *32*(9), 806-811. [http://dx.doi.org/10.1016/S1002-0721(14)60145-9]

[45] Nair, G.B.; Swart, H.C.; Dhoble, S.J. A review on the advancements in phosphor-converted light emitting diodes (pc-LEDs): Phosphor synthesis, device fabrication and characterization. *Prog. Mater. Sci.,* **2020**, *109*, 100622. [http://dx.doi.org/10.1016/j.pmatsci.2019.100622]

[46] Sun, J.; Sun, R.; Du, H. Improved moisture resistance of $SrSO_4$:Sm^{3+} phosphors coated with SiO_2. *Appl. Surf. Sci.,* **2012**, *258*(10), 4569-4573. [http://dx.doi.org/10.1016/j.apsusc.2012.01.029]

[47] Dutta, D.P.; Roy, M.; Tyagi, A.K. Dual function of rare earth doped nano Bi_2O_3: White light emission and photocatalytic properties. *Dalton Trans.,* **2012**, *41*(34), 10238-10248.

[http://dx.doi.org/10.1039/c2dt30653k] [PMID: 22791072]

[48] He, C.; Xia, Z.; Liu, Q. Microwave solid state synthesis and luminescence properties of green-emitting $Gd_2O_2S:Tb^{3+}$ phosphor. *Opt. Mater.,* **2015**, *42*, 11-16. [http://dx.doi.org/10.1016/j.optmat.2014.12.012]

[49] Lee, W.C.; Yang, H.K.; Kwak, M.; Moon, B.K. Microwave-assisted sintering synthesis and luminescence characteristics of Sr3SiO5: Eu^{2+} phosphors for Eu^{2+} concentrations. *Optik,* **2018**, *172*, 1205-1210. [http://dx.doi.org/10.1016/j.ijleo.2018.07.129]

[50] Mi, C.C.; Tian, Z.; Han, B.; Mao, C.; Xu, S. Microwave-assisted one-pot synthesis of water-soluble rare-earth doped fluoride luminescent nanoparticles with tunable colors. *J. Alloys Compd.,* **2012**, *525*, 154-158. [http://dx.doi.org/10.1016/j.jallcom.2012.02.095] [PMID: 22879690]

[51] Takeda, T.; Xie, R.J.; Suehiro, T.; Hirosaki, N. Nitride and oxynitride phosphors for white LEDs: Synthesis, new phosphor discovery, crystal structure. *Prog. Solid State Chem.,* **2018**, *51*, 41-51. [http://dx.doi.org/10.1016/j.progsolidstchem.2017.11.002]

[52] West, A.R. *Solid state chemistry and its application,* 2nd ed; Wiley and Sons, **2014**.

[53] Wu, H.; Sun, Z.; Gan, S.; Li, L. Effects of alkali metal as charge compensator on the luminescence properties of $ZnWO_4$: Eu^{3+} phosphors synthesized by solid-state reaction. *J. Photochem. Photobiol. Chem.,* **2019**, *368*, 258-262. [http://dx.doi.org/10.1016/j.jphotochem.2018.09.048]

[54] Zhou, Z.; Wang, N.; Zhou, N.; He, Z.; Liu, S.; Liu, Y.; Tian, Z.; Mao, Z.; Hintzen, H.T. High colour purity single-phased full colour emitting white LED phosphor $Sr_2V_2O_7$: Eu^{3+}. *J. Phys. D Appl. Phys.,* **2013**, *46*(3), 035104. [http://dx.doi.org/10.1088/0022-3727/46/3/035104]

[55] Zhang, Z.; Li, J.; Yang, N.; Liang, Q.; Xu, Y.; Fu, S.; Yan, J.; Zhou, J.; Shi, J.; Wu, M. A novel multi-center activated single-component white light-emitting phosphor for deep UV chip-based high color-rendering WLEDs. *Chem. Eng. J.,* **2020**, *390*, 124601. [http://dx.doi.org/10.1016/j.cej.2020.124601]

[56] Blasse, G.; Grabmaire, B.C. *Luminescence Materials*; Springer, **1994**. [http://dx.doi.org/10.1007/978-3-642-79017-1]

[57] Cornelis, R.R. *Luminescence: From theory to application*; Wiley, **2008**.

[58] Sun, L.; Devakumar, B.; Li, B.; Liang, J.; Guo, H.; Huang, X. Synthesis and characterization of $Ca_3Lu(GaO)_3(BO_3)_4$: Ce^{3+}, Tb^{3+} phosphors: Tunable-color emissions, energy transfer, and thermal stability. *RSC Advances,* **2018**, *8*(41), 23284-23293. [http://dx.doi.org/10.1039/C8RA03289K] [PMID: 35540160]

[59] Cho, J.; Park, J.H.; Kim, J.K.; Schubert, E.F. White light-emitting diodes: History, progress, and future. *Laser Photonics Rev.,* **2017**, *11*(2), 1600147. [http://dx.doi.org/10.1002/lpor.201600147]

[60] Wu, J.L.; Gundiah, G.; Cheetham, A.K. Structure–property correlations in Ce-doped garnet phosphors for use in solid state lighting. *Chem. Phys. Lett.,* **2007**, *441*(4-6), 250-254. [http://dx.doi.org/10.1016/j.cplett.2007.05.023]

[61] Chiu, Y-C.; Liu, W-R.; Chang, C-K.; Liao, C-C.; Yeh, Y-T.; Jang, S-M.; Chen, T-M. Ca_2PO_4Cl : Eu^{2+}: An intense near-ultraviolet converting blue phosphor for white light-emitting diodes. *J. Mater. Chem.,* **2010**, *20*(9), 1755-1758. [http://dx.doi.org/10.1039/b920610h]

[62] Yi-Chen, C.; Liu, W-R.; Chang, C-K.; Liao, C-C.; Yeh, Y-T.; Jang, S-M.; Chen, T-M. Intense blue-emitting $Ca_2PO_4Cl:Eu^{2+}$ phosphor for near-ultraviolet converting white light-emitting diodes. *J. Rare Earths,* **2010**, *28*, 250-253.

[http://dx.doi.org/10.1016/S1002-0721(10)60374-2]

[63] Barry, T.L. Fluorescence of Eu^{2+} activated phases in binary alkaline earth orthosilicate system. *J. Electrochem. Soc.*, **1968**, *115*(11), 1181-1184. [http://dx.doi.org/10.1149/1.2410935]

[64] Han, J.K.; Hannah, M.E.; Piquette, A.; Talbot, J.B.; Mishra, K.C.; McKittrick, J. Nano- and submicron sized europium activated silicate phosphors prepared by a modified co-precipitation method. *ECS J. Solid State Sci. Technol.*, **2012**, *1*(3), R98-R102. [http://dx.doi.org/10.1149/2.018203jss]

[65] Pavitra, E.; Seeta Rama Raju, G.; Oh, J.; Krishna Bharat, L.; Jeong, J.H.; Huh, Y.S.; Yu, J.S. $(BaSr)_2SiO_4:Eu^{2+}$ nanorods with enhanced luminescence properties as green-emitting phosphors for white LED applications. *Dyes Pigments,* **2017**, *142*, 447-456. [http://dx.doi.org/10.1016/j.dyepig.2017.03.065]

[66] Kyota, U.; Naoto, H.; Yoshinobu, Y.; Atsushi, N.; Takuya, N.; Hajime, Y. Luminescence properties of a red phosphor, $CaAlSiN_3:Eu^{2+}$, for white light-emitting diodes. *Electrochem. Solid-State Lett.*, **2016**, *9*, H22-H25.

[67] Wang, J.; Zhang, H.; Lei, B.; Xia, Z.; Dong, H.; Liu, Y.; Zheng, M.; Xiao, Y. Enhanced photoluminescence and phosphorescence properties of red $CaAlSiN_3:Eu^{2+}$ phosphor *via* simultaneous UV-NIR stimulation. *J. Mater. Chem. C Mater. Opt. Electron. Devices,* **2015**, *3*(17), 4445-4451. [http://dx.doi.org/10.1039/C5TC00236B]

[68] Yang, X.; Chen, J.; Chai, C.; Zheng, S.; Chen, C. Near ultraviolet excited white light emitting diode (WLED) based on the blue $LiCaPO_4$: Eu^{2+} phosphor. *Optik,* **2019**, *198*, 163238. [http://dx.doi.org/10.1016/j.ijleo.2019.163238]

[69] Zheng, J.; Cheng, Q.; Wu, S.; Guo, Z.; Zhuang, Y.; Lu, Y.; Li, Y.; Chen, C. An efficient blue-emitting $Sr_5(PO_4)_3Cl:Eu^{2+}$ phosphor for application in near-UV white light-emitting diodes. *J. Mater. Chem. C Mater. Opt. Electron. Devices,* **2015**, *3*(42), 11219-11227. [http://dx.doi.org/10.1039/C5TC02482J]

[70] Chen, J.; Liu, Y.; Mei, L.; Liu, H.; Fang, M.; Huang, Z. Crystal structure and temperature-dependent luminescence characteristics of $KMg_4(PO_4)_3:Eu^{2+}$ phosphor for white light-emitting diodes. *Sci. Rep.*, **2015**, *5*(1), 9673. [http://dx.doi.org/10.1038/srep09673] [PMID: 25855866]

[71] Zhou, J.; Yu, X.; Wang, T.; Zhou, D.; Qiu, J. A single-phased white-emitting $Ca_2Ga_2GeO_7:Dy^{3+}$ phosphor with different charge compensation ions. *J. Rare Earths,* **2017**, *35*(3), 241-246. [http://dx.doi.org/10.1016/S1002-0721(17)60906-2]

[72] Zhai, Y.; Li, X.; Liu, J.; Jiang, M. A novel white-emitting phosphor ZnWO4:Dy3+. *J. Rare Earths,* **2015**, *33*(4), 350-354. [http://dx.doi.org/10.1016/S1002-0721(14)60425-7]

[73] Li, C.; Zhang, C.; Hou, Z.; Wang, L.; Quan, Z.; Lian, H.; Lin, J. β-$NaYF_4$ and β-$NaYF_4:Eu^{3+}$ microstructures: Morphology control and tunable luminescence properties. *J. Phys. Chem. C,* **2009**, *113*(6), 2332-2339. [http://dx.doi.org/10.1021/jp8101628]

[74] Chang, C-K.; Chen, T-M. $Sr_3B_2O_6$: Ce^{3+}, Eu^{2+}: A potential single-phased white-emitting borate phosphor for ultraviolet light-emitting diodes. *Appl. Phys. Lett.*, **2007**, *91*(8), 081902. [http://dx.doi.org/10.1063/1.2772195]

[75] Huang, C-H.; Liu, W-R.; Chen, T-M. Single-phased white-light phosphors $Ca_6Gd(PO_4)_7:Eu^{2+}$, Mn^{2+} under near ultraviolet excitation. *J. Phys. Chem. C,* **2010**, *114*(43), 18698-18701. [http://dx.doi.org/10.1021/jp106693z]

[76] Zhang, C.; Lin, J. Defect-related luminescent materials: Synthesis, emission properties and applications. *Chem. Soc. Rev.,* **2012**, *41*(23), 7938-7961.

[http://dx.doi.org/10.1039/c2cs35215j] [PMID: 23019577]

[77] Green, W.H.; Le, K.P.; Grey, J.; Au, T.T.; Sailor, M.J. White phosphors from a silicate-carboxylate sol-gel precursor that lack metal activator ions. *Science,* **1997**, *276*(5320), 1826-1828.
[http://dx.doi.org/10.1126/science.276.5320.1826]

[78] Han, B.; Liu, B.; Dai, Y.; Zhang, J. Defect-induced luminescence of a self-activated borophosphate phosphor. *J. Electron. Mater.,* **2018**, *47*(8), 4550-4554.
[http://dx.doi.org/10.1007/s11664-018-6329-z]

[79] Lin, C.; Zhang, C.; Lin, J. Phase transformation and photoluminescence properties of nanocrystalline ZrO_2 powders prepared *via* the Pechini-type sol-gel process. *J. Phys. Chem. C,* **2007**, *111*(8), 3300-3307.
[http://dx.doi.org/10.1021/jp066615l]

[80] Lin, C.; Yu, M.; Cheng, Z.; Zhang, C.; Meng, Q.; Lin, J. Bluish-white emission from radical carbonyl impurities in amorphous Al_2O_3 prepared *via* the Pechini-type sol-gel process. *Inorg. Chem.,* **2008**, *47*(1), 49-55.
[http://dx.doi.org/10.1021/ic700652v] [PMID: 18069822]

CHAPTER 5

Electronic and Piezoelectric Properties of Non-metal Doped II-VI Monolayer Compounds

Lalmuan Chhana[1,2], **Zodinmawia**[1], **Ramesh Chandra Tiwari**[1], **Shivraj Gurung**[2] and **Lalhriat Zuala**[2,*]

[1] *Department of Physics, School of Physical Sciences, Mizoram University, Aizawl, Mizoram, India*

[2] *Physical Sciences Research Centre (PSRC), Department of Physics, Pachhunga University College, Mizoram University, Aizawl, Mizoram, India*

Abstract: The enhancement of nano-system properties, particularly low dimensional structures, is of great importance for future devices. Using spin-polarized Density Functional Theory (DFT), electronic and piezoelectric properties of II-VI monolayer (ZnO, ZnS, CdO and CdS) are studied. Variations of these properties are further studied under substitutional doping of non-metal atoms (boron and carbon). Doping with a B/C atom transforms all the monolayers into half-metallic ferromagnet, with changes arising mainly from *p*-orbitals of the dopant (B/C) atom. Reduction of band gap energy from its pristine structure is observed in all the doped cases. Observations predicted that the B-doped ZnO and ZnS monolayer showed negative structural stiffness and negative piezoelectric tensors, while C-doping remains stable with enhanced elastic as well as piezoelectric properties of the monolayer.

Keywords: Dynamic stability, Non-metal doping, Piezoelectric property, II-VI monolayer.

INTRODUCTION

The existence of two-dimensional material piezoelectricity, which is absent in its bulk form, has already been the focus of several theoretical predictions [1] and experimental reports [2]. In monolayer piezoelectric materials, crystal structure plays a crucial role in piezoelectricity and can lead to flexo-electricity [3]. Since the majority of the atoms in monolayer materials are surface-based, the impact of the surface on its physical properties cannot be disregarded. Thus, surface changes resulting in surface piezoelectricity and ferroelectricity can significantly modify the piezoelectric characteristics of monolayer materials [4 - 6]. For designing

* **Corresponding author Lalhriat Zuala:** Physical Sciences Research Centre (PSRC), Department of Physics, Pachhunga University College, Mizoram University, Aizawl, Mizoram, India; E-mail: hriatamail@gmail.com

Dibya Prakash Rai, Kingsley O. Obodo, & Jitendra Pal Singh (Eds.)

piezoelectricity, such universal surface characteristics of the monolayer materials are intriguing. In fact, some monolayer materials that are fundamentally non-piezoelectric due to the preservation of their inversion symmetry, even in their monolayer counterpart, can nonetheless be converted to this property through surface or structural changes. As a result, there are a lot more piezoelectric materials in two dimensions, and it is theoretically possible to develop materials with better piezoelectric capabilities.

Among the low dimensional materials, group II-VI monolayer compounds with wide band gaps are widely studied due to their versatility as a result of their tunable band gap nature. After the stabilities and possible realization of monolayers were verified experimentally [7], theoretical investigations of ZnO, ZnS, CdO, and CdS monolayers and their properties attracted huge attention among the II-VI monolayer compounds [8 - 15]. Later experimental reports of two monolayer thick ZnO films, in which the hexagonal ZnO sheets possess smaller buckling, confirmed the prediction of DFT calculations [16, 17]. Following the experimental confirmation of ZnO monolayer, a single layer of ZnS was also successfully grown on Ag (111) substrate [18]. Successful prediction of *ab initio* DFT calculation inspires further theoretical investigation on various properties of wide band gap II-VI semiconducting compounds. Reported calculations on CdO and CdS have also shown that these materials possess low formation energy in their two-dimensional structure with sp^2 hybridized orbitals [19, 20]. These studies motivate deeper insight into these monolayers and their properties and to find a way to enhance them by introducing dopant atoms.

The additional benefits of monolayer materials can also be vital for the design of piezoelectric monolayer devices in addition to these specific benefits, which include the availability of other piezoelectric materials and piezoelectricity manufacturing. Besides, monolayer materials are fundamentally flexible and strong. Furthermore, a limited number of atoms in these monolayers significantly enables atomistic-level simulations for a practical approach. Moreover, free charges in semi-conductive monolayer materials can be electrostatically controlled by gates without gates and interconnects. In this study, variation of piezoelectric properties of the proposed monolayers will be studied via induction of structural defects as a result of non-metal doping.

COMPUTATIONAL METHODS

For the study, a 4x4x1 supercell of each monolayer is taken for the pristine structure. A vacuum of 15 Å is inserted along the *z*-direction to avoid periodic interaction between monolayers. For each of the doped structures, one of the oxygen/sulfur atoms in a supercell is substituted with a non-metal (B/C) dopant atom. Structural optimization is performed until the forces on each atom become less than 0.01 eV/ Å with energy convergence criteria of 10^{-5} eV.

All the computations are performed with spin-polarized plane-wave density functional theory (DFT) as implemented in the Vienna ab initio simulation package (VASP) [21]. The ionic potentials are represented by Projector Augmented Wave Potentials (PAW) [22] with its electron exchange-correlation treated by Generalized Gradient Approximation (GGA) in the form of Perdew, Burke and Ernzerhof (PBE) for solids [23]. An energy cut-off of 400 eV is used for the plane-wave expansion of the electronic wave function. Brillouin zone integration is performed using 8 x 8 x 1 k-point mesh within the Monkhorst-Pack scheme [24], with the properties calculation sampled by denser 16 x 16 x 1 k-point mesh.

Elastic and piezoelectric calculations for zero external electric field can be expressed as:

$$\begin{bmatrix} \sigma_1 \\ \sigma_2 \\ \sigma_6 \end{bmatrix} = \begin{bmatrix} C_{11} & C_{12} & 0 \\ C_{12} & C_{11} & 0 \\ 0 & 0 & \frac{C_{11}-C_{12}}{2} \end{bmatrix} \begin{bmatrix} S_1 \\ S_2 \\ S_6 \end{bmatrix} \tag{1}$$

and

$$\begin{bmatrix} P_1 \\ P_2 - P_2^0 \end{bmatrix} = \begin{bmatrix} 0 & 0 & -d_1 \\ -d_1 & d_1 & 0 \end{bmatrix} \begin{bmatrix} S_1 \\ S_2 \\ S_6 \end{bmatrix} \tag{2}$$

respectively. In the above expressions, σ_1 and σ_2 are the normal stresses along *x* and *y* directions, respectively. S_1 and S_2 are the normal strains along the *x* and *y* directions, respectively. σ_6 and S_6 are the in-plane shear stress and shear strain, respectively. P20 is the spontaneous polarization in the *y*-direction. P_1 and P_2 are polarizations along *x* and *y* directions, respectively. C_{11} and C_{12} are two independent elastic constants, while d_1 is an independent piezoelectric constant.

Piezoelectric coefficients for monolayer system are related by the equation:

$$e_1 = d_1(C_{11} - C_{12}) \tag{3}$$

To analyze the relative stability of doped monolayer, formation energy is calculated using,

$$E_f = E_{D\text{-}ML} - E_{ML} + E_X - E_D \quad (4)$$

where $E_{D\text{-}ML}$ and E_{ML} are the total energies per supercell of the doped and pristine monolayer, respectively. E_X and E_D denote the energies of the isolated replaced atom and dopant atom, respectively.

RESULTS AND DISCUSSION

For initial calculation, the lattice parameters (a_0) were taken from the previously calculated results (theoretical or experimental), and structural optimization was performed. For ZnO ML a_0= 3.303 Å [17], a_0= 3.7995 Å for ZnS ML [25], a_0= 3.68 Å for CdO ML [19] and a_0= 4.24 Å CdS ML [20]. The calculated lattice parameters for ZnO ML, ZnS ML, CdO ML and CdS ML are 3.28 Å, 3.89 Å, 3.67 Å and 4.25 Å, respectively. The calculated bond lengths of Zn-O, Zn-S, Cd-O and CdS in their respective monolayer are 1.89 Å , 2.24 Å , 2.12 Å and 2.46 Å , respectively, which are shorter than their respective bulk counterpart of 2.01 Å , 2.36 Å , 2.39 Å and 2.58 Å . This decrease in the bond length as a result of reduced dimension may be attributed to the changes in hybridization from sp^3 to sp^2 in the monolayer structure. All the monolayers investigated in the present work are studied in their P1 structure with space group *P6m2,* as shown in Fig. (**1**). The optimized lattice parameters are slightly different than the previously reported values but fairly close and are given in Table **1**.

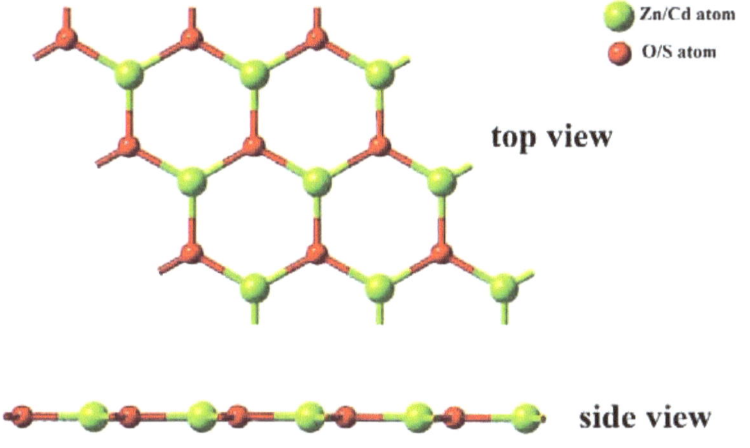

Fig. (1). Top view and side view of **a** graphene-like structure of II-VI monolayer compound.

Table 1. Structural parameters of monolayer ZnO, ZnS, CdO and CdS.

System	Lattice Parameter (a_0) (Å)		Bond Length (Å)	
	Previous results	Calculated results	Bulk (wurtzite)	Calculated results
ZnO ML	3.303[a]	3.28	2.01	1.89
ZnS ML	3.812[b]	3.89	2.36	2.24
CdO ML	3.68[c]	3.67	2.39	2.12
CdS ML	4.24[d]	4.25	2.58	2.46

[a] [17], [b] [25], [c] [19], [d] [20]

We also calculate the phonon dispersion curve to determine if the pristine monolayers maintain their crystal structure at ambient conditions, as shown in Fig. (**2**). Phonons are calculated by expanding the crystal Hamiltonian initially as a function of the displacements of ions from their equilibrium positions. Over the entire Brillouin zone, no imaginary phonon modes are present, confirming the dynamic stability of the proposed monolayer structures.

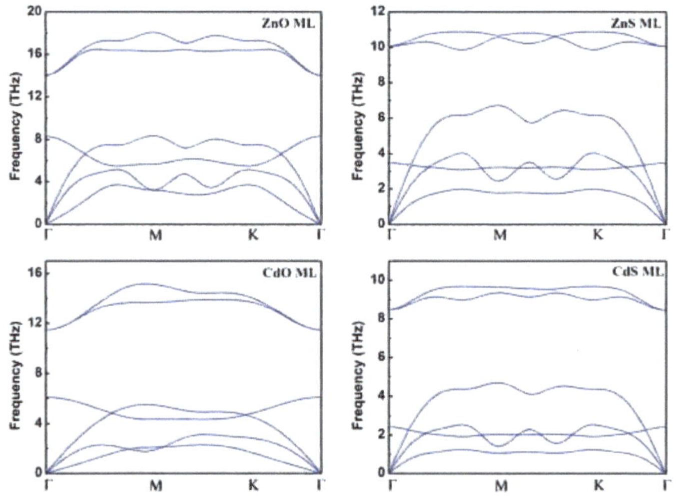

Fig. (2). Phonon dispersion curves of ZnO, ZnS, CdO and CdS monolayer.

Pristine Properties

Band structures and density of states (PDOS) are calculated from the optimized ZnO, ZnS, CdO and CdS monolayer structures and are shown in Fig. (**3**). The spin components of the electronic band structures and DOS are symmetric in all the structures, which indicates that they are non-magnetic semiconductors with a direct band gap of 1.65 eV, 2.64 eV, 0.69 eV and 1.82 eV for ZnO, ZnS, CdO and

CdS monolayer respectively, at the Γ point. The presence of Fermi energy closer to the valence band maximum indicates that the proposed monolayers possess intrinsic *p-type* behavior. These findings are consistent with other DFT results [10, 11, 26, 27]. Projected DOS (PDOS) is plotted in Fig. (**3**), showing the contribution and possible hybridization of the orbitals. It is evident from Fig. (**3a-d**) that S-*p* and O-*p* orbitals mainly contribute to the density of states near the band edge. Symmetry in the states confirms its non-magnetic behavior.

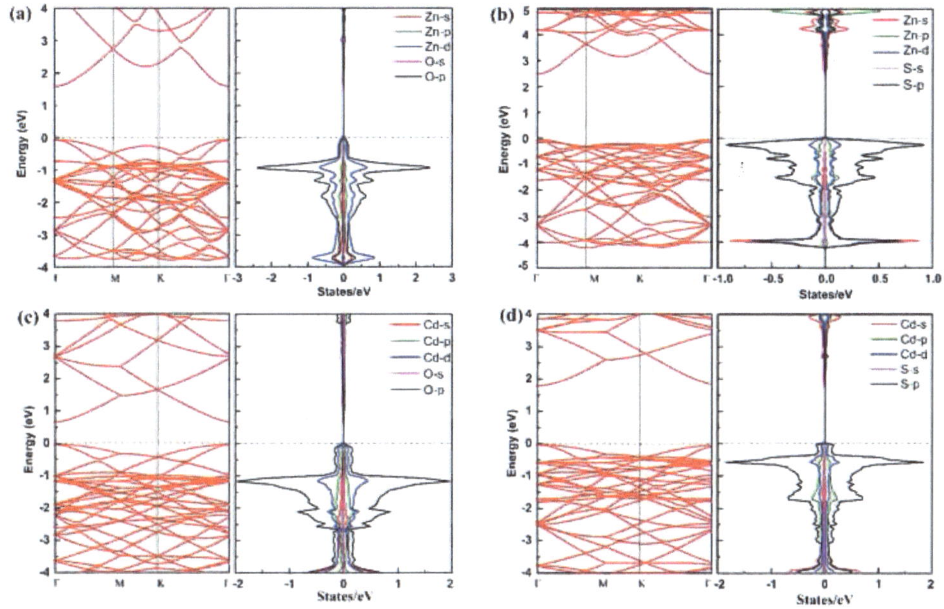

Fig. (3). Band structures and PDOS of monolayer **(a)** ZnO, **(b)** ZnS, **(c)** CdO and **(d)** CdS.

Elastic constants are crucial for the mechanical properties of materials. Elastic constants are found to be C_{11}=83.6 & C_{12}=54.39 for the ZnO monolayer and C_{11}=58.63 & C_{12}=28.63 for the ZnS monolayer. These findings are consistent with earlier studies [28]. Calculated elastic constants for the CdO monolayer are found to be C_{11}=60.94 & C_{12}=45.52 and also, C_{11}=44.34 & C_{12}=27.74 for the CdS monolayer.

All in-plane piezoelectric coefficients, e_1 and d_1, are calculated for all the proposed pristine monolayers and are given in Table **2**, along with the elastic coefficients. Calculated relaxed-ion piezoelectric stress coefficient (e_1) for ZnO, ZnS, CdO and CdS monolayer are found to be 2.44 x 10^{-10} C m^{-1}, 0.57 x 10^{-10} C m^{-1}, 3.13 x 10^{-10} C m^{-1} and 0.99 x 10^{-10} C m^{-1} respectively. Relatively, their respective piezoelectric strain coefficient (d_1) is calculated from e_1 and is found to be 8.35

pm V^{-1}, 1.91 pm V^{-1}, 20.30 pm V^{-1} and 5.96 pm V^{-1}. These findings are in good agreement with other studies [29, 30].

Table 2. Elastic constants (C_{11} and C_{12}) and piezoelectric coefficients (d_1 and e_1) of the proposed monolayers.

System	Elastic Constant		Piezoelectric Coefficient	
	C_{11}	C_{12}	e_1 (10^{-10} C m^{-1})	d_1 (pm V^{-1})
ZnO ML	83.6	54.39	2.44	8.35
ZnS ML	58.63	28.63	0.57	1.91
CdO ML	60.94	45.52	3.13	20.30
CdS ML	44.34	27.74	0.99	5.96

Non-metal Doping

Geometric optimization shows that all of the doped structures are perfectly planar. However, substituting smaller atoms at the O/S site changes the bond length around the dopant atom, as shown in Fig. (**4**). The bond length between Zn and B atoms is 1.98 Å in B-doped ZnO ML and 2.08 Å in B-doped ZnS ML. Cd-B bond length is also found to be 2.09 Å and 2.17 Å in B-doped CdO and CdS ML, respectively. By doping with the C atom, the Zn-C bond in doped ZnO and ZnS ML are found to be 1.90 Å and 1.97 Å , respectively, while respective Cd-C bond length in doped CdO and CdS monolayer are 2.03 Å and 2.10 Å . The formation energy of each of the doped structures is calculated and highlighted in Table **3**. along with the structural parameters. These calculations suggest that doping ZnO, ZnS, CdO and CdS monolayer with B/C are relatively stable.

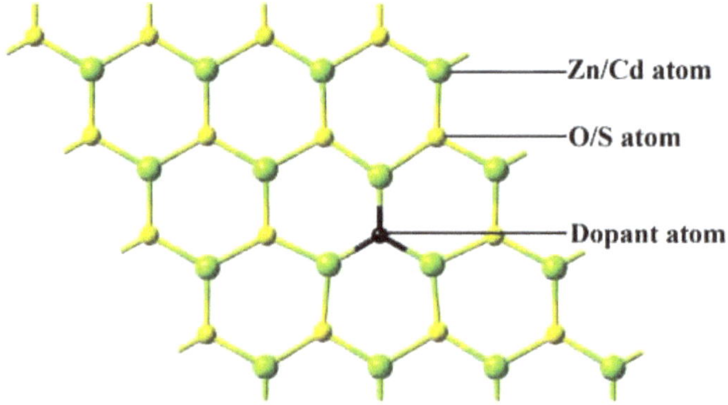

Fig. (4). Boron/Carbon doped II-VI monolayer.

Table 3. Bond length between Zn/Cd and dopant (B/C) atom and formation energies of the doped structures.

System	Bond Length (Å)		Formation Energy (eV)	
	Zn/Cd-B	Zn/Cd-C	B-doped	C-doped
ZnO ML	1.98	1.90	1.07	0.13
ZnS ML	2.08	1.97	2.84	0.46
CdO ML	2.09	2.03	5.12	2.96
CdS ML	2.17	2.10	2.27	0.08

In the ZnO monolayer, besides the introduction of half-metallic property, B-dopant significantly modified the energy bands. The band gap at the Γ-point is around 0.67 eV. The conduction band minimum here denotes the flat band, which arises due to the defect states created by B-atom. The energy gap corresponding between the top of the valence band and the bottom of the parabolic band is also found to be 1.23 eV. Similarly, the formation of deep-lying defect states in the semiconducting spin state is observed in the case of the B-doped ZnS monolayer, which is at 0.78 eV above the Fermi energy. In the B-doped CdO monolayer, the energy gap is 0.35 eV, while the energy band gap is found to be 0.65 eV in the B-doped CdS monolayer. Reduction of band gap energy from its pristine structure is observed in all the doped cases. This is due to the B atom having a significantly smaller atomic radius occupying the O/S site in the lattice, creating distortion and subsequently deforming the crystal structure around the dopant atom. This results in the reduction of a band gap, which agrees with reported studies on the effect of strain on monolayers [31, 32]. Band structures of B-doped ZnO, ZnS, CdO and CdS monolayer are presented in Fig. (**5**), with spin up (spin down) states represented by red (black) color.

Fig. (**6a-d**) shows the PDOS of atomic orbitals of B atom and its neighboring atoms in the B-doped ZnO, ZnS, CdO and CdS monolayers, respectively. The contribution of B-p orbitals, is quite significant in all the structures. Furthermore, the flat band line above the Fermi energy in the band structure corresponds to the unoccupied states of Zn-3*d*, O-2*p* and B-2*p* orbitals. In ZnS ML, the defect energy states in the electronic band structure are mainly due to the 2*p* orbitals of the dopant atom, suggesting weak hybridization between dopant 2*p* orbitals and host atomic orbitals. In the B-doped CdO ML system, Fig. (**6c**), a flat band line observed in the band structure arises from B-2*p*, Cd-4*d*, Cd-4*p*, and O-2*p* orbitals. PDOS shows that the spin-down states possess metallic character due to the B-2*p* orbital crossing the Fermi level. In CdS ML, a high density of states formed mainly by the unoccupied 2*p* orbitals of the dopant atom is observed in both spin orientations. Doping B atom in the proposed monolayers results in flat band lines, which, due to the atomic radius difference of B and O/S results in unoccupied defect energy states.

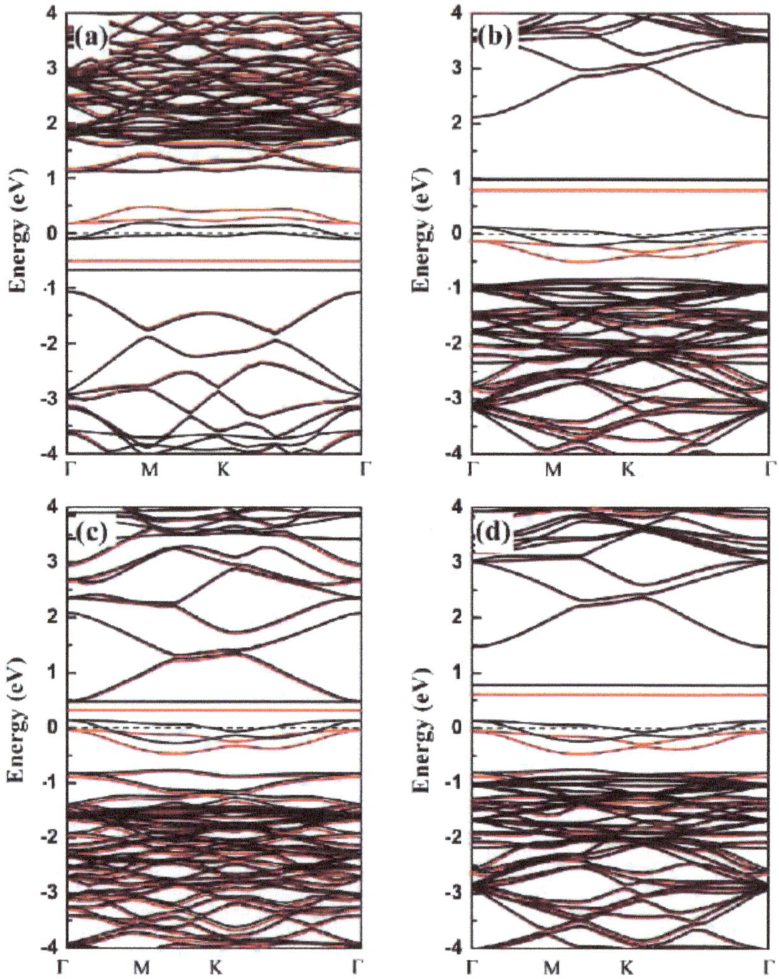

Fig. (5). Spin-polarized band structures of boron-doped monolayer.

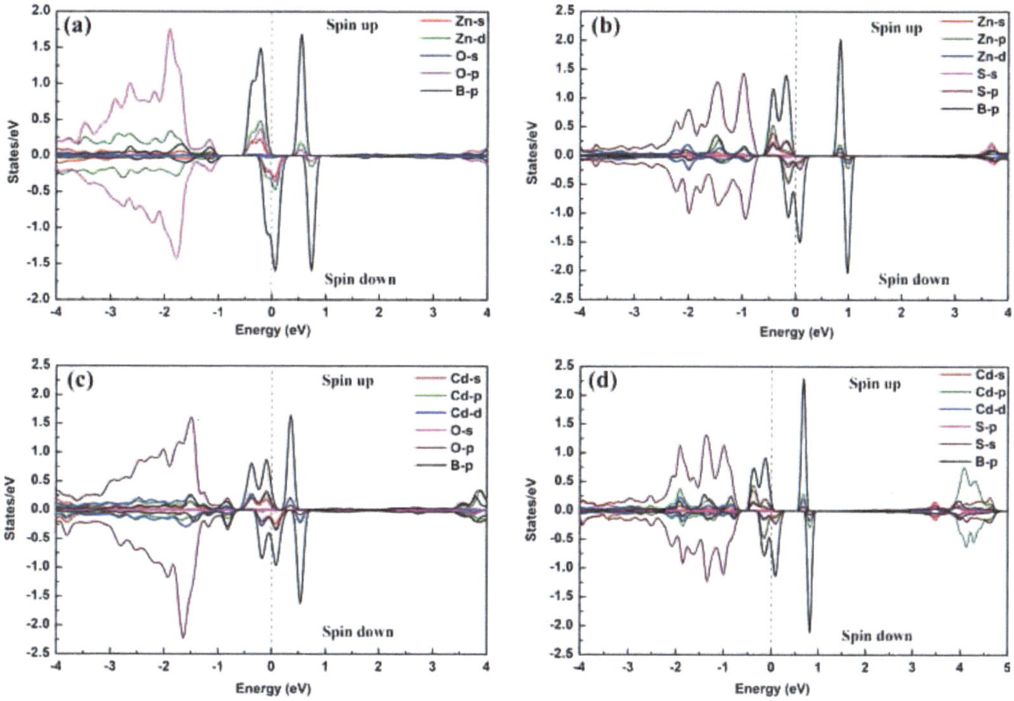

Fig. (6). Spin-polarized PDOS of boron-doped monolayer.

Band structures of carbon-doped ZnO, ZnS, CdO and CdS monolayer are presented in Fig. **(7a-d)**, respectively, with spin up (spin down) states represented by red (black) color. In C-doped ZnO ML, the energy gap is 1.53 eV, slightly smaller than its pristine counterpart (1.65 eV). Similar to B-doping, defect states are also observed in the band structure. Similarly, C-doping also results in the reduction of a band gap in ZnS ML. The energy band gap in a semiconducting channel is observed to be 2.41 eV, smaller than the pristine counterpart (2.64 eV). However, defect states in the semiconducting spin state of C-doped ZnS ML are lying just below the Fermi energy. C-doped CdO monolayer, on the other hand, shows an energy gap of 0.64 eV with a small change from its pristine counterpart. Relatively, the energy band gap in the spin up state in C-doped CdS ML is 1.67eV, which is almost comparable with the band gap energy of the host lattice. The introduction of C-atom in the band gap energy is not as drastic as B-doping. In fact, in the case Cd based monolayers in this study, the band gap almost remained unchanged. This may be attributed to the atomic radius being relatively closer to O/S in comparison with the previous B-atom. Also, doping C-atom all results in half-metallic ferromagnetism, with one of its spin states crossing the Fermi level in all the proposed monolayers. This is in agreement with other studies on strained monolayers [31, 32].

Fig. (**8a-d**) show the PDOS of atomic orbitals of the C atom and its neighboring atoms in the B-doped ZnO, ZnS, CdO and CdS monolayers, respectively. Like B-doping, a significant contribution of C-*p* orbitals is observed in all the structures. Orbitals contributions are in C-doped ZnO and ZnS ML is closely related to its pristine states as well as its B-doped counterpart. However, unlike the B-doped structure, no defect energy is observed in C-doped ZnO ML. PDOS of C-doped CdO ML, given in Fig. (**8d**), revealed that metallic behavior for spin down state arises due to C-2*p* orbitals. In C-doped CdS ML, the corresponding defect state formation is observed only in the spin-down channel and the empty defect states lie near the bottom of the CBM of the host lattice. The formation and location of empty defect states in the band gap are related to the number of electrons in the *p*-orbitals of isolated dopant atoms.

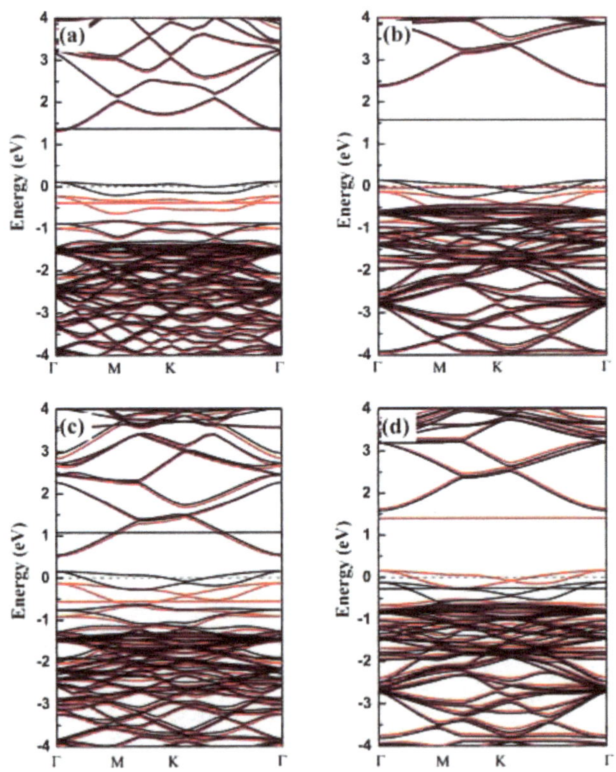

Fig. (7). Spin-polarized band structures of carbon-doped monolayer.

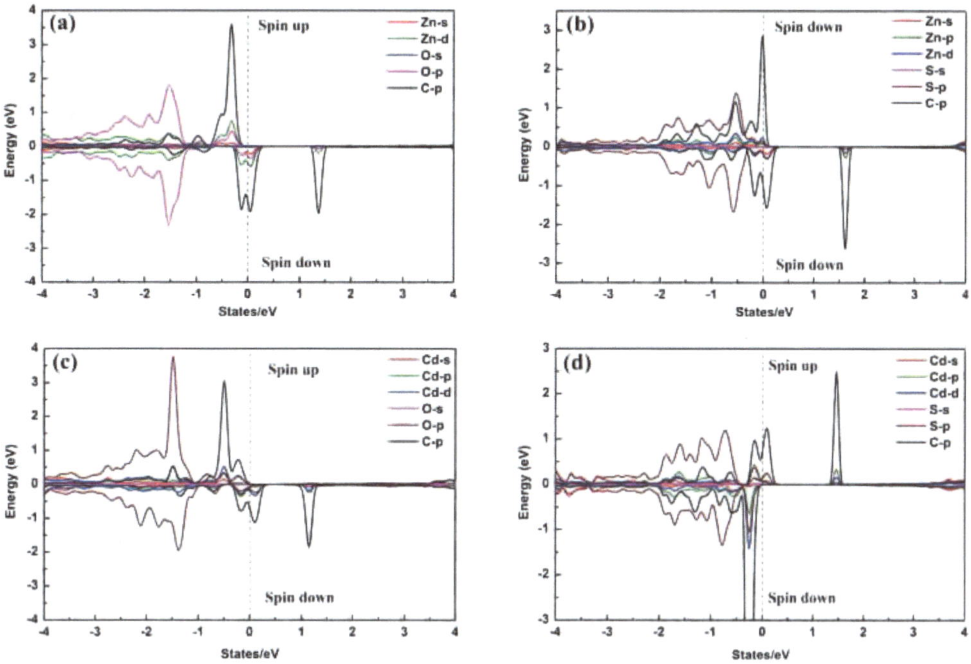

Fig. (8). Spin-polarized PDOS of carbon-doped monolayer.

Upon doping with B and C atoms, significant variation is observed in the elastic and piezoelectric properties of the monolayers. Negative stiffness is observed in the B-doped ZnS monolayer and, hence, also in its piezoelectric coefficient, d_1. While negative structural stiffness is absent on the B-doped ZnO monolayer, d_1 is found to be negative since $C_{12}>C_{11}$, as given in Table **4**. Carbon substitutional doping, on the other hand, results in a stable structure and enhanced piezoelectric tensors in comparison to its pristine counterpart. This may be attributed to C having a larger atomic radius than B, consequently offering less strain in the structure, as observed earlier from the electronic properties.

Negative structural stiffness is known to present a problem between force and displacement in deformed objects: applied force is in the opposite direction of displacement rather than in the same direction. In our study, doping of non-metal atom (B), which possesses different structural and electronic properties from its displaced atom (O/S), creates strain in the supercell. Such objects are known to possess stored energy and may be unstable unless they are restricted. Negative stiffness can be an interesting property since material damping can be observed with positive and negative stiffness. Negative structural stiffness is commonly observed in post-buckled elements and is, in fact, observed experimentally [14].

However, it is not always the case that partially constrained elastic material must exhibit a positive strain energy, which remains stable. Elastic objects under partial constraint may possess negative bulk modulus and yet still be stable. However, a negative elastic coefficient observed in doping, as a result of its strained structure in some cases, may reveal its instability. Nonetheless, the majority of the doped structures are still stable, and such observation may provide insight into tunable piezoelectric properties in the proposed materials.

Table 4. Elastic constants (C_{11} and C_{12}) and piezoelectric tensor, e_1 (x 10^{-10} Cm^{-1}) and d_1 (pm V^{-1}) B and C-doped ZnO, ZnS, CdO and CdS monolayer.

System	Boron Doping				Carbon Doping			
	C_{11}	C_{12}	e_1	d_1	C_{11}	C_{12}	e_1	d_1
ZnO ML	64.92	77.17	1.29	-10.53	73.1	64.96	1.62	19.90
ZnS ML	-10.8	108.46	10.47	-8.78	29.16	25.51	3.22	88.22
CdO ML	72.19	50.21	1.83	8.33	59.10	47.83	5.31	47.12
CdS ML	37.09	29.26	2.88	36.78	46.88	19.37	0.83	3.02

CONCLUSION

Theoretical investigation of II-VI monolayers (ZnO, ZnS, CdO, CdS) using Density Function Theory (DFT) is done to investigate their electronic and piezoelectric properties. After their dynamic stability is confirmed, spin-polarized calculation reveals that these pristine monolayers are non-magnetic and also mechanically stable. Upon doping a non-metal atom (B/C), all the monolayers transform into half-metallic ferromagnets with changes arising mainly from the *p*-orbitals of the dopant (B/C) atom. Reduction of band gap energy from its pristine structure is observed in all the doped cases. B-atom with a significantly smaller atomic radius, in comparison with C-atom, results in a larger variation of its structural as well as electronic properties. Larger structural distortion in B-doping further results in negative structural stiffness and negative piezoelectric coefficient of ZnO and ZnS monolayer. On the other hand, C-doping results in enhanced structural stiffness and piezoelectric properties of these monolayers. Our study suggests that B/C-atom doping is an effective way of tuning the electronic and piezoelectric properties of these II-VI monolayers.

ACKNOWLEDGEMENTS

Financial support for this research work was provided by SERB-DST, Govt. of India vide Grant No EEQ/2018/000854 dated 23rd May 2019 as a Major Research Project.

REFERENCES

[1] Duerloo, K.A.N.; Ong, M.T.; Reed, E.J. Intrinsic piezoelectricity in two-dimensional materials. *J. Phys. Chem. Lett.,* **2012**, *3*(19), 2871-2876. [http://dx.doi.org/10.1021/jz3012436]

[2] Wu, W.; Wang, L.; Li, Y.; Zhang, F.; Lin, L.; Niu, S.; Chenet, D.; Zhang, X.; Hao, Y.; Heinz, T.F.; Hone, J.; Wang, Z.L. Piezoelectricity of single-atomic-layer MoS2 for energy conversion and piezotronics. *Nature,* **2014**, *514*(7523), 470-474. [http://dx.doi.org/10.1038/nature13792] [PMID: 25317560]

[3] Hinchet, R.; Khan, U.; Falconi, C.; Kim, S-W. Piezoelectric properties in two-dimensional materials: Simulations and experiments. *Mater. Today,* **2018**, *21*(6), 611-630. [http://dx.doi.org/10.1016/j.mattod.2018.01.031]

[4] Mortazavi, B.; Shojaei, F.; Yagmurcukardes, M.; Shapeev, A.V.; Zhuang, X. Anisotropic and outstanding mechanical, thermal conduction, optical, and piezoelectric responses in a novel semiconducting BCN monolayer confirmed by first-principles and machine learning. *Carbon,* **2022**, *200*, 500-509. [http://dx.doi.org/10.1016/j.carbon.2022.08.077]

[5] Lv, Q.; Qiu, J.; Wen, Q.; Li, D.; Zhou, Y.; Lu, G. Large in-plane and out-of-plane piezoelectricity in 2D γ-LiMX2 (M=Al, Ga and In; X=S, Se and Te) monolayers. *Mater. Sci. Semicond. Process.,* **2023**, *154*, 107222. [http://dx.doi.org/10.1016/j.mssp.2022.107222]

[6] Kaur, A.; Sharma, S.; Nandi, P.; De Sarkar, A. Correlation between strain tunable piezoelectricity and Rashba effect in flexible Janus Ga2Ge2XY (X, Y = S, Se, and Te) monolayers with high carrier mobility. *Physica E,* **2023**, *154*, 115791. [http://dx.doi.org/10.1016/j.physe.2023.115791]

[7] Ivanovskii, A.L. Graphene-based and graphene-like materials. *Russ. Chem. Rev.,* **2012**, *81*(7), 571-605. [http://dx.doi.org/10.1070/RC2012v081n07ABEH004302]

[8] Chhana, L.; Vanchhawng, L.; Rai, D.P.; Tiwari, R.C.; Zuala, L. Comparative study of half-metallic ferromagnetic behaviour in ZnO monolayer doped with boron and carbon atoms. *Int. Nano Lett.,* **2021**, *11*(2), 113-123. [http://dx.doi.org/10.1007/s40089-021-00330-4]

[9] Chhana, L.; Tiwari, R.C.; Chettri, B.; Rai, D.P.; Gurung, S.; Zuala, L. Ab initio investigation of non-metal-doped ZnS monolayer. *Appl. Phys., A Mater. Sci. Process.,* **2021**, *127*(9), 729. [http://dx.doi.org/10.1007/s00339-021-04882-2]

[10] Lashgari, H.; Boochani, A.; Shekaari, A.; Solaymani, S.; Sartipi, E.; Mendi, R.T. Electronic and optical properties of 2D graphene-like ZnS: DFT calculations. *Appl. Surf. Sci.,* **2016**, *369*, 76-81. [http://dx.doi.org/10.1016/j.apsusc.2016.02.042]

[11] Zheng, H.; Li, X-B.; Chen, N-K.; Xie, S-Y.; Tian, W.Q.; Chen, Y.; Xia, H.; Zhang, S.B.; Sun, H-B. Monolayer II-VI semiconductors: A first-principles prediction. *Phys. Rev. B Condens. Matter Mater. Phys.,* **2015**, *92*(11), 115307. [http://dx.doi.org/10.1103/PhysRevB.92.115307]

[12] Lalrinkima, ; Lahriatzuala, ; Rai, D.P.; Srivastava, S. Strain dependence of electronic properties and effective masses of monolayer ZnO from density functional theory. *AIP Conf. Proc.,* **2019**, *2115*(1), 030093. [http://dx.doi.org/10.1063/1.5112932]

[13] Chaurasiya, R.; Dixit, A. Point defects induced magnetism in CdO monolayer: A theoretical study. *J. Magn. Magn. Mater.,* **2019**, *469*, 279-288. [http://dx.doi.org/10.1016/j.jmmm.2018.08.076]

[14] Hoat, D.M.; Naseri, M.; Vu, T.V.; Rivas-Silva, J.F.; Hieu, N.N.; Cocoletzi, G.H. Structural, electronic and optical properties of CdO monolayer and bilayers: Stacking effect investigations. *Superlattices Microstruct.,* **2020**, *145*, 106644. [http://dx.doi.org/10.1016/j.spmi.2020.106644]

[15] Atmani, E.H.; Bziz, I.; Fazouan, N.; Aazi, M. Comparative study by DFT method of structural, electronic and optical properties of monolayer, bilayer and bulk CdS. *Appl. Phys., A Mater. Sci. Process.,* **2021**, *127*(11), 878. [http://dx.doi.org/10.1007/s00339-021-05009-3]

[16] Weirum, G.; Barcaro, G.; Fortunelli, A.; Weber, F.; Schennach, R.; Surnev, S.; Netzer, F.P. Growth and surface structure of zinc oxide layers on a Pd(111) surface. *J. Phys. Chem. C,* **2010**, *114*(36), 15432-15439. [http://dx.doi.org/10.1021/jp104620n]

[17] Tusche, C.; Meyerheim, H.L.; Kirschner, J. Observation of depolarized ZnO(0001) monolayers: Formation of unreconstructed planar sheets. *Phys. Rev. Lett.,* **2007**, *99*(2), 026102. [http://dx.doi.org/10.1103/PhysRevLett.99.026102] [PMID: 17678236]

[18] Deng, X.; Sorescu, D.C.; Lee, J. Single-layer ZnS supported on Au(111): A combined XPS, LEED, STM and DFT study. *Surf. Sci.,* **2017**, *658*, 9-14. [http://dx.doi.org/10.1016/j.susc.2016.12.003]

[19] Zhuang, H.L.; Hennig, R.G. Computational identification of single-layer CdO for electronic and optical applications. *Appl. Phys. Lett.,* **2013**, *103*(21), 212102. [http://dx.doi.org/10.1063/1.4831972]

[20] Garg, P.; Kumar, S.; Choudhuri, I.; Mahata, A.; Pathak, B. Hexagonal planar CdS monolayer sheet for visible light photocatalysis. *J. Phys. Chem. C,* **2016**, *120*(13), 7052-7060. [http://dx.doi.org/10.1021/acs.jpcc.6b01622]

[21] Kresse, G.; Furthmüller, J. Efficient iterative schemes for *ab initio* total-energy calculations using a plane-wave basis set. *Phys. Rev. B Condens. Matter,* **1996**, *54*(16), 11169-11186. [http://dx.doi.org/10.1103/PhysRevB.54.11169] [PMID: 9984901]

[22] Blöchl, P.E. Projector augmented-wave method. *Phys. Rev. B Condens. Matter,* **1994**, *50*(24), 17953-17979. [http://dx.doi.org/10.1103/PhysRevB.50.17953] [PMID: 9976227]

[23] Perdew, J.P.; Burke, K.; Ernzerhof, M. Generalized gradient approximation made simple. *Phys. Rev. Lett.,* **1996**, *77*(18), 3865-3868. [http://dx.doi.org/10.1103/PhysRevLett.77.3865] [PMID: 10062328]

[24] Monkhorst, H.J.; Pack, J.D. Special points for Brillouin-zone integrations. *Phys. Rev., B, Solid State,* **1976**, *13*(12), 5188-5192. [http://dx.doi.org/10.1103/PhysRevB.13.5188]

[25] Es-Smairi, A.; Fazouan, N.; Atmani, E.H. Enhanced optical and thermoelectric properties of ZnS monolayer and stacked bilayer compared with bulk. *Mater. Res. Express,* **2019**, *6*(12), 125047. [http://dx.doi.org/10.1088/2053-1591/ab56fa]

[26] Chen, L.; Cui, Y.; Xiong, Z.; Zhou, M.; Gao, Y. Chemical functionalization of the ZnO monolayer: structural and electronic properties. *RSC Advances,* **2019**, *9*(38), 21831-21843. [http://dx.doi.org/10.1039/C9RA03484F] [PMID: 35518889]

[27] Xiao, W.-Z.; Wang, L.-L. Magnetic properties in CdS monolayer doped with first-row elements: A density functional theory investigation. *Phys. Status Solidi B ,* **2014**, *251*(6), 1257-1264.

[28] Tu, Z.C. First-principles study on physical properties of a single ZnO monolayer with graphene-like structure. *J. Comput. Theor. Nanosci.,* **2010**, *7*(6), 1182-1186. [http://dx.doi.org/10.1166/jctn.2010.1470]

[29] Sevik, C.; Çakır, D.; Gülseren, O.; Peeters, F.M. Peculiar piezoelectric properties of soft two-dimensional materials. *J. Phys. Chem. C,* **2016**, *120*(26), 13948-13953. [http://dx.doi.org/10.1021/acs.jpcc.6b03543]

[30] Alyörük, M.M. First principles studied tunable electronic and optical properties of 2D honeycomb ZnO monolayer engineered by biaxial strain and intrinsic vacancy. *Mater. Sci. Eng. B.,* **2016**, *254*, 114517.

[31] Miao, Y.; Wang, Z.; Zhao, H.; Chen, Q.; Wang, H.; Wan, M.; Chen, L.; He, K.; Wang, Q. First principles studied tunable electronic and optical properties of 2D honeycomb ZnO monolayer engineered by biaxial strain and intrinsic vacancy. *Mater. Sci. Eng. B,* **2020**, *254*, 114517. [http://dx.doi.org/10.1016/j.mseb.2020.114517]

[32] Chaurasiya, R.; Raman, R.; Tyagi, S.; Dixit, A. Strain modulated optoelectronic properties of CdO monolayer. *J. Electron. Mater.,* **2019**, *48*(6), 3963-3969. [http://dx.doi.org/10.1007/s11664-019-07160-3]

CHAPTER 6

A Theoretical Investigation on the New Quaternary MAX-phase Compounds $(Zr_{1-x}Ti_x)_3AlC_2$ (where x= 0-1)

Habib Rached[1,2,*] and **Ismail Ouadha**[1]

[1] *Magnetic Materials Laboratory, Faculty of Exact-Sciences, Djillali-Liabès University of Sidi-Bel-Abbès, Sidi-Bel-Abbès, Algeria*

[2] *Department of Physics, Faculty of Exact Sciences and Informatics, Hassiba Benbouali University of Chlef, Chlef, Algeria*

Abstract: The physical properties of the $(Zr_{1-x}Tix)_3AlC_2$ MAX-phase compounds have been studied using the first-principle plane-wave method in the framework of the DFT theory. The Perdew-Burke-Ernzerhof parametrization was chosen for the exchange correlation (XC) energy. The equilibrium ground-state properties of the named compounds were calculated and compared with the reported experimental and theoretical data. The stability of our compounds has been analyzed. The electronic structures were predicted, indicating all our compounds exhibit a metallic behavior. The mechanical stability and elastic moduli were evaluated from the elastic constants. The effect of temperature and pressure on Bulk modulus, Debye temperature and heat capacity have been investigated and discussed in detail.

Keywords: Electronic structures, First-principles calculations, Quaternary MAX-phases, Thermo-mechanical properties.

INTRODUCTION

The growing interest in materials science has led to a true technological revolution of our time. The need to develop new materials that can meet the rapidly changing and renewable needs of society is one of the major challenges in the field of engineering and new technologies. Materials science is concerned with the elaboration and characterization of materials of whatever types, and constantly searching for new materials that meet the increasing demands of technological applications. In fact, before choosing a material to be used in a particular technological application, the possibilities of its use must be first stud-

* **Corresponding author Habib Rached:** Magnetic Materials Laboratory, Faculty of Exact-Sciences, Djillali-Liabès University of Sidi-Bel-Abbès, Sidi-Bel-Abbès, Algeria and Department of Physics, Faculty of Exact Sciences and Informatics, Hassiba Benbouali University of Chlef, Chlef, Algeria; E-mail: habib_rached@yahoo.fr

Dibya Prakash Rai, Kingsley O. Obodo, & Jitendra Pal Singh (Eds.)

ied by determining its physical and chemical properties (structural, electronic, mechanical ... *etc.*) through both experimental and theoretical methods. For example, improving aircraft engines must include an increase in operating temperature, so the materials used should be lighter and more resistant to high temperatures. In recent decades, researchers have shown that even the strongest metal alloys become soft at temperatures above 1000°C. Scientists have developed a new class of materials known as ceramics, but unfortunately, it is characterized by fragility that limits its use despite undeniable progress in this field. Generally, brittleness, hardness and low usability are closely related to good resistance to high temperatures.

During the 1990s, the team Barsoum *et al.* [1] from Drexel University in Philadelphia discovered interesting properties of Ti_3SiC_2. In addition to being inexpensive, this material is distinguished by its hardness, lightness, useable, resistance to oxidation and thermal shock and ability to remain solid at temperatures above 1300°C in air. About fifty compounds with similar properties were synthesized and classified into a new family known as MAX phases, referring to their formulations.

MAX phases are ternary ceramic materials based on carbides or nitrides, which constitute a new class of nano lamellar materials combining the properties of metals and ceramics, which give them exceptional properties such as high elastic stiffness, high melting temperature, high thermal shock resistance and high electrical conductivity [2, 3]. Due to the exceptional and desirable properties of MAX phase materials is well-recognized that they are used in various new industrial applications such as in aerospace, defense, nuclear reactors, automotive, and medical [4 - 7].

Recently, in 2016, a successful experimental study of the new compound Zr_3AlC_2 of the MAX-phase family was reported [8]. After that, Zapata-Solvas *et al.*, have also synthesized experimentally solid solutions of MAX-phases $(Zr_{1-x}Ti_x)_3AlC_2$ for different x concentrations [4, 9, 10].

In addition to experimental work, there are many theoretical studies related to the properties of MAX phases, for example, elastic properties and electronic-structure calculations. Moreover, a theoretical study has significance in stimulating experimental research on the synthesis and application of the new MAX-phase compounds with better characteristics, reaching some results that are difficult to reach experimentally.

According to the theory of quantum mechanics, the electronic structures of materials can be theoretically determined by solving the Schrödinger equation, where defining the electronic structures of materials is fundamental to understanding and explaining experimental results obtained on their properties or even predicting their properties even if they have not yet been experimentally determined.

In order to expand and compare the previous works on Zr_3AlC_2 and Ti_3AlC_2 compounds, we have studied the structural, mechanical, electronic and thermodynamic properties of the uninvestigated MAX-phases $(Zr_{1-x}Ti_x)_3AlC_2$ using the first-principle full potential linearized augmented plane wave (FP-LAPW) method [13], in the framework of density functional theory (DFT) [11, 12] and simulated in the WIEN2k code [14].

GENERALITIES ON THE MAX-PHASES

The $M_{n+1}AX_n$ compounds are recognized as MAX-phase materials where M is an early transition metal, A is an A-group element, and X is C and/or N. During the last two decades, the MAX-phase materials have attracted intensive interest from scientists due to their combined best properties of metals and ceramics. The first study of this type of compound was done in the 1960s on the Ti_3SiC_2 powder to determine its specific properties [15]. However, interest in these phases remained limited for many years. Significantly renewed interest in the MAX phases since the mid-1990s after the intrinsic properties of the compounds became known, where Barsoum and El-Raghy reported on the synthesis and characterization of bulk and pure samples of Ti_3SiC_2 [1, 16, 17]. A major research effort has been carried out to synthesize pure MAX-phase compounds in the form of bulk or thin film [18, 19].

Several MAX-phases have been synthesized and many have been investigated for their interesting properties. Characterization and understanding of their physical, mechanical and chemical properties have fundamental and development goals for potential applications. Many recent articles focus on compounds such as Ti_3SiC_2 [20] and Ti_3AlC_2 [21], especially on their elastic and mechanical properties.

T. Laupauw *et al.* were the first team who experimentally synthesized a new 312-MAX-phase compound Zr_3AlC_2 [8]. Barsoum *et al.*, have also successfully synthesized the 413-MAX-phase compound Ti4AlN3 [22]. The MAX-phases have now been shown to include over 90 different compounds. The majority consists of 211 compounds, most of which were discovered by Nowotny [23]. There is a lot of work being done regarding the theoretical prediction of the stability of these types of compounds and there is still definitely an undiscovered chemical group [24]. Finally, it is interesting to note that it is possible to make a large number of solid solutions at sites M, A, and X, of which only a few have been synthesized thus far, so this will be particularly important to modify and improve the properties of these materials.

Chemical Elements

The MAX phases take their name from their chemical composition $\mathbf{M_{n+1}AX_n}$, where:

•**M** is a transition metal (Sc, Ti, Zr, Nb, …).

•**A** is an element of group A (Al, Si, P, …).

•**X** is either C and /or N.

The index n varies from 1 to 3, and the corresponding structures M_2AX, M_3AX_2 and M_4AX_3 are named 211, 312 and 413, respectively [18]. Fig. (**1**) shows the different possible elements used and gives the list of thermodynamically stable MAX phases.

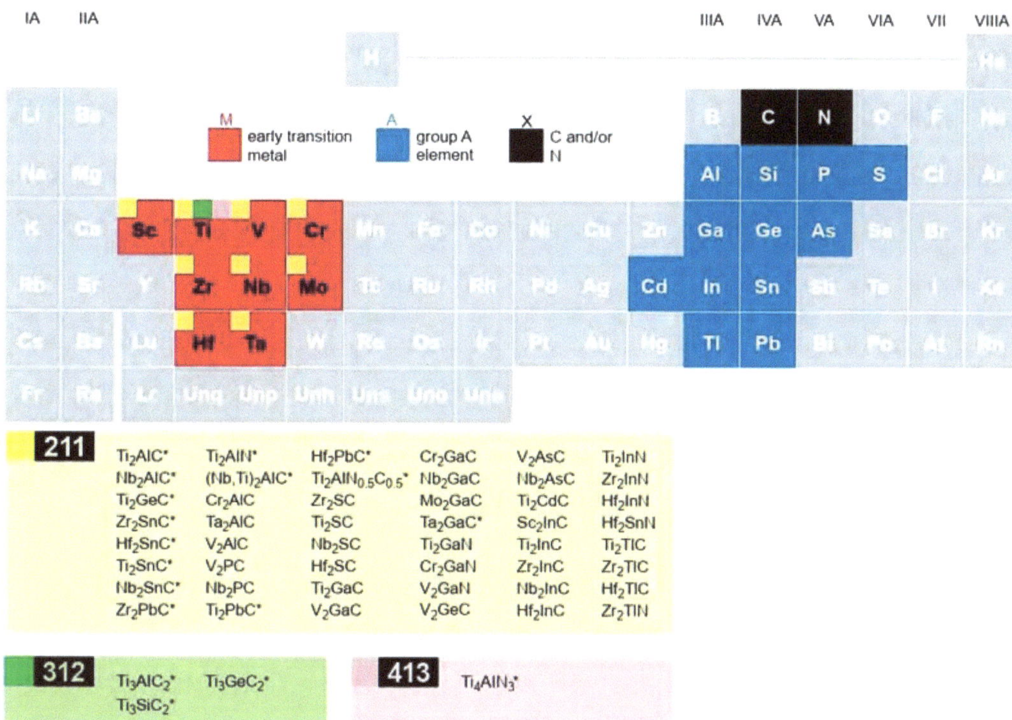

Fig. (1). Elements are known to make up the MAX phases. Red indicates M-elements, blue A-elements and grey X-elements, and a list of the different thermodynamically stable phases [2].

Crystal Structure

$M_{n+1}AX_n$ phases are hexagonal lamellar compounds that crystallize in the $P6_3/mmc$ space group. Fig. (**2**) represents the hexagonal unit cell 211, 312 and 413 MAX phase, respectively. The size of the unit cell increases along the c-axis with increasing values of n.

The MAX phase atoms are organized in alternate nanolaminate layers containing close-packed layers of M and X atoms – where the X atoms fill the octahedral sites between M layers – interleaved by pure A-element layers [18]. Wyckoff's positions for the atom sites in structures 211, 312, and 413 are listed in Table 1. The difference in MAX phases between the three structures is due to the number of M layers separating the A layers, so there are two layers in 211, three in 312 and four in 413 MAX phases. This difference, in particular, is a fundamental factor in the origin of the mechanical properties of the MAX phase.

Table 1. The atomic site positions of 211, 312 and 413 MAX phases.

Phases	Atoms	Wyckoff	*x*	*y*	*z*
211 M_2AX	M	4f	1/3	2/3	Z_M≈0.08
	A	2d	1/3	2/3	3/4
	X	2a	0	0	0
312 α-M_3AX_2	M_1	2a	0	0	0
	M_2	4f	2/3	1/3	Z_M≈0.14
	A	**2b**	0	0	1/4
	X	4f	1/3	2/3	Z_M≈0.07
312 β-M_3AX_2	M_1	2a	0	0	0
	M_2	4f	2/3	1/3	Z_M≈0.14
	A	**2d**	1/3	2/3	3/4
	X	4f	1/3	2/3	Z_M≈0.07
413 α-M_4AX_3	M_1	**4e**	0	0	Z_M≈0.16
	M_2	4f	1/3	2/3	Z_M≈0.05
	A	2c	1/3	2/3	1/4
	X_1	4f	2/3	1/2	Z_M≈0.10
	X_2	**2a**	0	0	0
413 β-M_4AX_3	M_1	**4f**	1/3	1/3	Z_M≈0.66
	M_2	4f	1/3	2/3	Z_M≈0.05
	A	2c	1/3	2/3	1/4
	X_1	**4e**	0	0	Z_M≈0.10

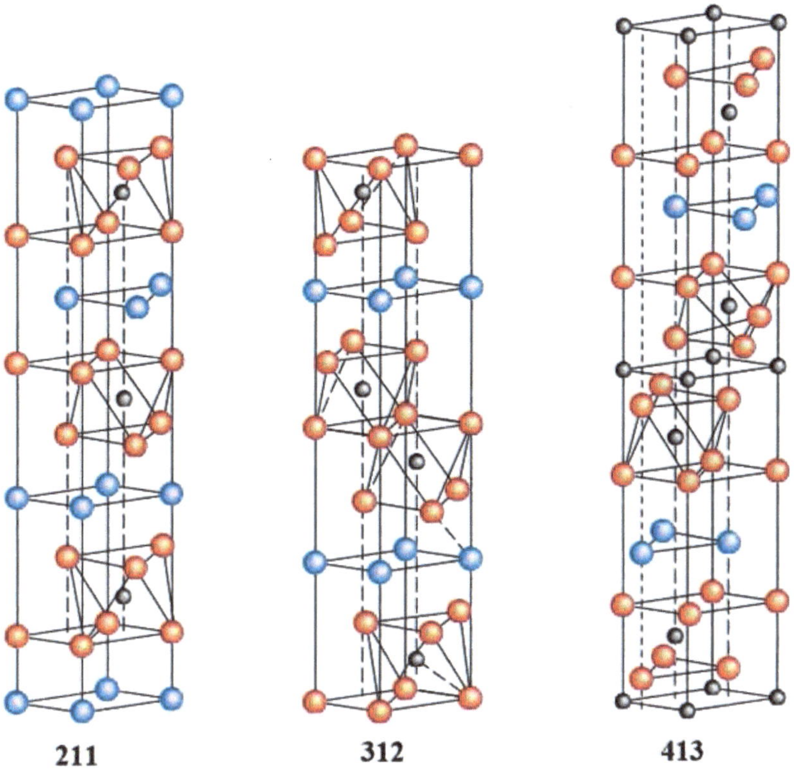

Fig. (2). Crystal structures of the three MAX-phase groups. (The TM atoms in red, the blue are the A-group elements, and C and/or N are in black.

The unit cells of MAX-phases containing M_6X octahedral, for example, Nb_6C, interleaved by A-element layers (such as Ga, Ge, As, or In). The edge-sharing octahedral building block of M_6X is similar to the binary MC, MN, carbides and nitrides, respectively. The M_6X layers are joined with respect to each other and disjointed by the A-layer, which acts as a mirror plane as shown in Fig. (**3**), which is of the Ti_3SiC_2 structure and taken with a high-resolution from transmission electron microscope (TEM) [2].

MAX Phases Properties

MAX phases have distinct properties of ceramic as well as certain properties of metals. The properties of the MAX-phase compounds are generally similar to those of their binary carbide and nitride counterparts of formula MX.

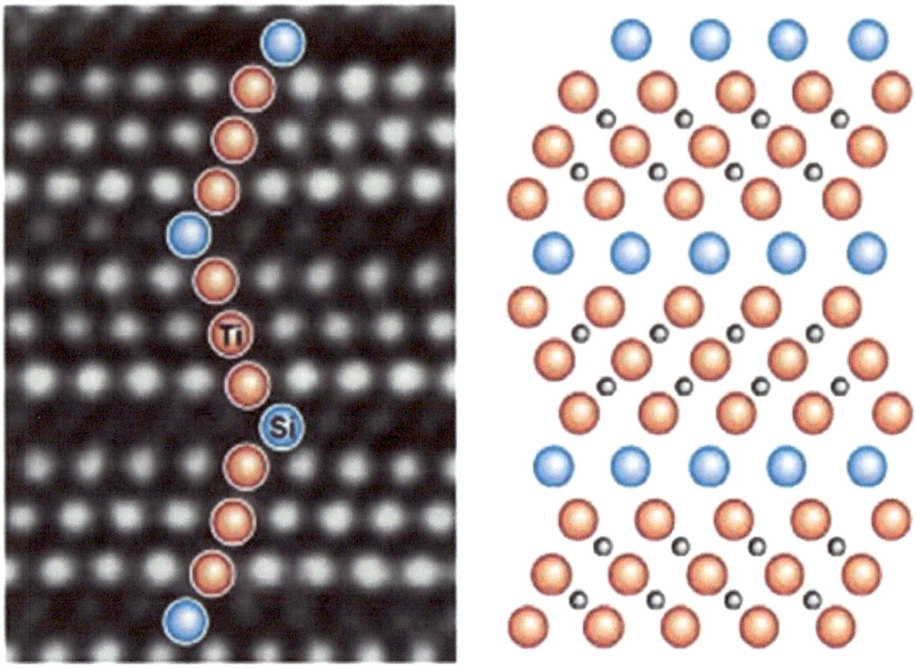

Fig. (3). The transmission electron microscope image of Ti_3SiC_2 [18].

Mechanical Properties

Mechanical properties of some MAX phases were reported by Barsoum [18] and by Zhang *et al.* [20]. In general, the MAX phase is very stiff, with values of Young's modulus and shear modulus at room temperature ranging from 178 to 362 and 80 to 142 GPa, respectively [25, 26]. This is especially true for the M_3AX_2 compounds where Young's modulus takes values greater than 300 GPa, and shear modulus is close to 120 and 200 GPa. For example, Young's modulus value for Ti_3AlC_2 is 300 GPa, while for Ti_2AlCit is only 200 GPa since the number of M-X bonds is greater for 312 compounds compared to 211 compounds and, therefore, is stiffer than it. The Poisson's ratio for all MAX phases is approximately 0.2. This value is much lower than that of pure Ti (0.3) and closer to the TiC value (0.19) [25]. Although many of the physical properties are similar between the MX and MAX phases, there are significant differences in mechanical properties. MX, unlike MAX-phases, are not machinable, do not tolerate damage, and is sensitive to thermal shock. The Debye temperature of MAX phases is relatively high and can be compared to ceramics instead of metals. The easy machinability of MAX phases at room temperature makes them candidates for many application areas [19, 25].

Electrical Properties

MAX phases are good metallic-type electrical conductors as their resistances are in a relatively narrow range of 0.2 to 0.7 μΩ.m at room temperature [27]. It should be noted that the resistivity of MAX-phases increases linearly with temperature as in the case of metals and its behavior can be described by:

$$\rho(T) = \rho_{300}\left(1 + \alpha(T - 300)\right) \; ; T > 100K \quad (1)$$

Where α is the temperature coefficient of resistivity, we note in Fig. (**4**) that the Ti_3SiC_2 has better conductivity compared to Ti_3AlC_2, and even the latter is a good conductor compared to pure Ti. Exceptionally, for Ti_4AlN_3, the resistivity of pure titanium and that of TiC are higher than those of compounds containing Ti.

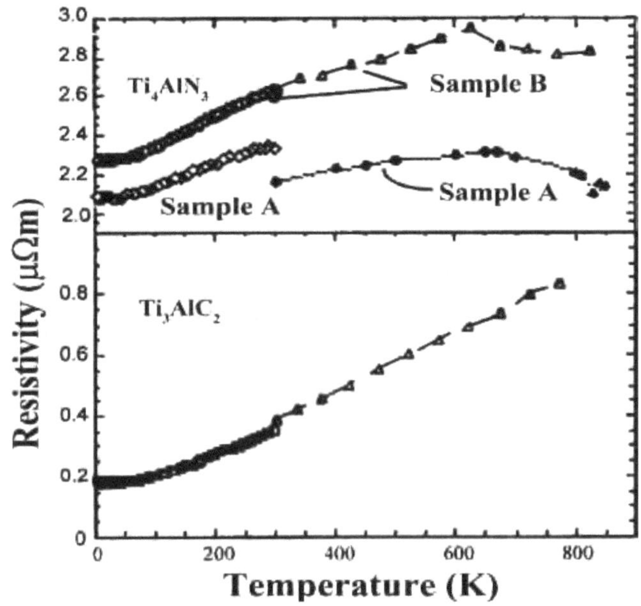

Fig. (4). Variation of the electrical resistivity with the temperature of some MAX phases [28, 29].

Thermal Properties

The MAX phases are good thermal conductors due to their good electrical conductivity, and their thermal conductivity ranges from 12 to 60 W/m.K at room temperature. Fig. (**5**) represents the variation in thermal conductivity of some MAX phases as a function of temperature.

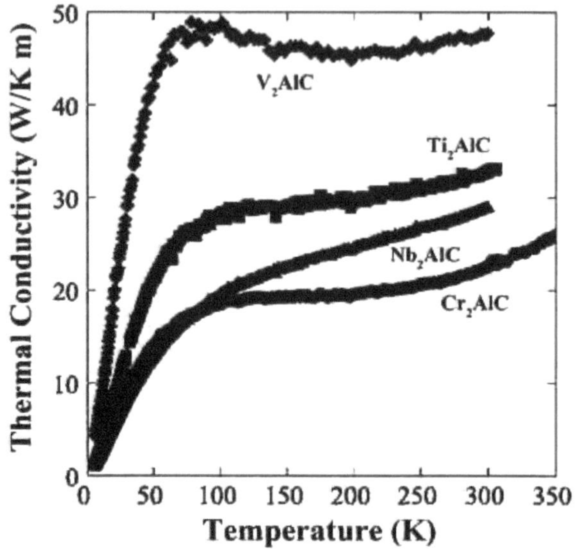

Fig. (5). The thermal conductivity of some MAX phases as a function of temperature [30].

MAX phases tend to have little anisotropy to thermal expansion; the coefficient of thermal expansion of the MAX phases is a little higher than that of the corresponding MX binary, which is, in general, between 5 and 10 μK^{-1} for the MAX phases [31]. More recently, it was discovered that Cr_2GeC has the highest thermal conductivity known to date, making it an excellent candidate for coating large areas of steel.

Also, Barsoum studied the evolution of the molar heat capacity C_p of the compounds Nb_2SnC, Ti_3SiC_2 and Ti_4AlN_3 in the range 300-1300K under constant pressure [18]. The results of the experiment are shown in Fig. (**6**), where the continuous lines are created using the Debye model [18].

The MAX phase doesn't melt but decomposes at high temperatures, producing transition metal carbide or nitride and the group A element depending on the reaction:

$$M_{n+1}AX_n \rightarrow M_{n+1}X_n + A \quad (2)$$

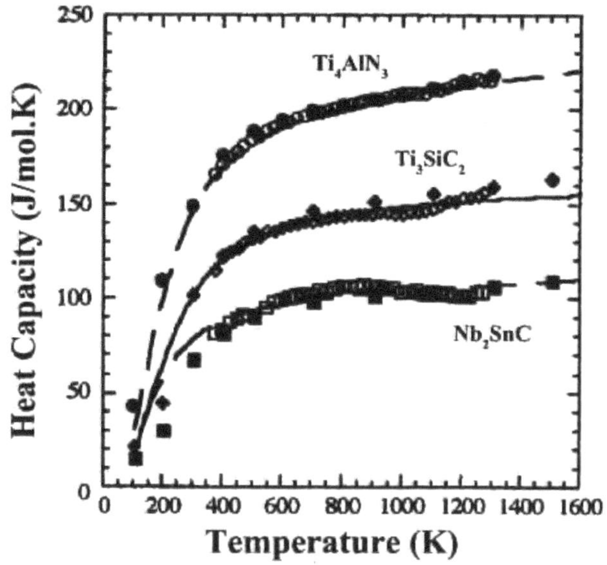

Fig. (6). The heat capacity C_P of Nb_2SnC, Ti_3SiC_2 and Ti_4AlN_3 compounds as a function of temperature [18].

This high-temperature decomposition usually results in the evaporation of element A. The Ti_3SiC_2 compound is considered a particularly heat-resistant material because it only decomposes at more than 2300°C. Thermal shock resistance is a very important factor in thermal properties, and it can be defined as the thermal gradient that ceramic can undergo without affecting mechanical properties. Despite the brittleness of MAX phases, they have high thermal shock resistance. For example, the Ti_3SiC_2 can withstand thermal shocks from 1400 °C to room temperature [18]. In Table **2**, we present some of the physical properties of MAX phase compounds.

Synthesis of MAX Phases

Herein, we briefly discuss the different methods of constructing MAX phases:

Synthesis of Massive Samples

The bulk MAX phases compounds are obtained from powder metallurgy. Two techniques are commonly used: Spark Plasma Sintering (SPS) [32] and Hot Isostatic Pressing (HIP) [1, 33]. The modifications of the experimental conditions (temperature, pressure, duration) give access to samples of different microstructures.

Synthesis of Thin Film

Development of thin films can be processed by two techniques: Physical vapor deposition (PVD) and chemical vapor deposition (CVD).

Table 2. Some physical properties of MAX-phase compounds.

Metallic Properties	Ceramic Properties
•Excellent electrical and thermal conductivity •Thermal shock resistance •Resistance to damage •Machinability at room temperature with traditional cutting tools •Low hardness •Relatively low thermoelectric power and Hall coefficient	•Very good resistance to acids and bases •Refractory decomposition at very high temperature (>1300 °C •Very good resistance to oxidation and corrosion •Thermal stability of the crystal structure and retention of most of their properties at high temperature •High stiffness and low density

- ***The PVD:*** employs purely the physical methods for generation vapor, which is condensed on substrate, such as thermal evaporation or sputtering.
- ***The CVD:*** the evolution occurs over the chemical reaction of species which form the chosen material. This procedure takes place at thermodynamical equilibrium and needs high temperatures.

Applications

The MAX phases have distinct properties because they combine the mineral and ceramic properties, and they can be considered good candidates for many applications. In addition to being good thermal and electrical conductors, they have good thermal and structural stability. Therefore, they are compounds that are used at high temperatures [18, 26]. We will cite below some examples of the use of these materials (Fig 7):

- They can be used to build internal combustion engines that operate at higher temperatures than is currently possible, making them more efficient.
- Used as a structural material for aerospace applications.
- Protective coating in thermal barriers or as protective layers on various components.
- Automotive and vehicle applications.
- Defense, medical and nuclear reactors, *etc.*

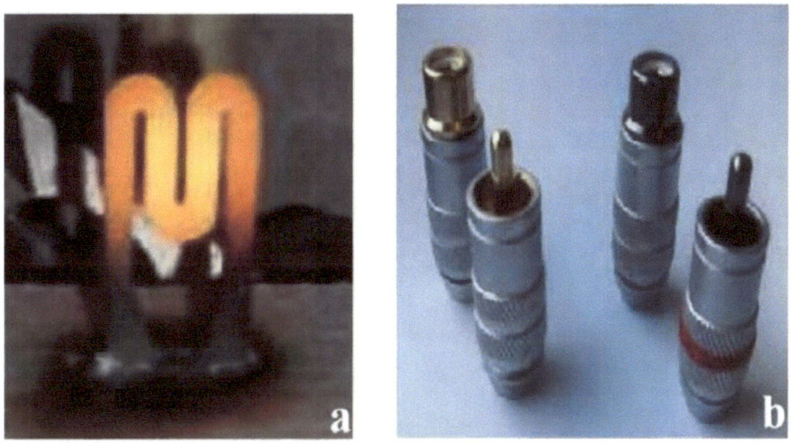

Fig. (7). (**a**): The Ti_2AlC in massive form is used as heating elements (~ 1400 ° C), (**b**): The replacement of the gold layer by Ti_3SiC_2 in the form of thin layers is carried out for electrical contact coating [18, 26].

RESULTS AND DISCUSSION OF THE FIRST-PRINCIPLE INVESTIGATIONS OF $(ZR1\text{-}XTIX)_3ALC_2$ MAX-PHASE COMPOUNDS (X=0-1)

Details of the Employed First-principle Method

In this work and in order to do our calculation, we have employed the full-potential-linearized-augmented-plane-wave (FP-LAPW) method [13] as implemented in the WIEN2k simulation code [34] and in the framework of the density functional theory (DFT) which has proven to be one of the perfect theory for the structural and electronic calculations of solids [35 - 50]. Further details of FP-LAPW are given in Refs [14, 51].

The generalized gradient approximation (GGA) parametrized by Perdew-Burke-Ernzerhof (PBE-96) is employed for the exchange-correlation functionals [52]. After the convergence tests, the optimized parameters are Rmt×Kmax equal to 8, the chosen R_{MT} is 1.96, 1.96, 2.38 and 1.74 au, for Zr, Ti, Al and C, respectively, while the G_{max} is equal to 14 (G_{max} defines the largest vector magnitude in the fourier expansion of the charge density [53]). The Monkorst-Pack method was employed for the first Brillouin zone (IBZ) using 1500 *k*-points. The charge convergence was set to 10^{-4}.

The 312 MAX-phase M_3AlC_2 compounds crystallize in a hexagonal system with the $P6_3/mmc$ (#194) space group [54, 55], as displayed in Fig. (**8**). Table **3** resumes the Wyckoff positions of the M_3AlC_2 compounds.

Fig. (8). The crystal structure of the M_3AlC_2 compound (where M=Zr and/or Ti) [56].

Table 3. The Wyckoff atomic positions of M_3AlC_2 compounds [56].

Material	Atom	Wyckoff Site	Positions
M_3AlC_2	M Ti, Zr	*2a*	(0,0,0) ; (0,0,1/2)
		4f	(1/3, 2/3, z_M); (2/3,1/3, z_M+1/2); (2/3,1/3,-z_M); (1/3,2/3,-z_M+1/2)
	Al	*2b*	(0,0,1/4); (0,0,3/4)
	C	*4f*	(1/3,2/3, z_C); (2/3,1/3, z_C+1/2); (2/3,1/3,-z_C); (1/3,2/3,-z_C+1/2)

Structural Properties

The structural properties of $(Zr_{1-x}Tix)_3AlC_2$ compounds were studied for x=0, 1/3, 1/2, 2/3 and 1). The behavior of a solid under pressure is described by the equation of state. We initialize by calculating the total energy as a function of the unit volume using the Perdew-Burke-Ernzerhof (PBE), in order to determine the equilibrium ground state of properties (lattice parameter a and c), the compressibility module B(GPa), the resulting data were adjusted by the Birch-Murnaghan state equation (EOS) [57, 58]. We have also used the 2D-optimize package developed by J. Morteza [59] which is implemented in the WIEN2K code.

$$E(V) = E_0 + \frac{9V_0B_0}{16}\left\{\left[\left(\frac{V_0}{V}\right)^{\frac{2}{3}} - 1\right]^3 B_0' + \left[\left(\frac{V_0}{V}\right)^{\frac{2}{3}} - 1\right]^2 \left[6 - 4\left(\frac{V_0}{V}\right)^{\frac{2}{3}}\right]\right\} \quad (3)$$

Where:

V_0 is the equilibrium volume of the unit cell.

E_0 is the corresponding total energy (the minimum of the total energy).

B_0 is the compressibility modulus, and B' is its pressure derivative, where:

$$B_0 = V(\frac{\partial^2 E}{\partial V^2})_{P=0} \tag{4}$$

Fig. (**9**) shows the change of the total energy as a function of the volume of a unit cell of our compounds. In Table **4**, we reported the obtained results and the available experimental and theoretical ones for comparison. It can be seen that the present results are in agreement with the corresponding experimental and theoretical results. Also, it's clear that when increasing the concentration, volumes of the unit cell decrease due to the atomic radius of Ti atom, which is less than that of the Zr atom. The values of the bulk modulus B increase when the x concentration increases. Thus, it achieves the inverse relationship between the volume and pressure.

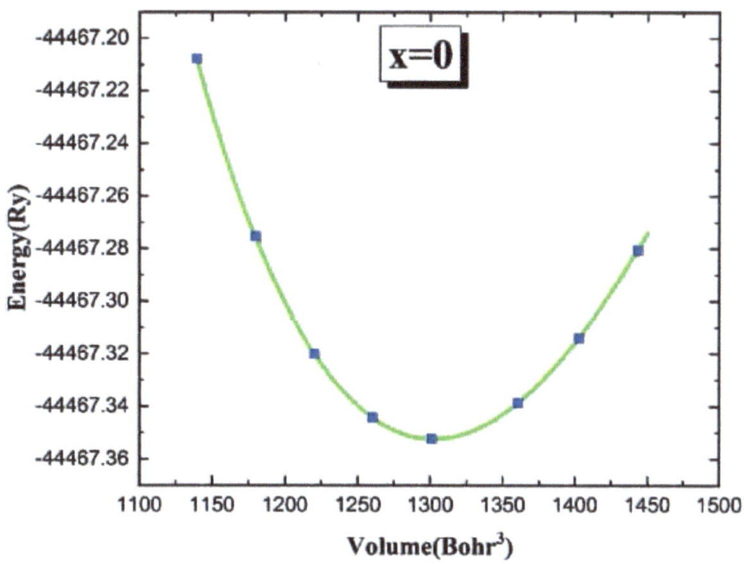

(Fig. 9) contd.....

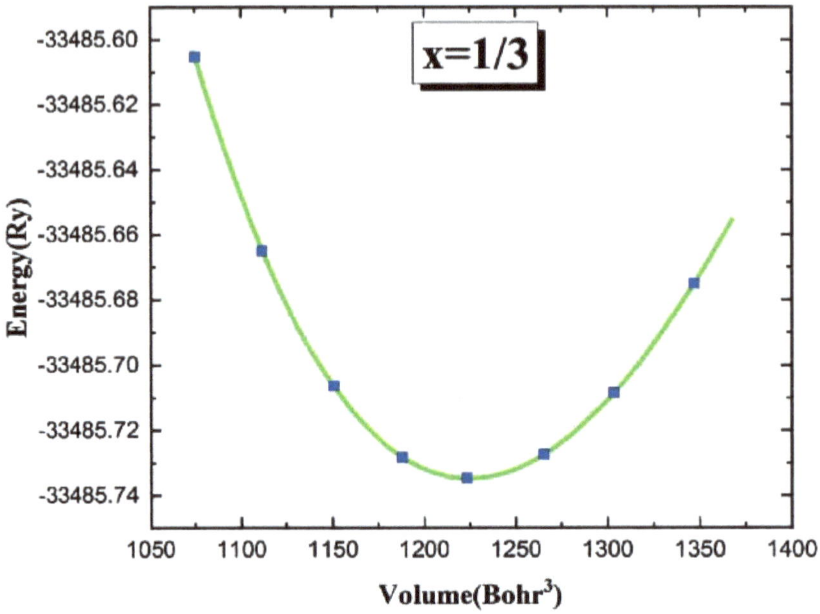

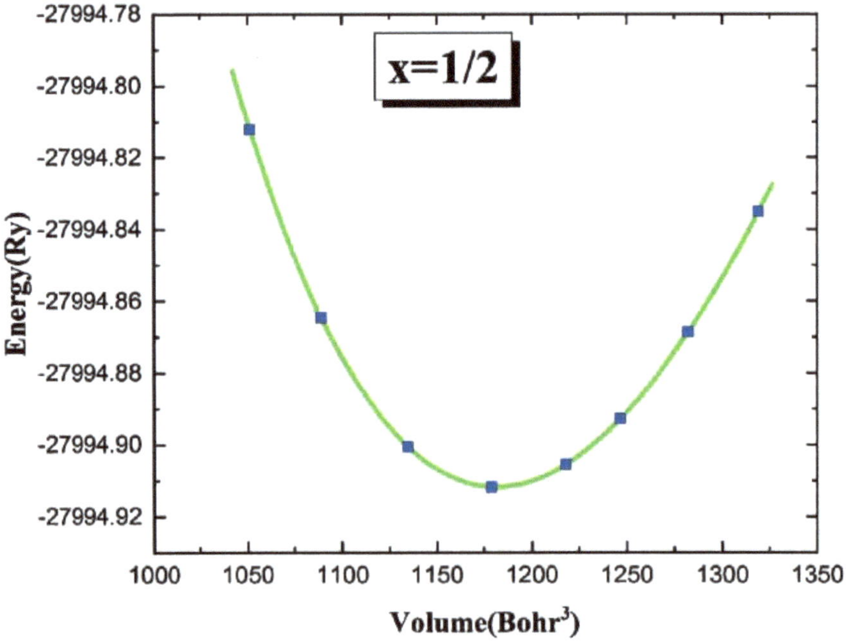

(Fig. 9) contd.....

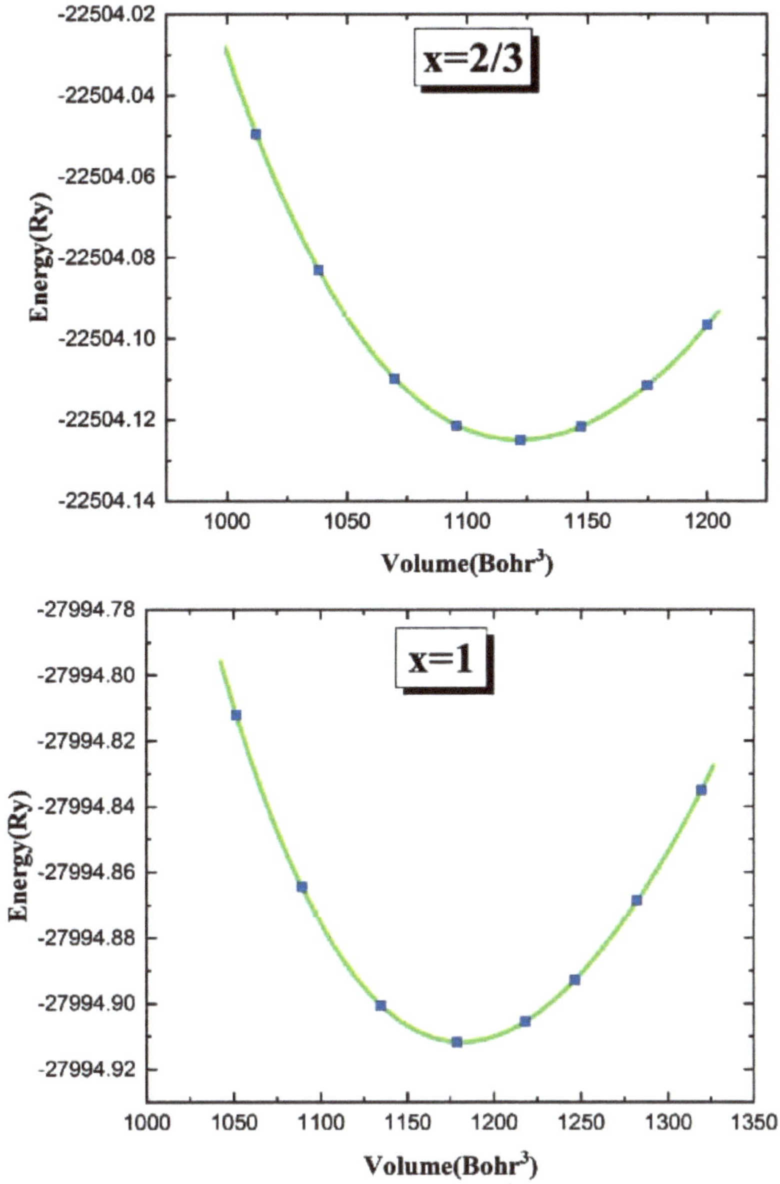

Fig. (9). Calculated the energy as a function of the volume for the $(Zr_xTi_{1-x})_3AlC_2$ compounds.

Table 4. The lattice parameter a, c/a ration, unit cell volume, and bulk modulus of $(Zr_{1-x}Ti_x)_3AlC_2$.

Concentration	a(Å)	c/a	V(Å³)	B *(GPa)*	B'	E (Ryd)	Remarks
0	**3.341** 3.333 3.335	**5.974** 5.986 5.985	**193.02** 191.95 192.27	**165.77** - 147.29	**4.26** - -		**Calc**[a] Expt[b] Calc[c]
1/3	**3.269** 3.267 3.240	**5.969** 5.991 5.999	**181.45** 180.97 176.75	**168.55** - 156.67	**3.96** - -		**Calc**[a]
1/2	**3.230** 3.232 3.197	**5.995** 6.002 6.012	**175.02** 175.45 170.11	**168.93** - 160.42	4.05 - -		**Calc**[a] Expt[d] Calc[c]
2/3	**3.184** 3.128 3.156	**5.987** 6.015 6.029	**166.16** 159.41 164.11	**172.39** - 159.06	**4.59** - -		**Calc**[a]
1	**3.077** 3.075 3.078	**6.057** 6.042 6.065	**152.78** 152.16 153.19	**185.26** - 158.23	**4.06** - -		**Calc**[a] Expt[e] Calc[c]

[a] **Present work.** [b] Ref [54]. [c] Ref [62]. [d] Ref [63]. [e] Ref [55].

The pressure effect on structural properties has been examined in the range from 0-25 GPa and illustrated in Figs. (**10**) and (**11**). It can be seen that when the pressure and concentration increase, the lattice parameters decrease.

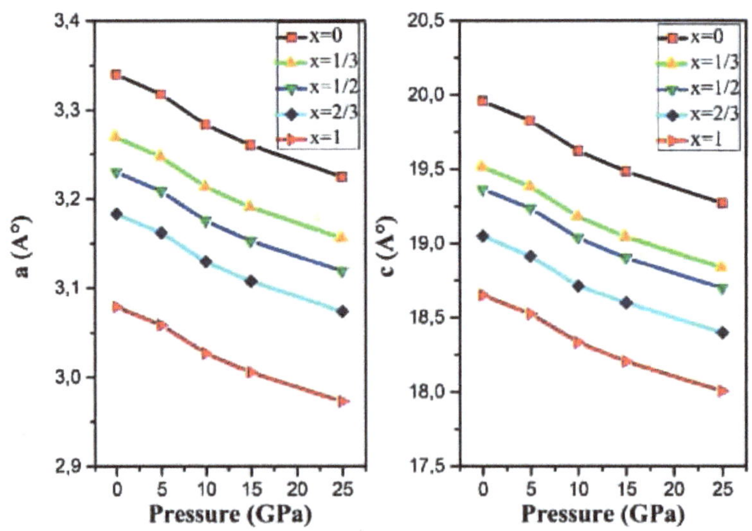

Fig. (10). The pressure effect on lattice parameters a and c of $(Zr_xTi_{1-x})_3AlC_2$ compounds.

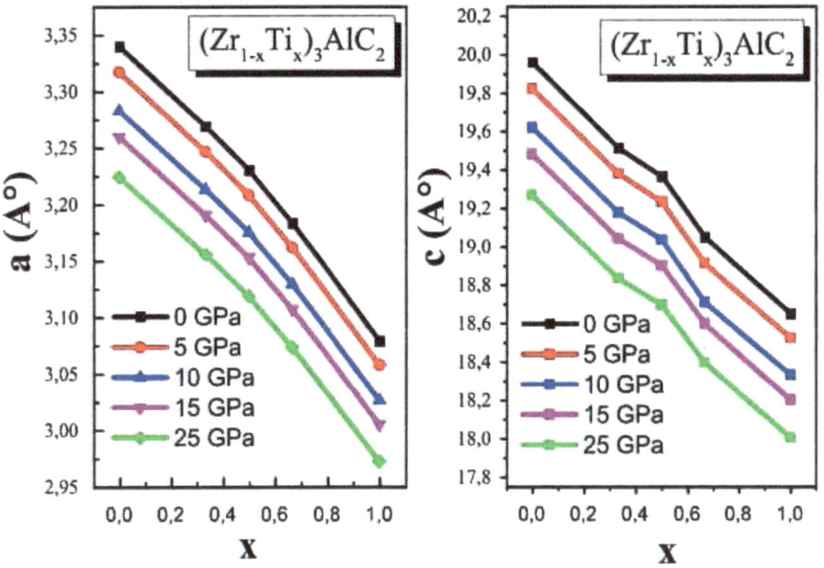

Fig. (11). The x concentrations affect the lattice parameters a and c of $(Zr_xTi_{1-x})_3AlC_2$ compounds.

- The formation energy is estimated to explore the thermodynamic stability of the compounds [60, 61]:

$$\Delta H_{f(\mathrm{Zr}_{1-x}\mathrm{Ti}_x)_3\mathrm{AlC}_2} = (E_{tot(\mathrm{Zr}_{1-x}\mathrm{Ti}_x)_3\mathrm{AlC}_2} - (3(1-x)E_{Zr} + 3xE_{Ti} + E_{Al} + 2E_C) \quad (5)$$

Where $\Delta H_{f(\mathrm{Zr}_{1-x}\mathrm{Ti}_x)_3\mathrm{AlC}_2}$ is the formation energy of $(\mathrm{Zr}_{1-x}\mathrm{Ti}_x)_3\mathrm{AlC}_2$, with (x = 0, 1/3,1/2, 2/3, 1), E_{tot} is the total energy per unit cell of the compounds, E (X :Zr, Ti, Al, and C) represents the total energies per atom of elements in their pure solid state. The calculated energies of the formation of the compounds are reported in Table 5. It is clear that all the values are negative, indicating that all the compounds are stable and can be synthesized.

Table 5. The calculated formation energy of $(Zr_{1-x}Ti_x)_3AlC_2$.

	Concentration				
	0	1/3	1/2	2/3	1
ΔHf (eV/f.u.)	-0.817	-0.753	-0.776	-0.729	-0.826

Mechanical Behavior

The elastic constants permit us to get information about the mechanical stability, stiffness, and anisotropy of a material. All solid objects are susceptible to deformation under the influence of applied external forces. The resistance against this deformation in crystals is greater than that of amorphous materials due to internal forces. Elasticity is the physical property of a material that returns the material to its original form and dimensions after removing the external forces that affect the material and lead to its deformation and consequently, the material has a preferred natural state. The behavior of a solid under an applied force F that stretches the solid elastically from L_0 to $L_0 + \delta L$ can be illustrated in Fig. (**12**). The force is divided between the chains of atoms that make up the solid, with each chain carrying a force δF.

The six elastic constants C_{11}, C_{12}, C_{13}, C_{33}, C_{44} and C_{66} are calculated using the Hex-elastic package of Jamal Morteza [65, 66]. The elastic constants are derived from the total energy with respect to the strain tensor ε_i, $E(V, \varepsilon_i)$:

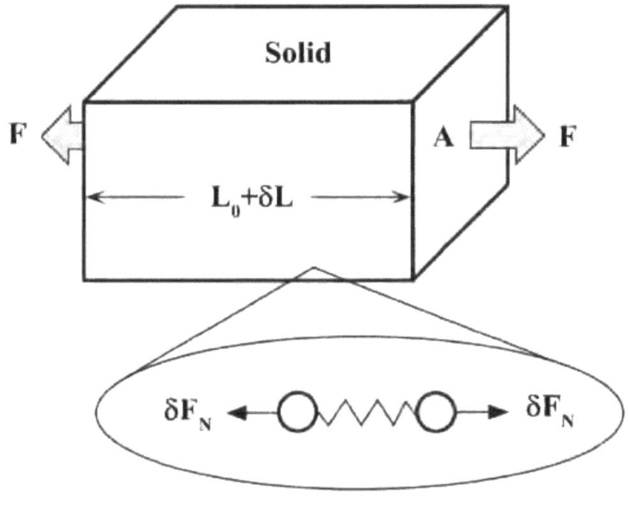

Fig. (12). The behavior of a solid material under an applied force *F* [64].

$$E(V, \varepsilon_i) = E_0(V_0, 0) + V_0\left(\sum_i \tau_i \xi_i \varepsilon_i + \frac{1}{2}\sum_{ij} c_{ij} \varepsilon_i \xi_i \varepsilon_j \xi_j\right) + O(\varepsilon^3) \quad (6)$$

Where; E_0 and V_0 are the energy and the volume at the equilibrium system. The ξi

JJFactor is equal to 1 if *i* is equal to 1, 2 or 3 and 2 if it is equal to 4, 5 or 6. In the equation 6τi are related to the strain on the crystal. The $E(V, \varepsilon_i)$ is distorted by applying six deformations as follows [56]:

$$D_1 = \begin{pmatrix} 1+\varepsilon & 0 & 0 \\ 0 & 1+\varepsilon & 0 \\ 0 & 0 & 1 \end{pmatrix} \tag{7}$$

$$D_2 = \begin{pmatrix} 1+\varepsilon & 0 & 0 \\ 0 & 1+\varepsilon & 0 \\ 0 & 0 & \dfrac{1}{(1+\varepsilon)^2} \end{pmatrix} \tag{8}$$

$$D_3 = \begin{pmatrix} 1 & 0 & 0 \\ 0 & 1 & 0 \\ 0 & 0 & 1+\varepsilon \end{pmatrix} \tag{9}$$

$$D_4 = \begin{pmatrix} \left(\dfrac{1+\varepsilon}{1-\varepsilon}\right)^{\frac{1}{2}} & 0 & 0 \\ 0 & \left(\dfrac{1-\varepsilon}{1+\varepsilon}\right)^{\frac{1}{2}} & 0 \\ 0 & 0 & 1 \end{pmatrix} \tag{10}$$

$$D_5 = \begin{pmatrix} 1 & 0 & \varepsilon \\ 0 & 1 & \varepsilon \\ \varepsilon & \varepsilon & 1+\varepsilon^2 \end{pmatrix} \tag{11}$$

$$D_6 = \begin{pmatrix} (1+\varepsilon^2)^{\frac{1}{2}} & \varepsilon & 0 \\ \varepsilon & (1+\varepsilon^2)^{\frac{1}{2}} & 0 \\ 0 & 0 & 1 \end{pmatrix} \tag{12}$$

Where $E(V, \varepsilon_i)$ can be obtained as:

$$E(V,\varepsilon) = E(V_0,0) + V_0\left((C_{11}+C_{12})\varepsilon^2 + O(\varepsilon^3)\right) \tag{13}$$

$$E(V,\varepsilon) = E(V_0,0) + V_0\left(\frac{C_{33}}{2}\varepsilon^2 + O(\varepsilon^3)\right) \tag{14}$$

$$E(V,\varepsilon) = E(V_0,0) + V_0\left((C_{zz})\varepsilon^2 + O(\varepsilon^3)\right) \tag{15}$$

$$E(V,\varepsilon) = E(V_0,0) + V_0\left((C_{11}-C_{12})\varepsilon^2 + O(\varepsilon^4)\right) \tag{16}$$

$$E(V,\varepsilon) = E(V_0,0) + V_0\left(4(C_{44})\varepsilon^2 + O(\varepsilon^3)\right) \tag{17}$$

$$E(V,\varepsilon) = E(V_0,0) + V_0\left(2(C_{66})\varepsilon^2 + O(\varepsilon^3)\right) \tag{18}$$

$$C_{zz} = C_{11} + C_{12} + 2C_{33} - 4C_{13} \tag{19}$$

Table **6** summarizes the obtained elastic. It can be seen that the elastic constants Cijsatisfy the mechanic stability conditions for hexagonal structure, which confirm the mechanical stability of the investigated compounds [67]:

Table 6. Calculated elastic constants C_{ij} (GPa) of $(Zr_{1-x}Ti_x)_3AlC_2$ compounds.

Compounds $(Zr_{1-x}Ti_x)_3AlC_2$	Ti content; *x*				
	0	1/3	1/2	2/3	1
C_{11} (GPa)	308.59	308.58	313.41	340.01	358.86
C_{12} (GPa)	89.33	89.35	90.86	90.62	99.95
C_{13} (GPa)	97.37	102.70	97.20	93.85	92.33
C_{33} (GPa)	318.24	328.91	331.38	355.97	366.22
C_{44} (GPa)	82.27	90.86	89.09	98.55	102.19
C_{66} (GPa)	109.63	109.62	111.28	124.69	129.45

$$\begin{cases} C_{11} > 0 \\ C_{33} > 0 \\ C_{44} > 0 \\ C_{11} - C_{12} > 0 \\ (C_{11} + C_{12})C_{33} > 2C_{13}^2 \end{cases} \tag{20}$$

The results show that C_{33} is larger than C_{11}, revealing that the *a* and *b*-axes are more compressible than the *c*-axis, explained in terms of the existence of strong covalent bonding in the [001] direction.

We can also notice that the C_{11} and C_{33} are considerably higher than all other ones, revealing an elastic anisotropy in these materials.

From the elastic constants, we have calculated the elastic moduli (the bulk modulus B and the shear modulus G). Based on the voigt-Reuss and Hill approaches, the bulk modulus B and the shear modulus G are given by the following expressions [68 - 71]:

$$B_V = \frac{2(C_{11} + C_{12}) + C_{33} + 4C_{13}}{9} \quad (21)$$

$$B_R = \frac{(C_{11} + C_{12})C_{33} - 2C_{13}^2}{C_{11} + C_{12} + 2C_{33} - 4C_{13}} \quad (22)$$

$$B_H = \frac{B_V + B_R}{2} \quad (23)$$

$$G_V = \frac{C_{11} + C_{12} + 2C_{33} - 4C_{13} + 12(C_{55} + C_{66})}{30} \quad (24)$$

$$G_R = \frac{5}{2}\frac{[(C_{11} + C_{12})C_{33} - 2C_{13}^2]C_{55}C_{66}}{3B_V C_{55}C_{66} + [(C_{11} + C_{12})C_{33} - 2C_{13}^2](C_{55}+C_{66})} \quad (25)$$

$$G_H = \frac{G_V + G_R}{2} \quad (26)$$

Young's modulus (*E*) and Poison's ratio ν, are also calculated using the following equations [72 - 74]:

$$E = \frac{9BG}{3B + G} \quad (27)$$

$$\nu = \frac{3B - E}{6B} \quad (28)$$

Table 7 summarizes the obtained values of the B, G, E andν; so, it can be said that:

- When increasing the x concentration, the values of B, G and E increase, which indicates that the hardness of these compounds increases with x.
- For all the compounds, the shear modulus G is smaller than the bulk modulus B.
- All values of the Poisson ratio are less than 0.26, revealing that compounds are brittle materials.
- The Ti_3AlC_2 compound presents a higher ability to resist deformation compared to others.

The Pugh B/G ratio is less than the critical value of 1.75, confirming the brittle nature of the studied materials [75].

Table 7. Calculated elastic moduli, B/G ratio, shear anisotropic factor, Cauchy pressure, and Vickers hardness HV (GPa).

Elastic moduli	0	1/3	1/2	2/3	1
B (GPa)	167.01	170.42	169.76	176.91	183.68
G (GPa)	97.33	101.27	102.11	114.05	119.10
E (GPa)	244.50	253.52	255.16	281.63	293.80
B/G	1.72	1.68	1.66	1.55	1.54
v	0.255	0.252	0.249	0.234	0.233
PxCauchy	15.10	11.84	8.11	-4.70	−9.86
PyCauchy	−20.30	-20.27	−20.42	-34.07	−29.50
A1	1.30	1.18	1.27	1.29	1.24
A2	0.75	0.45	0.80	0.45	0.79
A3	0.98	0.98	1.01	1.02	1.06
HV(Gpa)	12.48	13.20	13.52	16.11	16.74

The Poisson's ratio v is also calculated to analyze the chemical bonding nature [76]. The calculated values of the v indicate that the ionic bonding is more for Zr_3AlC_2 and $(Zr_{2/3}Ti_{1/3})_3AlC_2$ compounds and the ionic covalent mixed bonding for $(Zr_{1/2}Ti_{1/2})_3AlC_2$, while for $(Zr_{1/3}Ti_{2/3})_3AlC_2$ and Ti_3AlC_2 exhibit covalent bonding. These bonding natures are also confirmed using the Cauchy pressure [77].

The Cauchy pressures are estimated for the different directions and reported in Table 7.

$$\mathbf{P_x^{Cauchy} = C_{13} - C_{44}} \tag{29}$$

$$\mathbf{P_y^{Cauchy} = C_{12} - C_{66}} \tag{30}$$

The values for $(Zr_{2/3}Ti_{1/3})_3AlC_2$ and Ti_3AlC_2 compounds are negative, which confirms the more covalent character for these compounds, while the compounds Zr_3AlC_2, $(Zr_{1/3}Ti_{2/3})_3AlC_2$ and $(Zr_{1/2}Ti_{1/2})_3AlC_2$ have positive values of PxCauchy and negative values of PyCauchy which indicates the mixed nature of these compounds.

We have calculated the three independent elastic anisotropy factors A_1, A_2 and A_3 [78], which are important to study the degree of durability related mainly to micro-cracks in crystals:

$$A_1 = \frac{C_{11} + C_{12} + 2C_{33} - 4C_{13}}{6C_{44}} \quad (31)$$

$$A_2 = \frac{2C_{44}}{C_{11} + C_{12}} \quad (32)$$

$$A_3 = \frac{C_{11} + C_{12} + 2C_{33} - 4C_{13}}{3(C_{11} - C_{12})} \quad (33)$$

For an isotropic material, the A_1, A_2 and A_3 values must be equal to 1, otherwise it is an anisotropic material [79]. According to the results, all compounds are anisotropic materials.

Furthermore, the hardness is another important indicator of corrosion resistance of materials [80]. To evaluate this parameter, we have used the formula established by Chen [81]:

$$H_V = 2\left(\frac{G^3}{B^2}\right)^{0.585} - 3 \quad (34)$$

From the HV values as reported in Table **7**, it can be seen that the Zr_3AlC_2 compound has the smallest value among the three compounds, and Ti_3AlC_2 has the greatest value.

Electronic Properties

The electronic band structures of the studied compounds are illustrated in Fig. (**13**). It can be seen that there exists an overlap between the valence and conduction bands at the Fermi level E_F, showing a metallic behavior.

We have also calculated the total and partial state densities, TDOS and PDOS, respectively, as shown in Fig. (**14**).

The TDOS graphs confirm the metallic behavior for all the studied materials. The valence region can be divide into two mains regions:

- The first region between -5.0 eV to -1.5 eV, contain *p-d* hybridization of the transition element (Zr and Ti), with contribution of *s-Al* and *p*-C states.
- The second region between -1.5 eV to 0.0 eV, present also a *p-d* hybridization between the *d* state of transition metal and *p* state of Al atom.

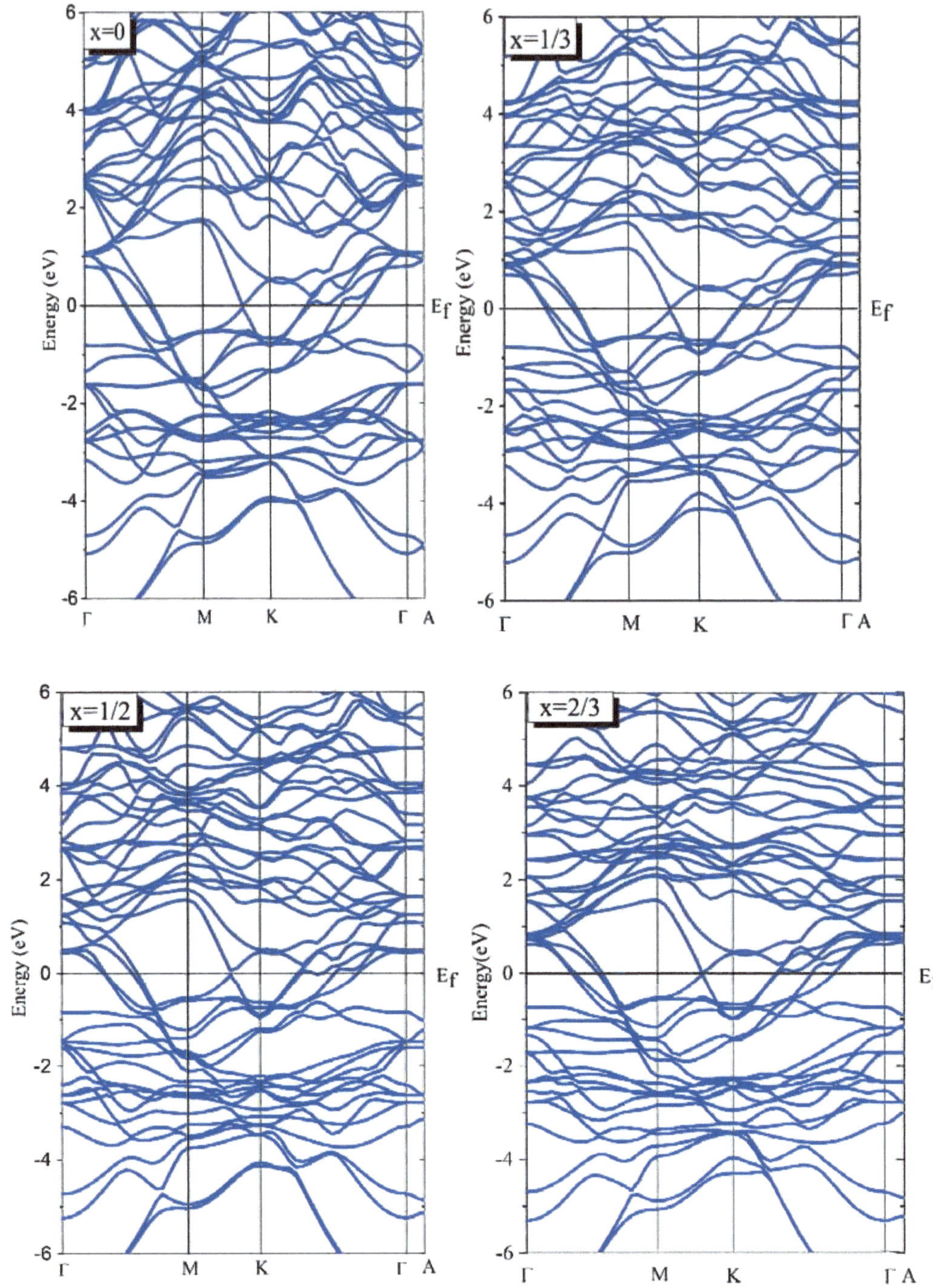

(Fig. 13) contd....

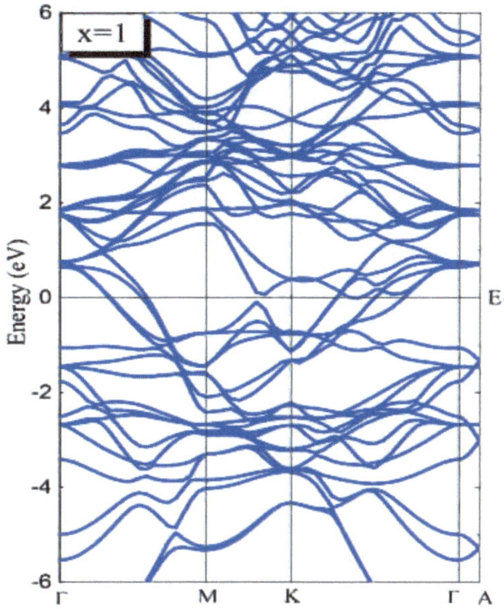

Fig. (13). The electronic band structures of $(Zr_{1-x}Ti_x)_3AlC_2$ compounds.

While, the conduction band contain a strong contribution of the *d* state of Zr element with a minor contribution of *s-p* states of Al and p states of C atoms.

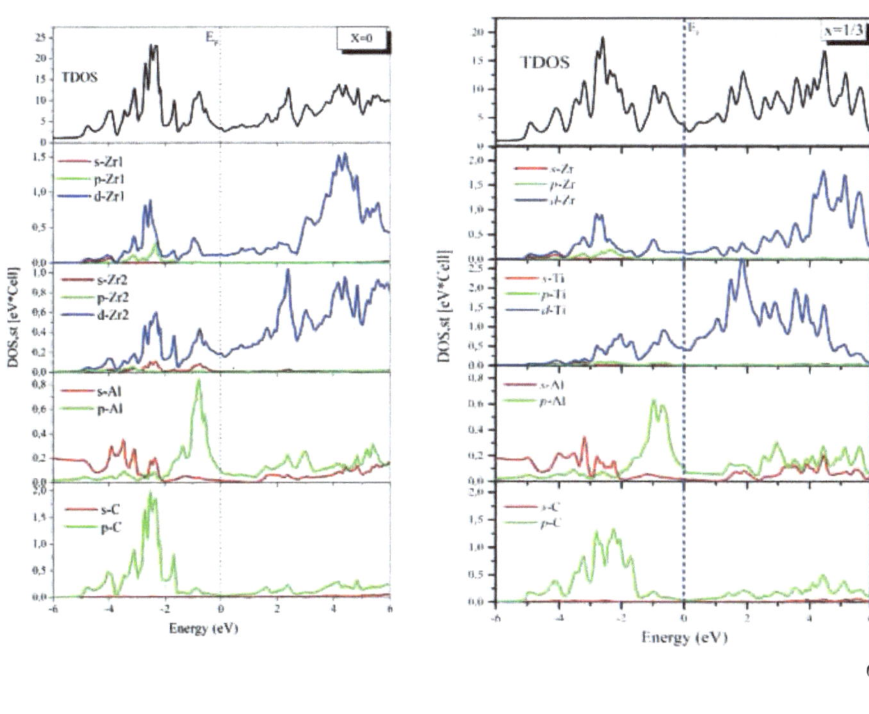

(Fig. 14) contd.....

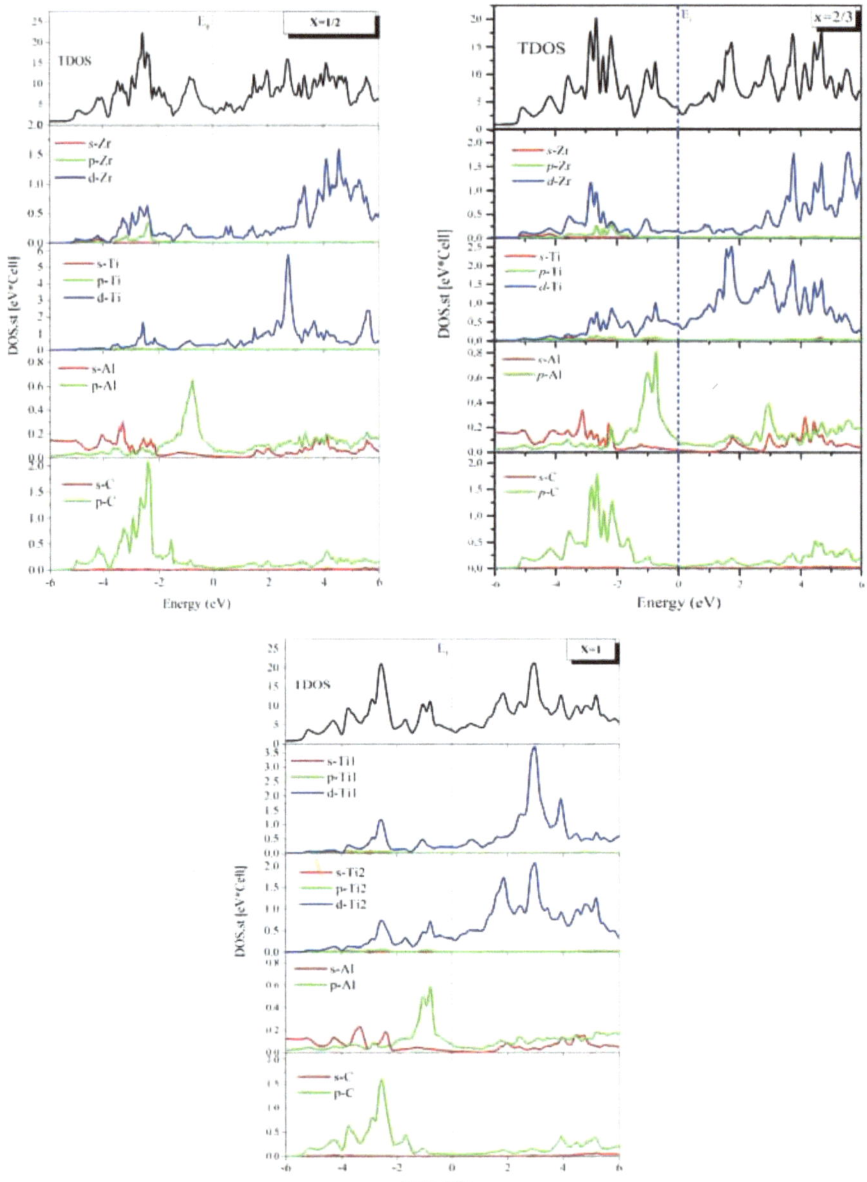

Fig. (14). The Total and Partial state densities (TDOS, PDOS).

Thermodynamic Properties

Thermodynamics is a branch of physics concerned with pressure and temperature and their relationship to energy and work [82]. The laws of thermodynamics are explained in terms of macroscopic constituents by statistical mechanics.

The Debye temperature θ_D and the average sound velocity v_m are calculated using the following relations [83, 84]:

$$\theta_D = \frac{h}{k_B}\left[\left(\frac{3n}{4\pi}\right)\frac{N_A\rho}{M}\right]^{\frac{1}{3}} \tag{35}$$

$$v_m = \left[\frac{1}{3}\left(\frac{2}{v_t^3} + \frac{1}{v_l^3}\right)\right]^{-\frac{1}{3}} \tag{36}$$

$$v_l = \left(\frac{3B + 4G}{3\rho}\right)^{\frac{1}{2}} \tag{37}$$

$$v_t = \left(\frac{G}{\rho}\right)^{\frac{1}{2}} \tag{38}$$

Where h, k_B, n N_A, ρ, M, v_m , v_l and v_t are Plank's constant, Boltzmann's constant, number of atoms per formula unit, Avogadro's number, density, molecular weight, average sound velocity, longitudinal and transverse sound velocities, respectively. The values of ρ, v_t, v_l ,v_m and θ_D are shown in the Table **8**.

Table 8. The calculated Density ρ, longitudinal, transverse and average sound velocity v_l, v_t and v_m as well as Debye temperature θ_D for $(Zr_{1-x}Ti_x)_3AlC_2$ compounds.

Concentration		ρ(g/cm3)	v_l Km/s	v_t Km/s	v_m Km/s	θ_D (K)
0	Present work	**5.95**	**7.29**	**4.17**	**4.64**	**546.8**
	[85]	5.61	12.74	4.57	5.19	613.0
1/3	Present work	**5.15**	**7.94**	**4.57**	**5.07**	**599.0**
	[85]	5.28	13.69	4.96	5.63	684.0
1/2	Present work	**4.93**	**7.88**	**4.55**	**5.05**	**615.8**
	[85]	5.07	14.16	5.14	5.83	718.0
2/3	Present work	**4.76**	**8.35**	**5.00**	**5.53**	**685.9**
	[85]	4.81	14.54	5.30	6.02	750.0
1	Present work	**4.22**	**9.01**	**5.31**	**5.89**	**749.9**
	[85]	4.22	15.48	5.64	6.40	815.0

We can note that:

- All the values of velocityvl, vt, vm and θD of $(Zr_{1-x}Ti_x)_3AlC_2$ are increase with x, while the density ρ is decreases with x.
- The longitudinal wave velocity is larger than the transverse one and both the longitudinal and transverse waves increase with increasing the value of x.
- The θ_D Value increases with increasing x, which corresponds to an increase in stiffness in the same sequence. This may indicate an increase in thermal conductivity when x is increased.

Many of the physical properties of an atomic crystal lattice can be successfully determined from the first principles within static approximation at zero temperature. However, the crystal lattice is in a dynamic, continuous state as the atoms vibrate around their equilibrium sites and hence, the material gains or loses heat by gaining or losing phonons [86].

By employing the quasi-harmonic Debye model as implemented in the Gibbs code [87], we have also evaluated the temperature and pressure effects on the Debye temperature, bulk modulus and heat capacity under temperatures by using the following formulas [88]:

$$\theta_D = \frac{\hbar}{K_B}\left[6\pi^2 V^{\frac{1}{2}} r\right]^{\frac{1}{3}} f(\nu)\sqrt{\frac{B_s}{M}} \tag{39}$$

$$f(\nu) = \left\{3\left[2\left(\frac{2}{3}\,\frac{1+\nu}{1-2\nu}\right)^{\frac{3}{2}} + \left(\frac{1}{3}\frac{1+\nu}{1-\nu}\right)^{\frac{3}{2}}\right]^{-1}\right\}^{\frac{1}{3}} \tag{40}$$

$$B_s \cong B_{static} = V\left(\frac{d^2E(V)}{dV^2}\right) \tag{41}$$

V is the equilibrium volume, M is the molecular mass, kB Is the Boltzmann constant and f (ν) the scaling function [89, 90], that depends on the Poisson's ratio ν of the isotropic material [91], while B_s is the adiabatic bulk modulus given by the static compressibility and the E (V) is the ground state energy.

The heat capacity C_V and C_P are also given by:

$$C_V = 3nk_B\left(4D\left(\frac{\theta_D}{T}\right)\frac{\frac{3\theta_D}{T}}{e^{\frac{\theta_D}{T}}-1}\right) \tag{42}$$

$$C_P = C_V(1+\alpha\gamma T) \tag{43}$$

$$D\left(\frac{\theta_D}{T}\right) = \frac{3}{(\theta_D/T)^3}\int_0^{\theta_D/T}\frac{x^3}{e^x-1}dx \tag{44}$$

$$\alpha = \frac{\gamma C_V}{B_T V} \tag{45}$$

$$\gamma = -\frac{d\,ln\theta_D(V)}{d\,lnV} \tag{46}$$

$D\left(\frac{\theta_D}{T}\right)$, n, γ, α denote the Debye integral, number of atoms per unit-cell, Gruneisen parameter, and volumetric thermal expansion, respectively.

Figs. **(15)**, **(16)** and **(17)** give respectively the pressure and temperature effects on the bulk modulus B for Zr_3AlC_2, $(Zr_{0.5}Ti_{0.5})_3AlC_2$ and Ti_3AlC_2, respectively. It can be seen that the values of B decrease slowly when the temperature increases under constant pressure, but they increase rapidly when the pressure increases under constant temperature. It is noted from the figure that the relationship between bulk modulus and pressure/temperature is nearly linear at fixed temperature/pressure.

The pressure and temperature effects on Debye temperature θ_D for Zr_3AlC_2, $(Zr_{0.5}Ti_{0.5})_3AlC_2$ and Ti_3AlC_2 compounds are given in Figs. **(18)**, **(19)** and **(20)**, respectively. It can be noted that the θ_D decrease slowly when the temperature increases under constant pressure, but they increase rapidly when the pressure increases under constant temperature.

From Figs. **(15-20)**, we point out that the change in B and θ_D are topologically similar for all our compounds, which is harmonious with the trend of volume.

The pressure and temperature effects on heat capacity C_P and C_V are also investigated and reported in Figs. **(21-26)**.

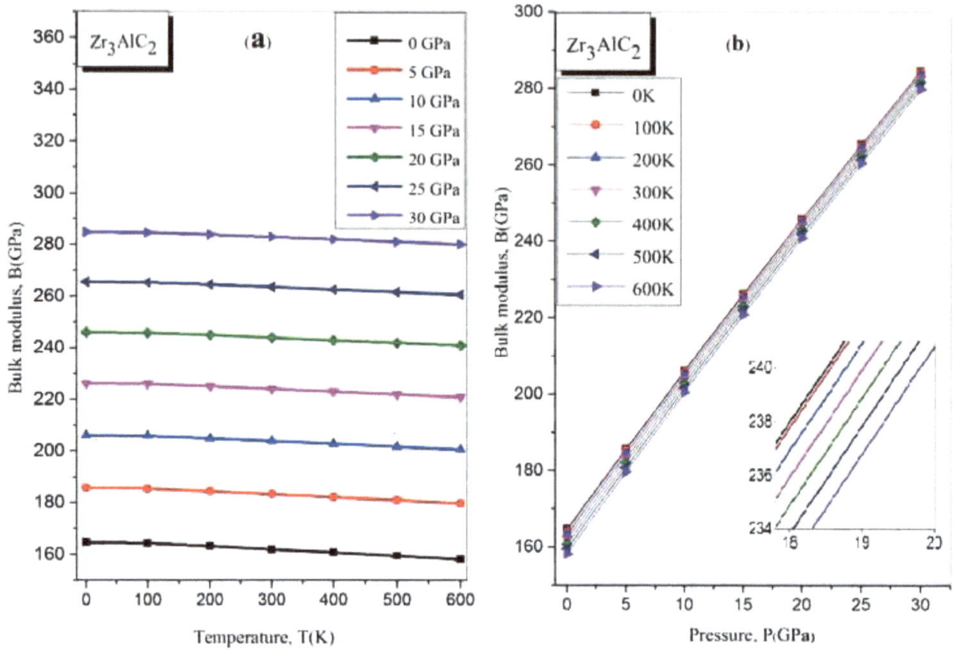

Fig. (15). The pressure and temperature effects on bulk modulus of Zr_3AlC_2 compound.

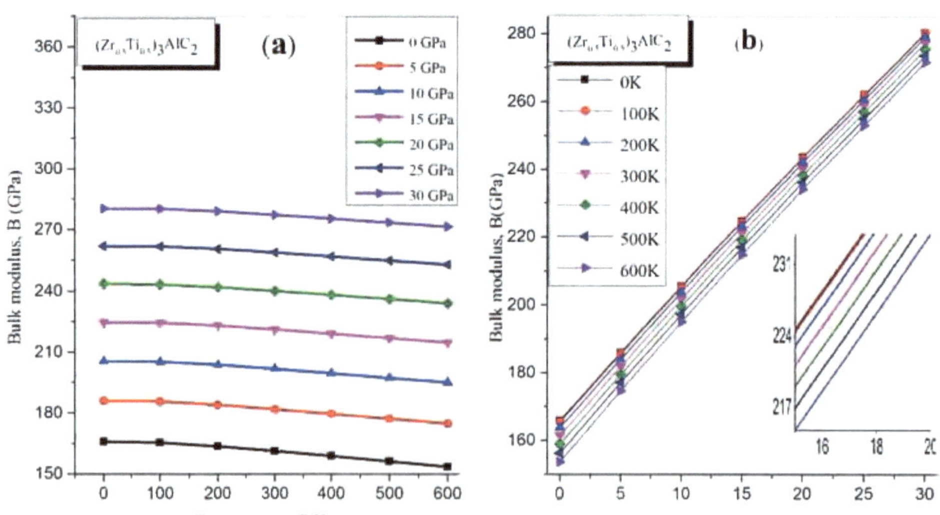

Fig. (16). The pressure and temperature effects on bulk modulus of $(Zr_{0.5}Ti_{0.5})_3AlC_2$ compound.

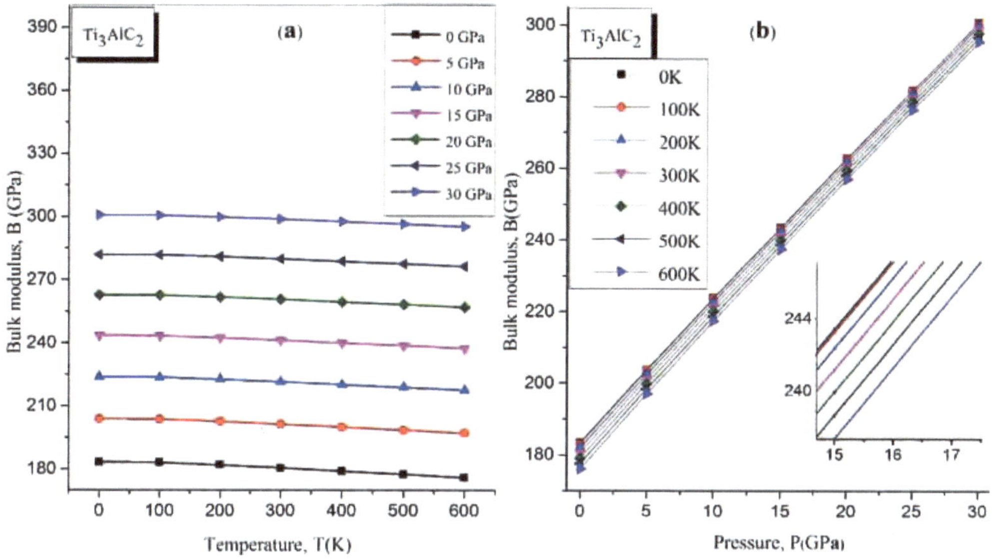

Fig. (17). The pressure and temperature effects on bulk modulus of Ti_3AlC_2 compound.

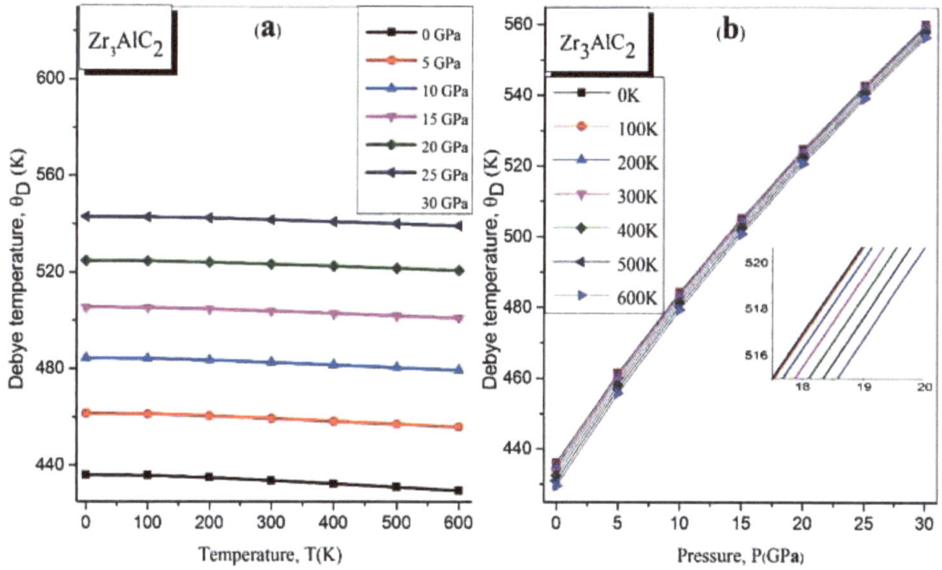

Fig. (18). The pressure and temperature effects on Debye temperature of Zr_3AlC_2 compound.

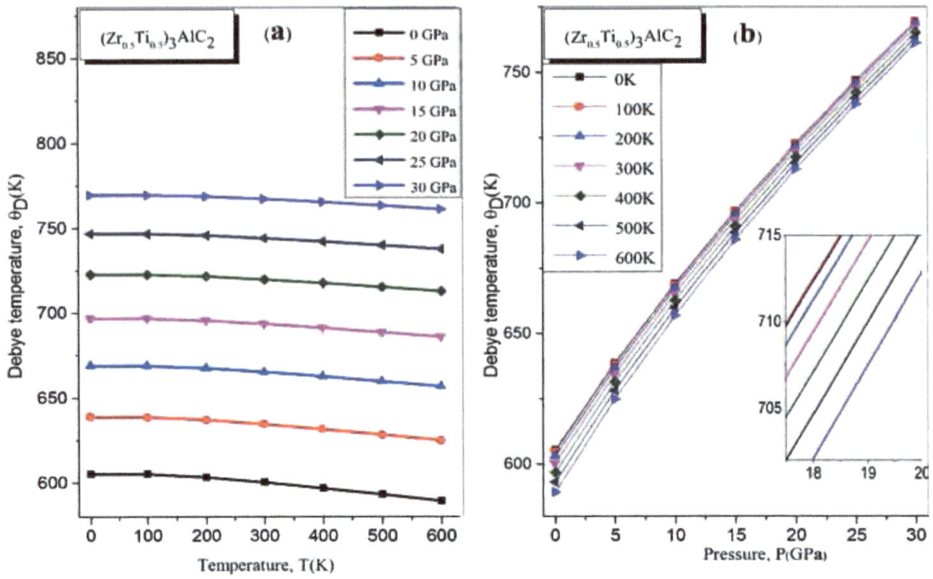

Fig. (19). The pressure and temperature effects on Debye temperature of $(Zr_{0.5}Ti_{0.5})_3AlC_2$ compound.

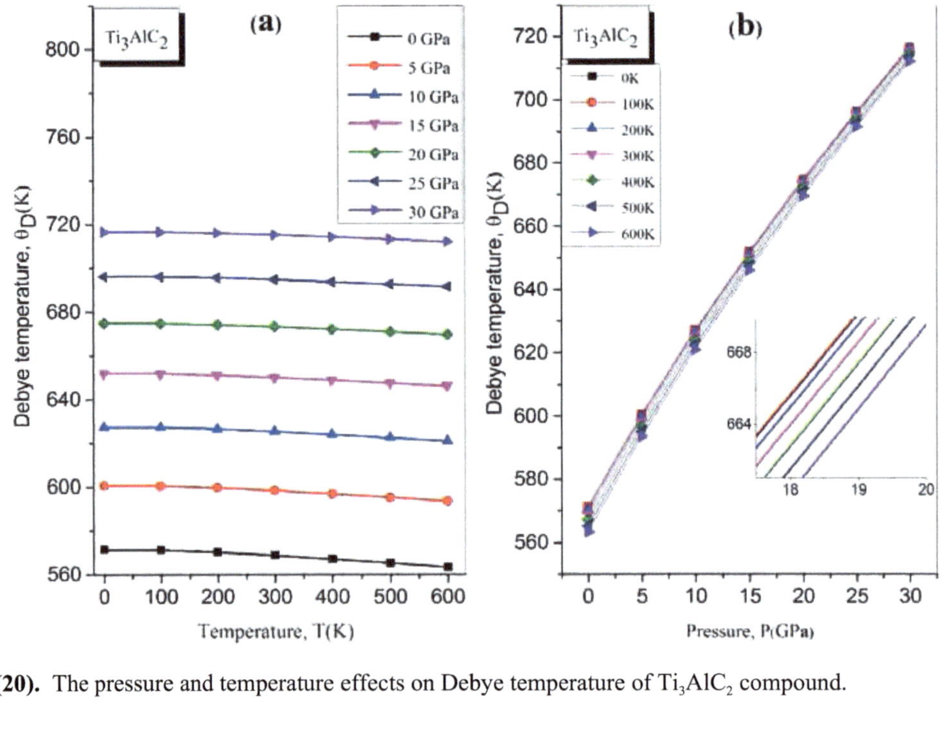

Fig. (20). The pressure and temperature effects on Debye temperature of Ti_3AlC_2 compound.

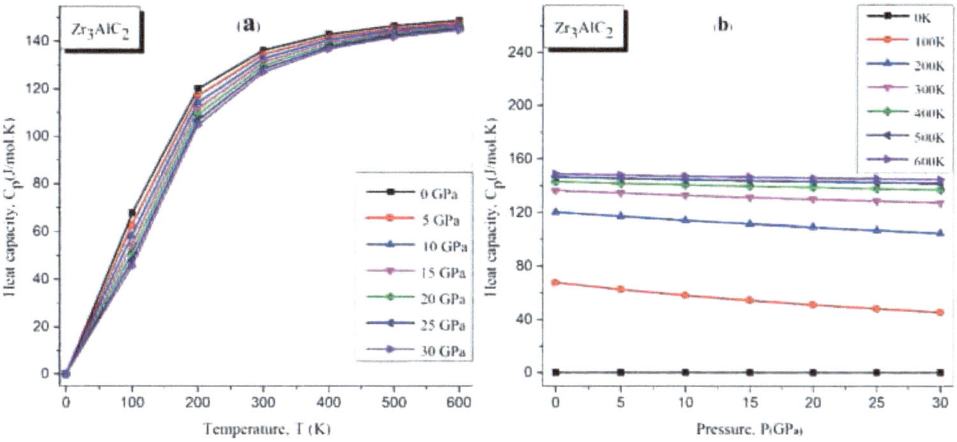

Fig. (21). The pressure and temperature effects on heat capacity C_P of Zr_3AlC_2 compound.

The change in C_P and C_V are mainly similar for all the compounds. Below 200 K the C_P and C_V increase rapidly with increasing temperature, while before 200 K, they continue to increase but weakly. Further, the C_P and C_V decrease slowly under constant temperature by increasing the pressure except when the temperature is equal to zero, all values remain to zero.

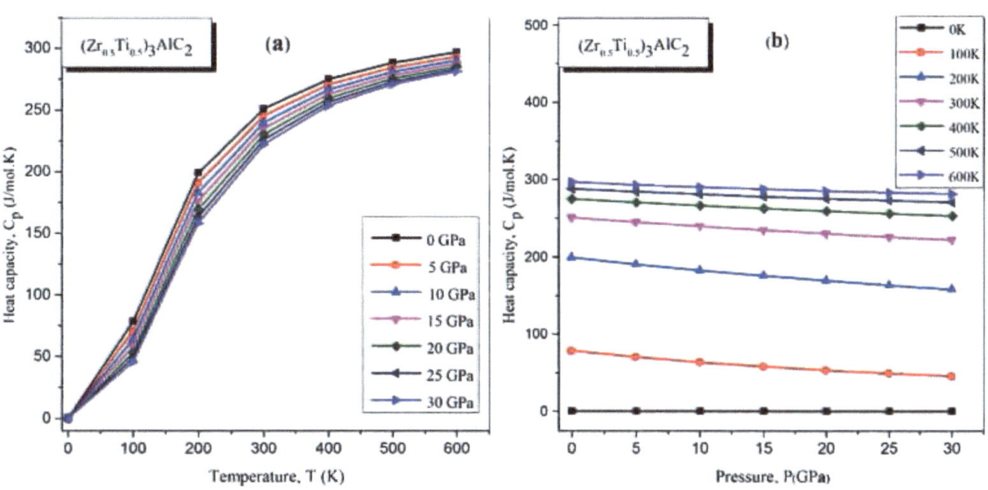

Fig. (22). The pressure and temperature effects on heat capacity C_P of $(Zr_{0.5}Ti_{0.5})_3AlC_2$ compound.

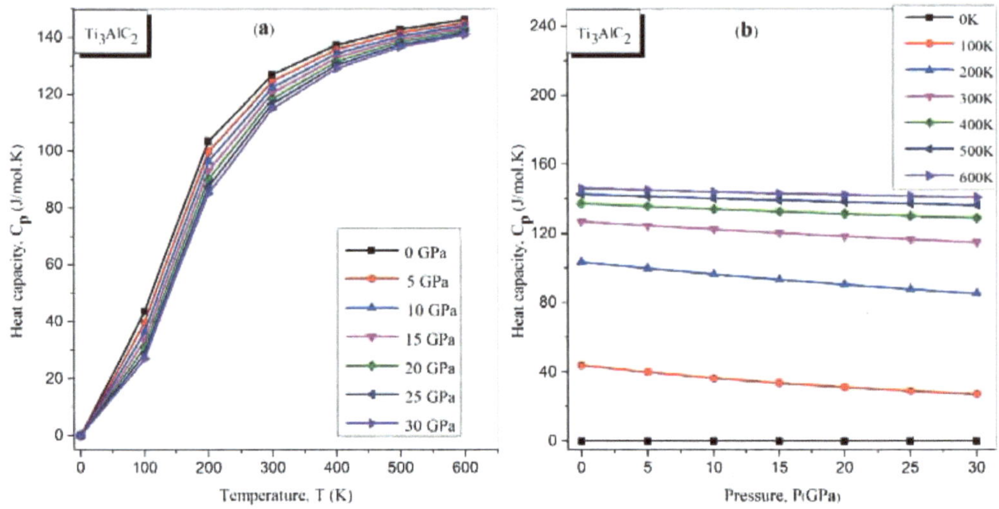

Fig. (23). The pressure and temperature effects on heat capacity C_P of Ti_3AlC_2 compound.

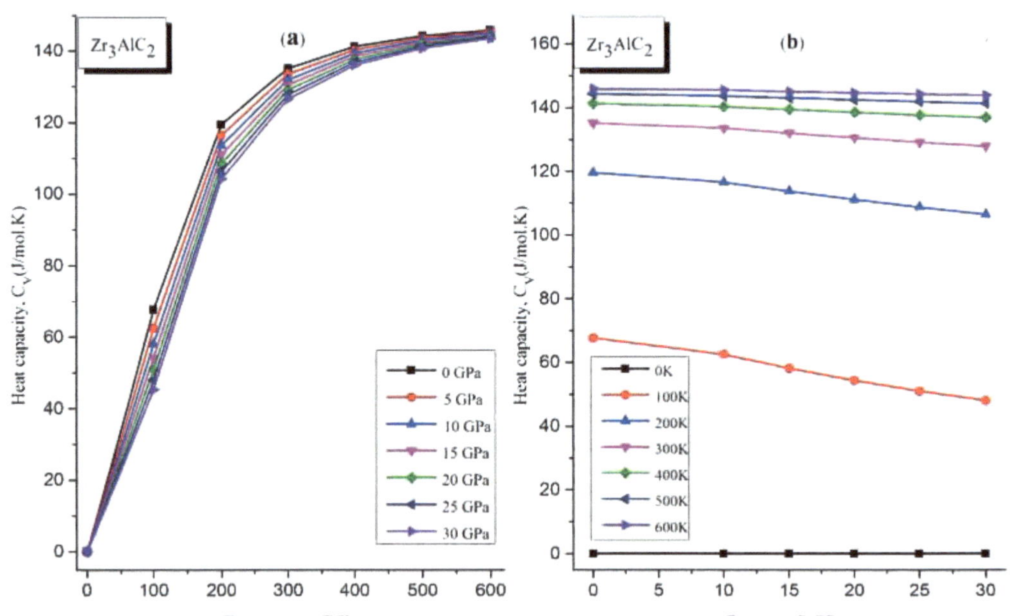

Fig. (24). The pressure and temperature effects on heat capacity C_V of Zr_3AlC_2 compound.

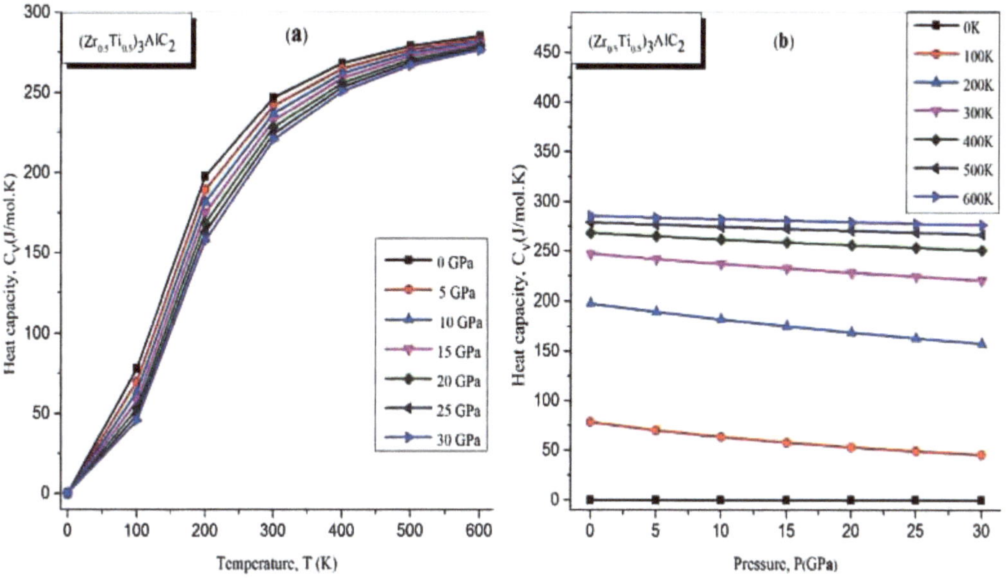

Fig. (25). The pressure and temperature effects on heat capacity C_V of $(Zr_{0.5}Ti_{0.5})_3AlC_2$ compound.

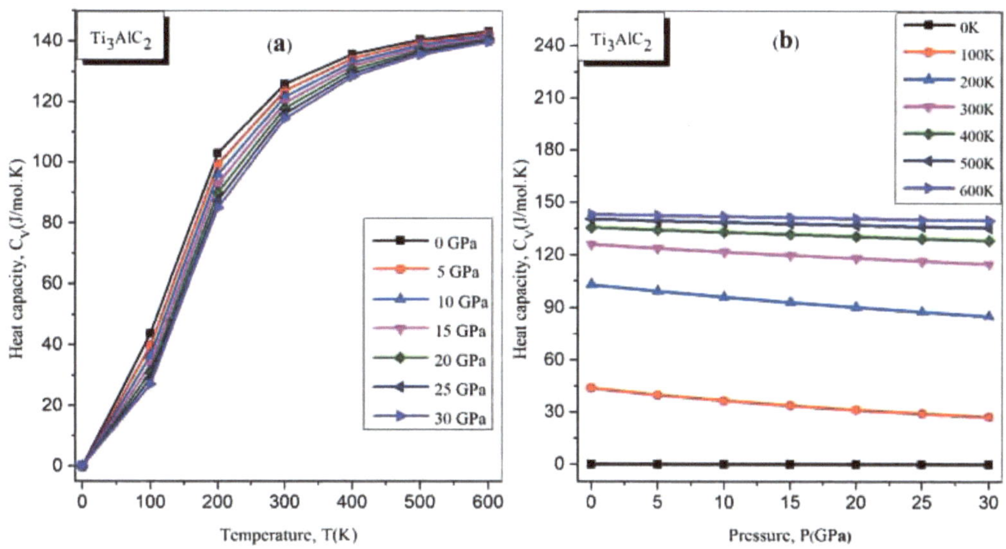

Fig. (26). The pressure and temperature effects on heat capacity C_V of Ti_3AlC_2 compound.

CONCLUSION

Herein, we have reported on various physical properties of the 312 MAX-phase compounds $(Zr_{1-x}Ti_x)_3AlC_2$ using the DFT first-principles calculations. The equilibrium ground-state of properties was well estimated and compared with other works. The thermodynamic stability was analyzed by calculating the formation energy, which shows that all compounds are quite stable and can be synthesized. The obtained elastic constants have confirmed the mechanical stability of the $(Zr_{1-x}Ti_x)_3AlC_2$ compounds. The elastic moduli and macroscopic mechanical behavior were studied and discussed in detail. The electronic structures revealed that the studied compounds show a metallic behavior with the presence of *p-d* and *s-p* hybridization. Furthermore, we have also studied the pressure and temperature effects on the bulk modulus, Debye temperature and heat capacity.

REFERENCES

[1] Barsoum, M.W.; El-Raghy, T. Synthesis and Characterization of a Remarkable Ceramic: Ti $_3$ SiC $_2$. *J. Am. Ceram. Soc.*, **1996**, *79*(7), 1953-1956. [http://dx.doi.org/10.1111/j.1151-2916.1996.tb08018.x]

[2] Barsoum, M.; El-Raghy, T. The MAX phases: Unique new carbide and nitride materials. *Am. Sci.*, **2001**, *89*(4), 334-343. [http://dx.doi.org/10.1511/2001.28.334]

[3] Yoo, H.I.; Barsoum, M.W.; El-Raghy, T. Ti_3SiC_2 has negligible thermopower. *Nature,* **2000**, *407*(6804), 581-582. [http://dx.doi.org/10.1038/35036686] [PMID: 11034196]

[4] Zapata-Solvas, E.; Christopoulos, S.R.G.; Ni, N.; Parfitt, D.C.; Horlait, D.; Fitzpatrick, M.E.; Chroneos, A.; Lee, W.E. Experimental synthesis and density functional theory investigation of radiation tolerance of Zr $_3$ (Al $_{1□x}$S i $_x$)C $_2$MAX phases. *J. Am. Ceram. Soc.,* **2017**, *100*(4), 1377-1387. [http://dx.doi.org/10.1111/jace.14742]

[5] El-Raghy, T.; Barsoum, M.W.; Zavaliangos, A.; Kalidindi, S.R. Processing and mechanical properties of Ti_3 SiC_2: II, effect of grain size and deformation temperature. *J. Am. Ceram. Soc.,* **1999**, *82*(10), 2855-2860. [http://dx.doi.org/10.1111/j.1151-2916.1999.tb02167.x]

[6] Barsoum, M.W.; Farber, L.; El-Raghy, T. Dislocations, kink bands, and room-temperature plasticity of Ti_3SiC_2. *Metall. Mater. Trans., A Phys. Metall. Mater. Sci.,* **1999**, *30*(7), 1727-1738. [http://dx.doi.org/10.1007/s11661-999-0172-z]

[7] Tallman, D.J.; Hoffman, E.N.; Caspi, E.N.; Garcia-Diaz, B.L.; Kohse, G.; Sindelar, R.L.; Barsoum, M.W. Effect of neutron irradiation on select MAX phases. *Acta Mater.,* **2015**, *85*, 132-143. [http://dx.doi.org/10.1016/j.actamat.2014.10.068]

[8] Lapauw, T.; Halim, J.; Lu, J.; Cabioc'h, T.; Hultman, L.; Barsoum, M.W.; Lambrinou, K.; Vleugels, J. Synthesis of the novel Zr_3 AlC_2 MAX phase. *J. Eur. Ceram. Soc.,* **2016**, *36*(3), 943-947. [http://dx.doi.org/10.1016/j.jeurceramsoc.2015.10.011]

[9] Hadi, M.A.; Panayiotatos, Y.; Chroneos, A. Structural and optical properties of the recently synthesized (Zr_3−x Ti x)AlC_2 MAX phases. *J. Mater. Sci. Mater. Electron.,* **2017**, *28*(4), 3386-3393. [http://dx.doi.org/10.1007/s10854-016-5933-z]

[10] Hadi, M.A.; Roknuzzaman, M.; Chroneos, A.; Naqib, S.H.; Islam, A.K.M.A.; Vovk, R.V.; Ostrikov, K. Elastic and thermodynamic properties of new (Zr_3−Ti)AlC_2 MAX-phase solid solutions. *Comput. Mater. Sci.,* **2017**, *137*, 318-326. [http://dx.doi.org/10.1016/j.commatsci.2017.06.007]

[11] Hohenberg, P.; Kohn, W. Inhomogeneous electron gas. *Phys. Rev.,* **1964**, *136*(3B), B864-B871. [http://dx.doi.org/10.1103/PhysRev.136.B864]

[12] Kohn, W.; Sham, L.J. Self-consistent equations including exchange and correlation effects. *Phys. Rev.,* **1965**, *140*(4A), A1133-A1138. [http://dx.doi.org/10.1103/PhysRev.140.A1133]

[13] Slater, J.C. *Adv. Quantum Chem.,* **1994**, *1*, 5564.

[14] Blaha, P.; Schwarz, K.; Madsen, G.K.H.; Kvasnicka, D.; Luitz, J. WIEN2K, an augmented plane wave +local orbitals program for calculating crystal properties. In: *TU Wien*; , **2001**; 28, .

[15] Nowotny, H. *Prog. Solid State Chem.,* **1970**, *2*, 27. [http://dx.doi.org/10.1016/0022-4596(70)90028-9]

[16] Barsoum, M.W.; Farber, L.; Levin, I.; Procopio, A.; El-Raghy, T.; Berner, A. High-resolution transmission electron microscopy of Ti_4 AlN_3, or Ti_3 Al_2 N_2 revisited. *J. Am. Ceram. Soc.,* **1999**, *82*(9), 2545-2547. [http://dx.doi.org/10.1111/j.1151-2916.1999.tb02117.x]

[17] Hu, C.; Li, F.; Zhang, J.; Wang, J.; Wang, J.; Zhou, Y. Nb_4AlC_3: A new compound belonging to the MAX phases. *Scr. Mater.,* **2007**, *57*(10), 893-896. [http://dx.doi.org/10.1016/j.scriptamat.2007.07.038]

[18] Barsoum, M.W. The MN+1AXN phases: A new class of solids. *Prog. Solid State Chem.,* **2000**, *28*(1-4), 201-281. [http://dx.doi.org/10.1016/S0079-6786(00)00006-6]

[19] Eklund, P.; Beckers, M.; Jansson, U.; Högberg, H.; Hultman, L. The M+1AX phases: Materials science and thin-film processing. *Thin Solid Films,* **2010**, *518*(8), 1851-1878. [http://dx.doi.org/10.1016/j.tsf.2009.07.184]

[20] Zhang, H.B.; Bao, Y.W.; Zhou, Y.C. *J. Mater. Sci. Technol.,* **2009**, *25*, 1-38. [http://dx.doi.org/10.1179/174328409X397247]

[21] Wang, X.H.; Zhou, Y.C. Layered machinable and electrically conductive Ti2AlC and Ti3AlC2 ceramics: A review. *J. Mater. Sci. Technol.,* **2010**, *26*(5), 385-416. [http://dx.doi.org/10.1016/S1005-0302(10)60064-3]

[22] Barsoum, M.W.; El-Raghy, T.; Procopio, A. Characterization of Ti^4AlN^3. *Metall Mater Trans A,* **2000**, *31*, 333-337.

[23] Nowotny, V. Structural chemistry of some transition metal compounds with the elements C, Si, Ge, Sn. *Prog. Solid State Chem.,* **1971**, *5*, 27-70. [http://dx.doi.org/10.1016/0079-6786(71)90016-1]

[24] Wang, J.; Zhou, Y. Recent progress in theoretical prediction, preparation, and characterization of layered ternary transition-metal carbides. *Annu. Rev. Mater. Res.,* **2009**, *39*(1), 415-443. [http://dx.doi.org/10.1146/annurev-matsci-082908-145340]

[25] Barsoum, M.W.; Radovic, M. Elastic and mechanical properties of the MAX phases. *Annu. Rev. Mater. Res.,* **2011**, *41*(1), 195-227. [http://dx.doi.org/10.1146/annurev-matsci-062910-100448]

[26] Sun, Z.M. Progress in research and development on MAX phases: A family of layered ternary compounds. *Int. Mater. Rev.,* **2011**, *56*(3), 143-166. [http://dx.doi.org/10.1179/1743280410Y.0000000001]

[27] Barsoum, M.W. *Encyclopedia of Materials Science and Technology*; Elsevier: Amsterdam, **2006**, pp. 1-1.

[28] Finkel, P.; Barsoum, M.W.; Hettinger, J.D.; Lofland, S.E.; Yoo, H.I. Low-temperature transport properties of nanolaminates Ti_3 AlC_2 and Ti_4 AlN_3. *Phys. Rev. B Condens. Matter,* **2003**, *67*(23), 235108.
[http://dx.doi.org/10.1103/PhysRevB.67.235108]

[29] Hettinger, J.D.; Lofland, S.E.; Finkel, P.; Palma, J.; Harrell, K.; Gupta, S.; Ganguly, A.; El-Raghy, T. Electrical transport, thermal transport, and elastic properties of M2AlC(M=Ti, Cr, Nb, and V). *Phy. Rev. B ,* **2005**, *72*, 115120.

[30] Barsoum, M. W. *MAX Phases: Properties of Machinable Carbides and Nitrides.* ; Wiley VCH-Verlag Gmb H&Co. KGaA, **2013**.

[31] Barsoum, M.W. *Ceramic Sience and Technology*; Edited by R. Riedel and I-W.Chen. Wiley-VCH Verlag, **2010**, 2, .

[32] El Saeed, M.A.; Deorsola, F.A.; Rashad, R.M. Influence of SPS parameters on the density and mechanical properties of sintered Ti_3SiC_2 powders. *Int. J. Refract. Hard Met.,* **2013**, *41*, 48-53.
[http://dx.doi.org/10.1016/j.ijrmhm.2013.01.016]

[33] Wang, L.; Zhang, J.; Jiang, W. Recent development in reactive synthesis of nanostructured bulk materials by spark plasma sintering. *Int. J. Refract. Hard Met.,* **2013**, *39*, 103-112.
[http://dx.doi.org/10.1016/j.ijrmhm.2013.01.017]

[34] Blaha, P.; Schwarz, K.; Tran, F.; Laskowski, R.; Madsen, G.K.H.; Marks, L.D. WIEN2k: An APW+lo program for calculating the properties of solids. *J. Chem. Phys.,* **2020**, *152*(7), 074101.
[http://dx.doi.org/10.1063/1.5143061] [PMID: 32087668]

[35] Bourachid, I.; Djamel , Rached; Rached, H.; Bentouaf, A.; Rached, Y.; Caid, M.; Abidri, B.; Djamel Rached, H. Magneto-electronic and thermoelectric properties of V-based Heusler in ferrimagnetic phase. *Appl. Phys., A Mater. Sci. Process.,* **2022**, *128*, 493.
[http://dx.doi.org/10.1007/s00339-022-05641-7]

[36] Rached, Y.; Caid, M.; Rached, H.; Merabet, M.; Benalia, S.; Al-Qaisi, S.; Djoudi, L.; Rached, D. Theoretical insight into the stability, magneto-electronic and thermoelectric properties of XCrSb (X: Fe, Ni) half-heusler alloys and their superlattices. *J. Supercond. Nov. Magn.,* **2022**, *35*(3), 875-887.
[http://dx.doi.org/10.1007/s10948-021-06131-2]

[37] Rached, H. Prediction of a new quaternary Heusler alloy within a good electrical response at high temperature for spintronics applications: DFT calculations. *Int. J. Quantum Chem.,* **2021**, *121*(12), e26647.
[http://dx.doi.org/10.1002/qua.26647]

[38] Belkacem, A.A.; Rached, H.; Caid, M.; Rached, Y.; Rached, D.; Mahmoud, N.T.; Benkhettou, N. The stability analysis and efficiency of the new MAX-phase compounds M3GaC2 (M: Ti or Zr): A first-principles assessment. *Results Phys.,* **2022**, *38*, 105621.
[http://dx.doi.org/10.1016/j.rinp.2022.105621]

[39] Rached, H.; Bendaoudia, S.; Rached, D. Investigation of iron-based double perovskite oxides on the magnetic phase stability, mechanical, electronic and optical properties *via* first-principles calculation. *Mater. Chem. Phys.,* **2017**, *193*, 453-469.
[http://dx.doi.org/10.1016/j.matchemphys.2017.03.006]

[40] Azzouz-Rached, A.; Rached, H.; Ouadha, I.; Rached, D.; Reggad, A. The vanadium-doping effect on physical properties of the Zr2AlC MAX phase compound. *Mater. Chem. Phys.,* **2021**, *260*, 124189.
[http://dx.doi.org/10.1016/j.matchemphys.2020.124189]

[41] Azzouz-Rached, A.; Haque Babu, M.M.; Rached, H.; Hadji, T.; Rached, D. Prediction of a new Sn-based MAX phases for nuclear industry applications: DFT calculations. *Mater. Today Commun.,* **2021**, *27*, 102233.

[http://dx.doi.org/10.1016/j.mtcomm.2021.102233]

[42] Azzouz-Rached, A.; Hadi, M.A.; Rached, H.; Hadji, T.; Rached, D.; Bouhemadou, A. Pressure effects on the structural, elastic, magnetic and thermodynamic properties of Mn_2AlC and Mn_2SiC MAX phases. *J. Alloys Compd.,* **2021**, *885*, 160998. [http://dx.doi.org/10.1016/j.jallcom.2021.160998]

[43] Rabah, I.E.; Rached, H.; Rabah, M.; Rached, D.; Benkhettou, N. A theoretical analysis of physical properties and half-metallic stability under pressure effect of the ScNiCrZ (Z=Ga, Al, In) heusler alloys. *SPIN,* **2021**, *11*(1), 2150007. [http://dx.doi.org/10.1142/S2010324721500077]

[44] Al-Qaisi, S.; Mushtaq, M.; Alomairy, S.; Vu, T.V.; Rached, H.; Haq, B.U.; Mahmood, Q.; Al-Buriahi, M.S. First-principles investigations of Na_2CuMCl_6 (M = Bi, Sb) double perovskite semiconductors: Materials for green technology. *Mater. Sci. Semicond. Process.,* **2022**, *150*, 106947. [http://dx.doi.org/10.1016/j.mssp.2022.106947]

[45] Tagrerout, A.; Rached, H.; Drief, M.; Guermit, Y.; Al-Qaisi, S.; Caid, M.; Rached, D. An extensive computational report on the quinary alloys Cu2Zn1−xCdxSnS4 for the solar cell systems: DFT simulation. *Comput. Condens. Matter.,* **2022**, *31*, e00670.

[46] Caid, M.; Rached, Y.; Rached, D.; Cheref, O.; Rached, H.; Benalia, S.; Merabet, M. Electronic structure of short-period ZnSe/ZnTe superlattices based on DFT calculations. *Condens. Matter Phys.,* **2022**, *25*(1), 13701. [http://dx.doi.org/10.5488/CMP.25.13701]

[47] Ould Moussa, M.; Kouidri, S.; Rached, H.; Haid, S.; Beloufa, N.; Ouadha, I.; El Keurti, M.; Meliani, S.; Ben Slimane, A. Theoretical investigation on the optoelectronic properties of $ZrxSi_1$-xO_2 tetragonal hypothetical alloys from zircon family. *Appl. Phys., A Mater. Sci. Process.,* **2022**, *128*(3), 231. [http://dx.doi.org/10.1007/s00339-022-05362-x]

[48] Samira, Belkhouane.; Ali, Bentouaf.; Habib, Rached.; Zahira Amel, Bouyakoub. Electronic structure, magnetic and structural properties of binary cubic C15 Laves phases PrX2 (X = Co and Fe): a first-principles study. *Appl. Phys., A Mater. Sci. Process.,* **2021**, *127*, 835. [http://dx.doi.org/10.1007/s00339-021-04980-1]

[49] Sadok Cherif, H.; Bentouaf, A.; Bouyakoub, Z.A.; Rached, H.; Aïssa, B. Computational determination of structural, electronic, magnetic and thermodynamic properties of Co2HfZ (Z = Al, Ga, Si and Sn) full Heusler compounds for spintronic applications. *J. Alloys Compd.,* **2022**, *894*, 162503. [http://dx.doi.org/10.1016/j.jallcom.2021.162503]

[50] Berrahal, M.; Bentouaf, A.; Rached, H.; Mebsout, R.; Aissa, B. Investigation of ruthenium based full-heusler compound for thermic, spintronics and thermoelectric applications: DFT computation. *Mater. Sci. Semicond. Process.,* **2021**, *134*, 106047. [http://dx.doi.org/10.1016/j.mssp.2021.106047]

[51] Reggad, A.; Lardjani, R.; Baghdad, R.; Bouhafs, B. The effect of the correlation and exchange interactions on the electronic and magnetic properties of the hexagonal NiS using the onsite exact exchange/hybrid functionals. *Physica B,* **2017**, *526*, 89-95. [http://dx.doi.org/10.1016/j.physb.2017.07.002]

[52] Perdew, J.P.; Burke, K.; Ernzerhof, M. Generalized gradient approximation made simple. *Phys. Rev. Lett.,* **1996**, *77*(18), 3865-3868. [http://dx.doi.org/10.1103/PhysRevLett.77.3865] [PMID: 10062328]

[53] Berri, S.; Maouche, D.; Zerarga, F.; Medkour, Y. Ab initio study of the structural, electronic, elastic and magnetic properties of Cu_2GdIn, Ag_2GdIn and Au_2GdIn. *Physica B,* **2012**, *407*(17), 3328-3334. [http://dx.doi.org/10.1016/j.physb.2012.04.012]

[54] Lapauw, T.; Halim, J.; Lu, J.; Cabioc'h, T.; Hultman, L.; Barsoum, M.W.; Lambrinou, K.; Vleugels, J. Synthesis of the novel Zr_3 AlC_2 MAX phase. *J. Eur. Ceram. Soc.,* **2016**, *36*(3), 943-947. [http://dx.doi.org/10.1016/j.jeurceramsoc.2015.10.011]

[55] Pietzka, M.A.; Schuster, J.C. Summary of constitutional data on the Aluminum-Carbon-Titanium system. *J. Phase Equilibria,* **1994**, *15*(4), 392-400. [http://dx.doi.org/10.1007/BF02647559]

[56] Ouadha, I.; Rached, H.; Azzouz-Rached, A.; Reggad, A.; Rached, D. Study of the structural, mechanical and thermodynamic properties of the new MAX phase compounds (Zr1-xTix)3AlC2. *Condens. Matter,* **2020**, *23*(4), e00468.

[57] Hu, C.; Li, F.; Zhang, J.; Wang, J.; Wang, J.; Zhou, Y. Nb4AlC3: A new compound belonging to the MAX phases. *Scr. Mater.,* **2007**, *57*(10), 893-896. [http://dx.doi.org/10.1016/j.scriptamat.2007.07.038]

[58] Barsoum, M.W.; El-Raghy, T. Synthesis and characterization of a remarkable ceramic: Ti_3SiC_2. *J. Am. Ceram. Soc.,* **1996**, *79*(7), 1953-1956. [http://dx.doi.org/10.1111/j.1151-2916.1996.tb08018.x]

[59] Available from: http://www.wien2k.at/

[60] Surucu, G. Investigation of structural, electronic, anisotropic elastic, and lattice dynamical properties of MAX phases borides: An Ab-initio study on hypothetical MAB (M = Ti, Zr, Hf; A = Al, Ga, In) compounds. *Mater. Chem. Phys.,* **2018**, *203*, 106-117. [http://dx.doi.org/10.1016/j.matchemphys.2017.09.050]

[61] Surucu, G.; Kaderoglu, C.; Deligoz, E.; Ozisik, H. Investigation of structural, electronic and anisotropic elastic properties of Ru -doped WB 2 compound by increased valence electron concentration. *Mater. Chem. Phys.,* **2017**, *189*, 90-95. [http://dx.doi.org/10.1016/j.matchemphys.2016.12.036]

[62] Hadi, M.A.; Panayiotatos, Y.; Chroneos, A. Structural and optical properties of the recently synthesized (Zr3−x Ti x)AlC2 MAX phases. *J. Mater. Sci. Mater. Electron.,* **2017**, *28*(4), 3386-3393. [http://dx.doi.org/10.1007/s10854-016-5933-z]

[63] Zapata, E.; Hadi, M.A.; Horlait, D.; Parfitt, D.C.; Thibaud, A.; Chroneos, A.; Lee, W.E. *J. Am. Ceram. Soc.,* **2017**, *100*, 3393-3401. [http://dx.doi.org/10.1111/jace.14870]

[64] Kasap, S.O. *Principles of electronic materials and devices,* 3rd ed; The McGraw-Hill, **2006**.

[65] Ashcroft, N.W.; Mermin, N.D. *Solid State Physics*; Saunders College: Philadelphia, **1976**.

[66] Reshak, A.H. Moreteza jamal. *Int. J. Electrochem. Sci.,* **2013**, *8*, 12252-12263. [http://dx.doi.org/10.1016/S1452-3981(23)13261-5]

[67] Rached, H.; Rached, D.; Benalia, S.; Reshak, A.H.; Rabah, M.; Khenata, R.; Bin Omran, S. First-principles study of structural stabilities, elastic and electronic properties of transition metal monocarbides (TMCs) and mononitrides (TMNs). *Mater. Chem. Phys.,* **2013**, *143*(1), 93-108. [http://dx.doi.org/10.1016/j.matchemphys.2013.08.020]

[68] Wang, J.; Wang, J.; Zhou, Y.; Hu, C. Phase stability, electronic structure and mechanical properties of ternary-layered carbide Nb4AlC3: An ab initio study. *Acta Mater.,* **2008**, *56*(7), 1511-1518. [http://dx.doi.org/10.1016/j.actamat.2007.12.003]

[69] Hill, R. The elastic behaviour of a crystalline aggregate. *Proc. Phys. Soc. A,* **1952**, *65*(5), 349-354. [http://dx.doi.org/10.1088/0370-1298/65/5/307]

[70] Voigt, W. *Textbook of crystal physics*; Teubner Verlag, Leipzig., **1928**, pp. 95-100.

[71] Reuss, A. Calculation of the yield point of mixed crystals based on the plasticity condition for single crystals. *Z. Angew. Math. Mech.,* **1929**, *9*(1), 49-58. [http://dx.doi.org/10.1002/zamm.19290090104]

[72] Tu, Y.; Wang, Y. First-principles study of the elastic properties of and alloys. *Solid State Commun.,* **2011**, *151*(3), 238-241.

[http://dx.doi.org/10.1016/j.ssc.2010.11.023]

[73] Vaitheeswaran, G.; Kanchana, V.; Svane, A.; Delin, A. Elastic properties of $MgCNi_3$ —a superconducting perovskite. *J. Phys. Condens. Matter,* **2007**, *19*(32), 326214. [http://dx.doi.org/10.1088/0953-8984/19/32/326214]

[74] TaO, X.; Yang, J.; Xi, L.; Ouyang, Y. *J. Solid State Chem.,* **2012**, *194*, 179-187.

[75] Pugh, S.F. XCII. Relations between the elastic moduli and the plastic properties of polycrystalline pure metals. *Lond. Edinb. Dublin Philos. Mag. J. Sci.,* **1954**, *45*(367), 823-843. [http://dx.doi.org/10.1080/14786440808520496]

[76] Yildirim, A.; Koc, H.; Deligoz, E. First-principles study of the structural, elastic, electronic, optical, and vibrational properties of intermetallic Pd_2Ga. *Chin. Phys. B,* **2012**, *21*(3), 037101. [http://dx.doi.org/10.1088/1674-1056/21/3/037101]

[77] Jamal, M.; Sarvestani, N.K.; Yazdani, A.; Reshak, A.H. Mechanical and thermodynamical properties of hexagonal compounds at optimized lattice parameters from two-dimensional search of the equation of state. *RSC Advances,* **2014**, *4*(101), 57903-57915. [http://dx.doi.org/10.1039/C4RA09358E]

[78] Ledbetter, H.M. *J. Phys. Chem.,* **1977**, *6*(4), 1109-1135.

[79] Biskri, Z.E.; Rached, H.; Bouchear, M.; Rached, D. Computational study of structural, elastic and electronic properties of lithium disilicate ($Li_2Si_2O_5$) glass-ceramic. *J. Mech. Behav. Biomed. Mater.,* **2014**, *32*, 345-350. [http://dx.doi.org/10.1016/j.jmbbm.2013.10.029] [PMID: 24411692]

[80] Ding, A.L.; Li, C.M.; Wang, J.; Ao, J.; Li, F.; Chen, Z-Q. Anisotropy of elasticity and minimum thermal conductivity of monocrystal M_4AlC_3 (M = Ti, Zr, Hf). *Chin. Phys. B,* **2014**, *23*(9), 096201. [http://dx.doi.org/10.1088/1674-1056/23/9/096201]

[81] Chen, X.Q.; Niu, H.; Li, D.; Li, Y. Modeling hardness of polycrystalline materials and bulk metallic glasses. *Intermetallics,* **2011**, *19*(9), 1275-1281. [http://dx.doi.org/10.1016/j.intermet.2011.03.026]

[82] Kaufman, M. *Principles of Thermodynamics*; Marcel Dekker, Inc: New York, **2002**. [http://dx.doi.org/10.1201/9780203909768]

[83] Elahmar, M.H.; Rached, H.; Rached, D.; Khenata, R.; Murtaza, G.; Bin Omran, S.; Ahmed, W.K. Structural, mechanical, electronic and magnetic properties of a new series of quaternary Heusler alloys CoFeMnZ (Z=Si, As, Sb): A first-principle study. *J. Magn. Magn. Mater.,* **2015**, *393*, 165-174. [http://dx.doi.org/10.1016/j.jmmm.2015.05.019]

[84] Benkabou, M.; Rached, H.; Abdellaoui, A.; Rached, D.; Khenata, R.; Elahmar, M.H.; Abidri, B.; Benkhettou, N.; Bin-Omran, S. Electronic structure and magnetic properties of quaternary Heusler alloys CoRhMnZ (Z = Al, Ga, Ge and Si) *via* first-principle calculations. *J. Alloys Compd.,* **2015**, *647*, 276-286. [http://dx.doi.org/10.1016/j.jallcom.2015.05.273]

[85] Hadi, M.A.; Roknuzzaman, M.; Chroneos, A.; Naqib, S.H.; Islam, A.K.M.A.; Vovk, R.V.; Ostrikov, K. Elastic and thermodynamic properties of new (Zr3−Ti)AlC2 MAX-phase solid solutions. *Comput. Mater. Sci.,* **2017**, *137*, 318-326. [http://dx.doi.org/10.1016/j.commatsci.2017.06.007]

[86] Haddadi, K. *Study of the structural, elastic and electronic properties of antiperovskite compounds of type XNCa3, doctoral thesis, faculty of Science.,* **2013**,

[87] Blanco, M.A.; Francisco, E.; Luaña, V. GIBBS: isothermal-isobaric thermodynamics of solids from energy curves using a quasi-harmonic Debye model. *Comput. Phys. Commun.,* **2004**, *158*(1), 57-72. [http://dx.doi.org/10.1016/j.comphy.2003.12.001]

[88] McLellan, A.G. *The Classical Thermodynamics of Deformable Materials*; Cambridge University

Press: Cambridge, **1980**, p. 165.

[89] Francisco, E.; Recio, J.M.; Blanco, M.A. A. Martín pend as. *J. Phys. Chem.,* **1998**, *102*, 1595. [http://dx.doi.org/10.1021/jp972516j]

[90] Francisco, E.; Blanco, M.A.; Sanjurjo, G. Atomistic simulation of Sr F 2 polymorphs. *Phys. Rev. B Condens. Matter,* **2001**, *63*(9), 094107. [http://dx.doi.org/10.1103/PhysRevB.63.094107]

[91] Poirier, J.P. *Introduction to the Physics of the Earth's Interior*; Cambridge University Press, **1991**.

CHAPTER 7

Surface Segregation in Pt_3Nb and Pt_3Ti using Density Functional Theory-based Methods

Kingsley O. Obodo[1,2,3], **Lalrin Kima**[4,5], **Adedapo S. Adeyinka**[6] and **Dibya Prakash Rai**[4,*]

[1] *ICTP-East African Institute for Fundamental Research, Kigali, Rwanda*

[2] *Center for Space Research, North-West University Potchefstroom, South Africa*

[3] *National Institute of Theoretical and Computational Sciences, Johannesburg, South Africa*

[4] *Department of Physics, Mizoram University, Aizawl, India*

[5] *Department of Physics, Pachhunga University College, Mizoram University, Aizawl, India*

[6] *Research Centre for Synthesis and Catalysis, Department of Chemical Sciences, University of Johannesburg, Johannesburg, South Africa*

Abstract: First-principles DFT calculations were used to investigate surface segregation processes in ordered Pt_3X (where X=Nb, Ti) alloys. Using pristine Pt (111) surface as a reference, the effect of surface segregation on the adsorption energy of O_2 atoms in Pt_3X alloys was evaluated. Our results showed that surface segregation due to direct exchange is only feasible for the Pt_3Nb alloy (E_{segr} = - 0.3833 eV) but not for its Ti analogue (E_{segr} = 0.516 eV). In contrast, for both Pt_3X alloys, surface segregation due to antisite migration and leading to the formation of a Pt-skin or overlayer, favouring oxygen atom adsorption, an essential step in ORR, is possible. Interestingly, reverse migration of X atoms from the bulk to replace Pt atoms on the surface is an endothermic process and is thus very unlikely. Analysis of the surface segregation energy for configurations involving a direct exchange of Pt atoms located beyond the third layer in the slab model with Nb atoms at the surface indicates the formation of pristine bulk like Pt (111) surface from Pt_3Nb surface is unlikely. The energy of adsorption for the O-atom on pristine Pt (111) surface shows that the presence of minute quantities of dopant Nb atoms in the sub-surface layer could enhance its suitability for ORR. Comparison of O-atom adsorption energy on the various surface segregation models of Pt_3X alloys to that of pristine Pt (111) surface shows that the formations of a Pt-skin or overlayer on the Pt_3Nb surface due to surface segregation change the O-atom adsorption energy on this surface to 0.34 eV which is just 0.14 eV higher than the optimal value of 0.20 eV. Our results also show that the binding of an oxygen atom to the fcc Pt site in Pt_3Ti is lower in energy compared to its binding on a pristine Pt (111) surface. In comparison, the binding of an oxygen atom to the fcc Pt site in Pt_3Ti is of the same magnitude as that of the pristine Pt (111) surface.

* **Corresponding author Dibya Prakash Rai:** Department of Physics, Mizoram University, Aizawl, India; E-mail: dibyaprakashrai@gmail.com

Dibya Prakash Rai, Kingsley O. Obodo, & Jitendra Pal Singh (Eds.)

Keywords: Density functional theory, Platinum, Surface segregation, ORR.

INTRODUCTION

Developing improved polymer electrolyte membrane fuel cells (PEMFC) is an active area of research within the industry and academia. In industry, manufacturing long-lasting and durable PEMFC at a meager cost (reduction of platinum usage) is one of the key drivers of research in this area [1 - 3]. The PEMFC could provide a viable means of on-board energy for different applications such as transportation, stationary generators, etc. This industrial development is only possible if specific academic questions are tackled. Some interesting questions currently addressed within the academic domain are the improvement of the fuel cell catalyst, segregation issues in alloyed materials, reaction kinetics, nanoparticle optimization, *etc*. [1 - 4].

One key reaction carried out by the fuel cell membrane is the oxygen reduction reaction (ORR), which involves the reduction of water to produce oxygen and energy. The ORR reaction proceeds slowly, even in the presence of platinum, the archetypal catalyst for this process. Several Pt alloys have been investigated as catalysts for this reaction, and some of them have displayed enhanced catalytic activity. However, their long-term durability and stability are still in question due to the possibility of surface segregation in the alloys. Hence, this work investigates the segregation of Pt_3Nb and Pt_3Ti-based alloy catalysts in ORR. Surface segregation involves the movement of atoms from the bulk or vice versa due to the prevailing chemical environment, which might result in a difference between the chemical composition of the surface region and the bulk region. Surface segregation could significantly affect the overall properties of materials overall properties encompassing crucial aspects such as wetting, electrical contact, adsorption, oxidation, crystal growth, corrosion, friction and wear, and catalysis. This effect could occur in both ordered and disordered alloys [5 - 8].

Various Pt alloys, including late transition metals such as Ni, Co, Cr, and Fe, as well as partially dealloyed core-shell catalysts derived from Pt-Cu nanoparticles, have undergone extensive investigation [1 - 11]. This chapter introduces a novel collection of Pt alloys and intermetallic compounds that exhibit promising properties for the oxygen reduction reaction (ORR). The ORR is a challenging reaction to catalyze because the catalyst material must possess both stability under highly corrosive conditions at a fuel cell cathode and chemical activity to activate O_2 while also being capable of releasing oxygen from the surface in the form of H_2O. O_2 activation typically involves proton and electron transfers to form adsorbed OOH before breaking the O-O bond [12]. Therefore, the catalyst must moderately stabilize OOH. After dissociation, the catalyst surface forms adsorbed

O and OH, and these species should not bind too strongly to facilitate rapid H_2O desorption. Our focus lies on metallic ORR catalysts comprised of alloys or intermetallic compounds, hereafter referred to as alloys. Specifically, systems that form Pt overlayers or "skins" on the surface are explored. These two metallic elements are the only ones stable under the high potentials and acidic conditions of a PEMFC while also exhibiting surface chemical properties that closely align with the optimum requirements for the ORR. The ultimate objective is to discover Pt_3X alloy meeting the following criteria: (1) they should form Pt or Pd overlayers with greater ORR activity than Pt, and (2) they should demonstrate maximum stability [9 - 15].

The approach involves employing density functional theory (DFT) calculations initially to identify intriguing candidates, followed by experimental testing. While the primary goal of this approach is computational electrocatalyst discovery, it also serves as a valuable exploration of the current understanding of the ORR. In the realm of catalysis, nothing attests to the robustness and accuracy of a theoretical framework more convincingly than its ability to identify new active materials.

Surface segregation can manifest in ordered and disordered alloys, driven by three primary physical mechanisms in solid solutions or disordered binary alloys. Firstly, for alloys with heat of solution negative, the majority component enriches the surface towards maximizing the bulk component mixing. Next, when there is a significant difference in atomic sizes between the components, the smaller component segregates to the surface to reduce strain energy. The component with lower surface energy, in its pure elemental form, segregates to the alloy surface. Lastly, surface segregation in ordered alloys can occur due to the generation and movement of antisite defects caused by the off-stoichiometric effect. Surface segregation in disordered alloys is always exothermic and results in a decrease in the system's configurational entropy. In contrast, surface segregation in ordered binary alloys would increase the system configurational entropy, and it can be exothermic or endothermic. The prediction of surface segregation in binary alloys is long been a subject of research [13 - 19] utilizing theoretical models such as tight-binding theory (TBT), embedded atom method (EAM), density functional theory (DFT), and statistical mechanics methods to evaluate energy and sample different configurations [18 - 22].

Previous studies have primarily focused on surface segregation in disordered alloy configurations. In this work, we employ a similar computational approach to predict the feasibility of surface segregation in ordered Pt_3Nb and Pt_3Ti alloys. The surface structure and composition of bimetallic alloys greatly influence their catalytic activities. Therefore, surface segregation models can be used to design and synthesize highly active catalysts. For example, platinum (Pt) is an efficient catalyst in various chemical reactions but is also a precious metal. To develop catalysts with reduced or low Pt content and excellent performance, studies on Pt-based bimetallic alloys are ongoing by replacing some Pt content with less expensive metals. Some studies have found that surface segregation enhances the activities of certain Pt binary alloys in oxygen reduction reactions (ORRs). For instance, surface segregation results in a 100% Pt outermost layer and a sublayer composed of 50% Pt (*i.e.*, 50% Ni) in the Pt_3Ni (111) surface. The presence of Ni atoms beneath the Pt monolayer shifts the *d* electron band of Pt, resulting in higher catalytic activity for ORRs [23 - 29]. To design Pt alloy catalysts using surface segregation, it is crucial to establish the correlation between surface composition and catalytic activities through experimental examination or theoretical determination of d electron band structures [30 - 34].

In this study, we first simulate the surface segregation process to predict the composition of the $Pt_3Nb(111)$ and $Pt_3Ti(111)$ surfaces. Then, using first-principles DFT calculations, we calculate the electronic structure and O adsorption on the Pt-segregated $Pt_3Nb(111)$ and $Pt_3Ti(111)$ surfaces. This investigation provides insights into how surface segregation can modify the catalytic behaviour of Pt_3Nb and Pt_3Ti alloys. Pt-Ti alloys exhibit a high negative heat of solution due to bonding interactions in the *d*-orbitals of Pt and Ti atoms being strong, resulting in high Pt_3Ti chemical stability. Furthermore, Pt_3Ti alloy demonstrates excellent catalytic performance in ORRs. Carbon-supported $Pt_{75}Ti_{25}$ nanoparticles are found to exhibit twofold improved catalytic activities or a shift of 20 mV towards a higher potential in the electroreduction of molecular oxygen compared to the benchmark Pt/C catalyst [35]. Using the low energy electron diffraction (LEED) structural analysis, the surface of the $Pt_3Ti(111)$ single crystal can exhibit two distinct terminations, depending on the crystal preparation, using a reproducible sputtering/annealing method. These terminations are distinguished by the presence of either one monolayer of Pt (Pt-$Pt_3Ti(111)$) or two Pt layers covering the underlying bulk crystal (2Pt-$Pt_3Ti(111)$) [36]. Experimental observations using ultraviolet photoemission spectroscopy also suggest the formation of a pure Pt outermost surface layer in annealed Pt_3Ti polycrystals, indicated by the movement of the *d*-band center away from the Fermi level [37]. It's worth noting that the surface segregation process in ordered Pt_3Ti crystals is different from what is seen in disordered Pt alloys [38].

Theoretical calculations have provided insights into surface segregation phenomena in most Pt alloys, such as Pt_3Co, Pt_3Ni, and Pt_3Fe, where the Pt atoms in the sublayer exchange with the second-element atoms in the outermost layer, explaining the observed segregation patterns [39-41]. To further investigate the surface segregation process in the (111) surface of ordered Pt_3X (X = Nb and Ti) crystals, we employ the density functional theory (DFT) method in this study. By utilizing DFT, we aim to gain a deeper understanding of the surface composition and structure of these alloys. Furthermore, we explore the adsorption energies of oxygen on various surfaces to model the impact of surface segregation on the catalytic activity of these Pt alloys in the oxygen reduction reaction (ORR). By examining the adsorption energies, we can evaluate the potential enhancement or alteration in catalytic activity resulting from surface segregation in Pt_3Nb and Pt_3Ti alloys. This comprehensive investigation using the DFT method allows us to shed light on the intricate surface segregation process in ordered Pt_3X crystals and its implications for catalytic performance in the ORR.

COMPUTATIONAL METHOD

In this study, we employed density functional theory (DFT) using the grid-based projector augmented wave (GPAW) code within the atomic-simulation environment (ASE) to model the bulk and surface structures of different Pt3X systems (where X refers to Nb and Ti atoms) [42, 43]. The Revised-Perde--Burke-Ernzerhof (RPBE) functional, a parameterization of the generalized gradient approximation (GGA), is utilized to approximate the exchange-correlation function. The Brillouin zone was sampled using a (3x3x1) Monkhorst-Pack grid of K points, with a grid spacing of 0.18 [44 - 47]. The behaviour of the ionic cores is described using ultrasoft pseudopotentials, and the valence states in Kohn-Sham formalism are expanded using the local combination of atomic orbitals with a kinetic energy of 340 eV [48]. Hellmann-Feynman forces acting on the atoms were set to 0.03 eV/Å.

To evaluate the relaxed geometries, we employed a two-step process. Firstly, we conducted a full geometry optimization of the bulk structure in the plane wave mode. Secondly, we created (111) surfaces for Pt_3X (where X refers to Nb and Ti) from the bulk structure and only optimized the atomic positions. We modeled the Pt_3X surfaces with different numbers of layers using a slab. We calculated the energies of surface segregation *via* direct exchange or anti-site migration, as well as oxygen adsorption at various sites on these surfaces, to evaluate the relevance of these systems for ORR.

RESULTS AND DISCUSSION

Surface Segregation in Pt_3X (111) Surface

To evaluate the surface segregation of Pt/X atoms to the outermost layer of the Pt_3X (111) surface, we carried out a direct exchange by swapping the positions of the X/Pt atoms on the surface with the corresponding Pt/X atoms in the interior. To examine the feasibility of this path, the segregation energy, $E_{(\text{segregation})}$, was computed. $E_{(segregation)}$ represents the difference in energy between two surface models: a bulk-terminated $Pt_3X(111)$ surface model and a Pt-segregated surface model. The latter was created by swapping the X (Nb or Ti) atom in the outermost layer with a subsurface Pt atom from the bulk-terminated $Pt_3Ti(111)$ surface. A negative value of $E_{(segregation)}$ indicates that Pt/X surface segregation is energetically favourable, while a positive value suggests the opposite. Therefore, the sign of $E_{(segregation)}$ provides insight into the feasibility of Pt/X segregation on the surface.

In Antisite migration, the Pt/X atoms segregation towards the outermost layer of the Pt_3X (111) surface by moving Pt/X antisites to the surface is considered. An antisite is a type of off-stoichiometric structural defect that occurs in ordered alloys, where an atom of one species occupies the site of the other species. In this study, we considered a Pt antisite (denoted as PtX) as an X-atom that substitutes a Pt atom in the ordered Pt_3Ti crystal (Table **(1d)** and Table **(2d)**). An X antisite (represented as Ti_{Pt}), on the other hand, refers to a Pt atom substituting an X atom (Table **(1e)** and Table **(2e)**). To evaluate the surface segregation energy in the case of Antisite migration, we calculated the energy difference between two Pt_3X (111) surface slab models. The first model contained a Pt/X antisite defect located at the surface region, specifically at the first, second, or third atomic layer. The second model contained an antisite defect in the bulk region, specifically at the fourth atomic layer. The energy difference between these models represents the energy required for the Pt/X atoms to migrate to the surface, and a negative value indicates an exothermic process.

Table 1. Calculated Pt surface segregation energy for the $Pt_3Nb(111)$ surface (where the grey and green balls represent Pt and Nb-atoms, respectively).

-	Bulk	-	Direct Exchange	$E_{(Segregation)}$ eV	-	Antisite	$E_{(Segregation)}$ eV
(a)	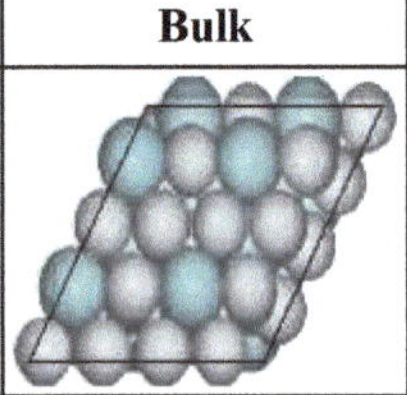	(b)	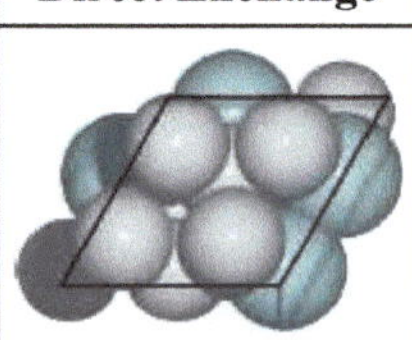	-0.38331	(d)	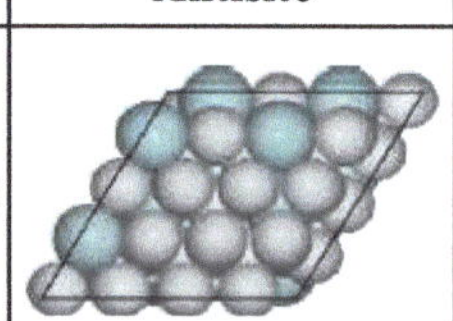	-1.09687
	-	(c)	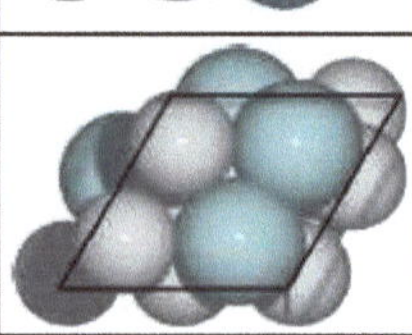	1.943373	(e)	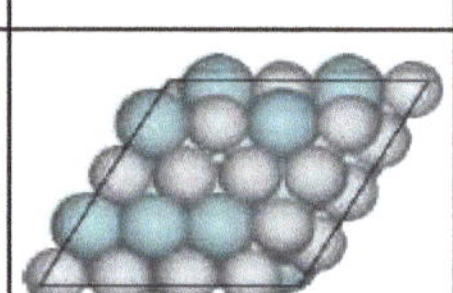	0.991

Direct Exchange and Antisite Migration

Tables **1** and **2** show the bulk, direct exchange and antisite migration configurations for the $Pt_3Nb(111)$ and $Pt_3Ti(111)$ surfaces, respectively, along with their corresponding surface segregation energies in eV. Table **(1a)** shows the bulk structure of $Pt_3Nb(111)$ surface structure, **(b)** and **(c)** indicate the direct exchange configurations and the antisite configurations are presented in **(d)** and **(e).** Table **(2a)** shows the bulk structure of $Pt_3Ti(111)$ surface structure; **(b)** and **(c)** shows the direct exchange configurations; **(d)** and **(e)** the antisite configurations.

Table 2. Calculated Pt surface segregation energy for the $Pt_3Ti(111)$ surface using the first-principles DFT method (where the grey and green balls represent Pt and Ti-atoms, respectively).

-	Bulk	-	Direct Exchange	$E_{(Segregation)}$ eV	-	Antisite	$E_{(Segregation)}$ eV
(a)	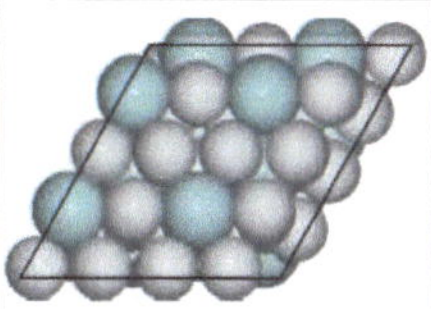	(b)	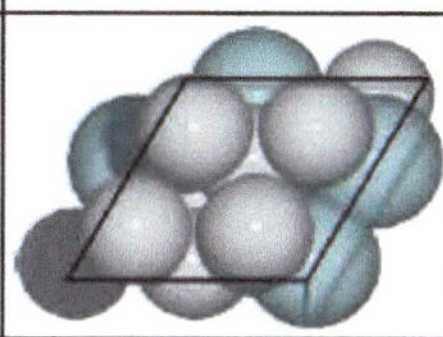	0.516	(d)	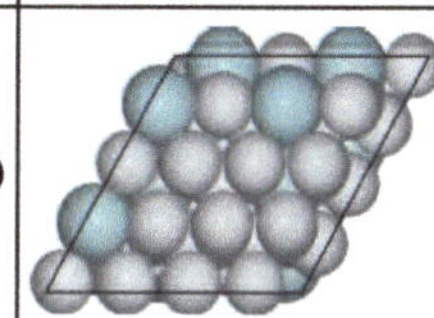	-1.236
	-	(c)	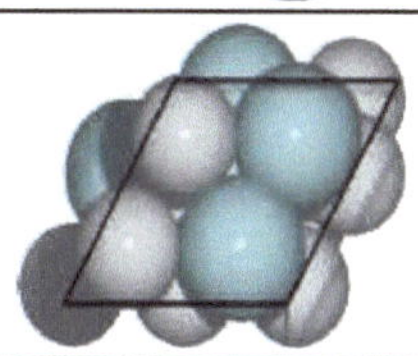	2.810	(e)	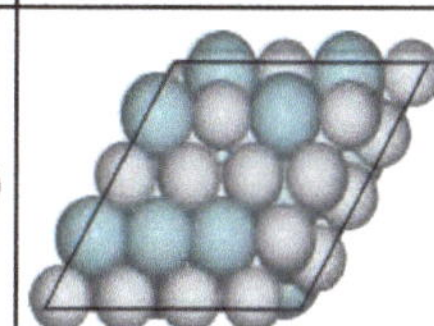	0.433

The calculated surface segregation energies, as presented in Tables **1** and **2**, show that the direct exchange of the Ti and Nb atoms (X_{Pt}) from the sub-surface to the top layer is highly endothermic with the energy of 2.8096eV and 1.9434eV, respectively. The calculated direct exchange energy of Pt_{Ti} is 0.516eV, and that of Pt_{Nb} is -0.3833eV. This implies that direct exchange is not favourable for Pt_3Ti-ordered alloys. However, our calculated results agree with previously published data. However, the exothermic character of the direct exchange energy of Pt_{Nb} implied that for the Pt_3Nb-based alloy, there is a possibility of having this defective kind of structure.

Further investigation on the ease of direct exchange of Pt_{Nb} is shown in Table **3**. Various configurations for six layers of the Pt_3Nb slab are considered. Table **3, 1** revealed the exchange of Pt in the top layer with Nb from the third layer. The segregation energy is positive. Configurations **2**, **3** and **4** were all positive and highly endothermic. They involve the direct exchange of Pt atoms located further lower in the slab with Nb atoms to create two layers of Pt(111) surface. The results of our theoretical approach suggest that the creation of pristine bulk like Pt(111) surface from $Pt_3Nb(111)$ surface is unlikely.

Table **4** presents the results of segregation energy for antisite migration. Considering Pt_3Ti ordered alloy, the formation of Pt_{Ti} (*i.e.*, the movement of extra

Pt atoms from the bulk structure towards the outermost layer of the $Pt_3Ti(111)$ surface) is energetically favourable, gaining about 1.236eV. However, the Ti_{Pt} antisite migration energy is about 0.433eV, an endothermic process. A similar trend is observed for the $Pt_3Nb(111)$ surface, where the Pt_{Nb} is exothermic with a calculated segregation energy of -1.096eV and Nb_{Pt} is endothermic with a calculated segregation energy of 0.991eV. Off-stoichiometric effect in the Pt_3X-based alloys would result in the segregation of Pt_X atoms towards the outermost surface layer (when the Pt atom concentration is greater than 75 at. %) or the X_{Pt} atoms to the inside of the bulk (when the Pt atom concentration is less than 75 at. %).

Table 3. Calculated Pt_3Nb direct exchange for different configurations.

-	Direct Exchange	$E_{(Segregation)}$ eV	-	Direct Exchange	$E_{(Segregation)}$ eV
1	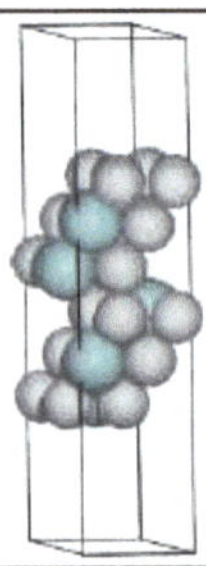	0.17049	3	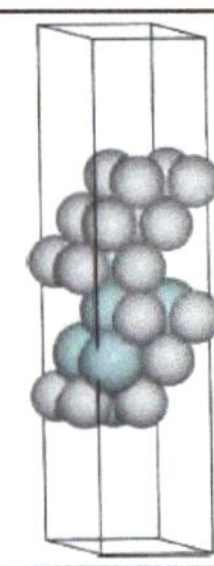	3.01152
2	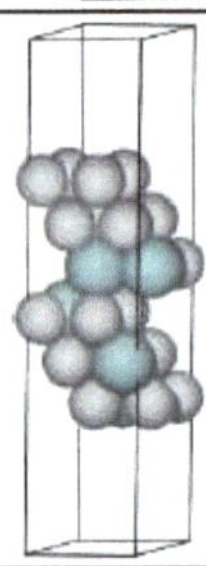	2.30465	4	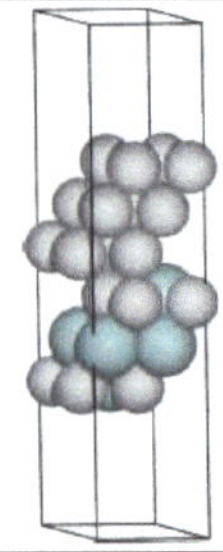	2.90730

Table 4. Calculated segregation energy for the antisite migration and direct exchange on Pt_3X (111) surface.

Antisite Migration				
Position of antisite	Pt_{Ti}	Ti_{Pt}	Pt_{Nb}	Nb_{Pt}
First layer	-1.2361	0.4335	-1.0969	0.9912
Second layer	0.1598	-0.0223	0.0808	0.0142
Direct Exchange				
Position of atoms before exchange	Pt_{Ti}	Ti_{Pt}	Pt_{Nb}	Nb_{Pt}
Second layer	0.5160	2.8096	-0.3833	1.9434

DFT-based studies on stoichiometric ordered Pt_3X system show that Pt_3Ti would not directly exchange its surface Ti atoms with Pt from the bulk, but the corresponding direct exchange mechanism is favourable in Pt_3Nb alloys. Hence, it is highly unlikely for the segregation of Pt to the outermost surface layer to occur *via* the direct exchange route in Pt_3Ti alloy but very likely in Pt_3Nb alloys. We considered the effect of off-stoichiometry on the surface segregation of Pt_3X alloys for DFT calculations.

The segregation energies of antisite defects in both the surface and bulk regions of $Pt_3Ti(111)$ surface were examined. We found that the off-stoichiometric configuration (Pt 75 at. %) of Pt_3X (111) would result in the migration of the extra Pt atoms to the surface. In contrast, when the Pt_3Ti crystal contains an excess of X atoms compared to the stoichiometric value (25 at. %), the additional Pt/Ti atoms migrate to the second layer of the (111) surface. Hence, the migration of Pt_X defects would result in the formation of an overlayer on the Pt_3X system.

Oxygen Adsorption on Pristine Pt (111) Surface

The adsorption energy of oxygen on the Pt (111) surface depends on the unit cell size and coverage. Table **5** (see **1** and **3**) shows the integrated oxygen adsorption energy for Pt (111) surface with 0.0625 and 0.25ML oxygen coverage, respectively. Their adsorption energy is slightly different due to the coverage effect.

Table **5** (see **2** and **3**) shows the integrated and differential adsorption energy of 4O-atom on Pt (111) clean surface. The differential E_{ads} is the energy required to add another adsorbate on the surface. In contrast, the integrated E_{ads} is the average energy needed to add a certain number of adsorbates to the surface.

The differential E_{ads} of oxygen on the Pt metallic surface is smaller than the integrated E_{ads}. This is attributed to the Coulombic effect and surface effects. This effect is more pronounced at higher coverage. Table **5** (see **6** and **7**) has one Nb dopant at the top and sub-surface, respectively. The sub-surface structure is about 1.47eV, more stable compared to the top-surface structure. The presence of the O-atom at the surface does not change the trend, with the sub-surface still more stable with about 1.31eV compared to the top-surface structure. Also, the oxygen

adsorption energy of the sub-surface configuration is lower compared to the top-surface and bulk-terminated Pt (111) surface.

The implication is that O-adsorption shifts to an optimal position in Table **5** (see **6**) configuration compared to pristine Pt. The complete Nb sub-surface shown in Table **5** (see **8**) results in a relatively weaker system that would bind O-atom compared to Pt. Therefore, the minute dopant in the sub-surface configuration is favourable in modifying the energetics. Noteworthy: the optimal adsorption energy of oxygen is about +0.2 eV.

Table 5. Adsorbed oxygen on the top surface of Pt (111) structures and their oxygen adsorption energy E_{ads} (where E_{ads} is referenced to Pt (111) E_{ads} in eV)

-	Clean Surface	$E_{(ads)}$ eV	-	Doped Surface	$E_{(ads)}$ eV
1	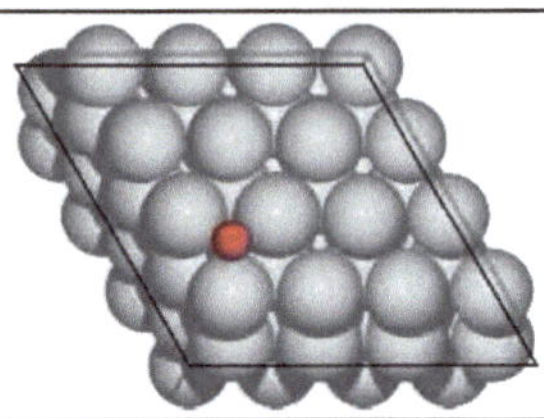	-0.83372	4	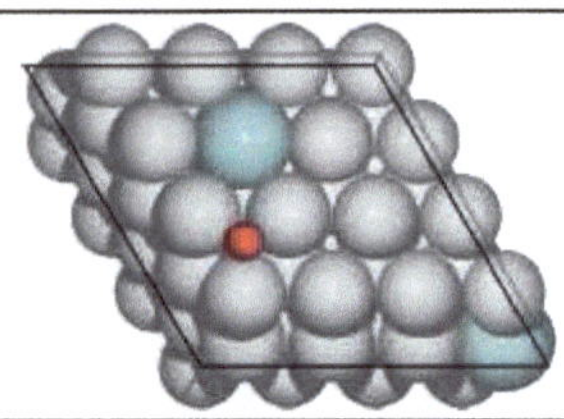	-0.842
2	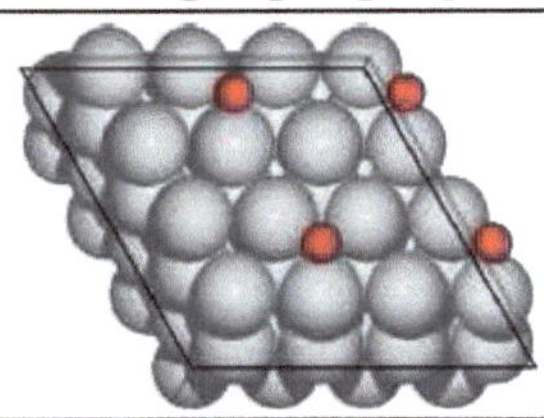	-0.81877	5	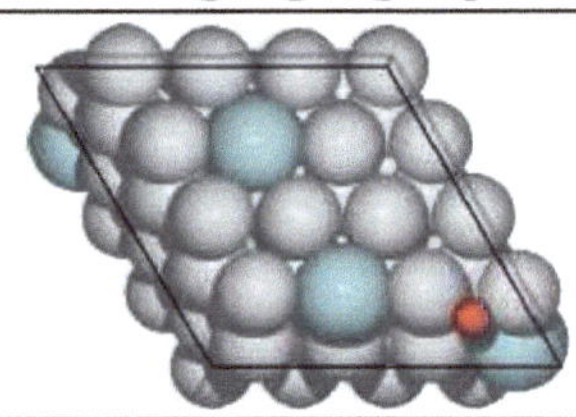	-0.77649
3	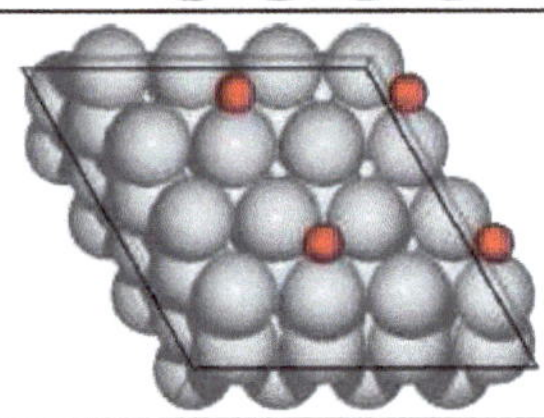	-0.79613	6	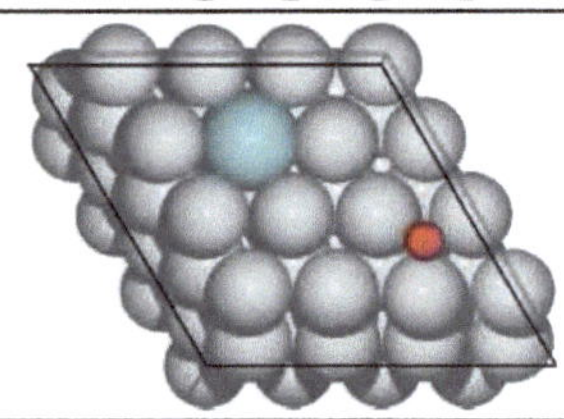	-0.89776
	-	-	7	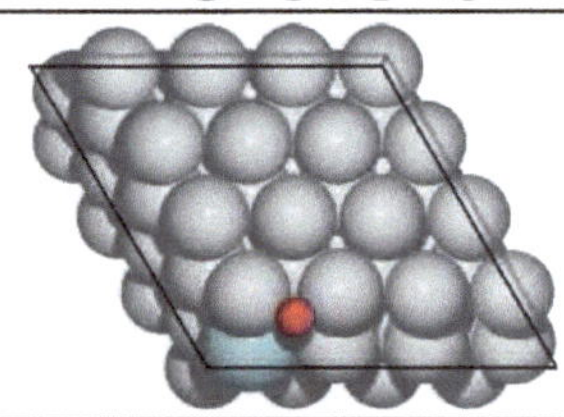	-0.73786
	-	-	8	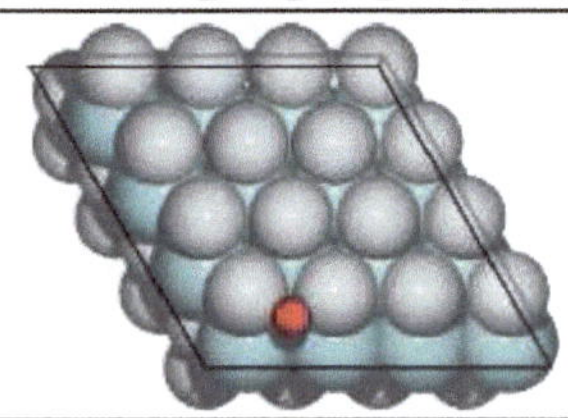	-0.07289

Oxygen Adsorption Energy on Pt_3X (111) Surface

Skin: Five-layer slab of Pt_3X is used to describe this system, as shown in Table **6**

(see **1** and **2**) shows the top and side view, respectively, with an oxygen atom adsorbed on the FCC site. The topmost layer is either Pt/X - over layer.

Table 6. (a) Schematic of the ball and stick model and (b) Calculated adsorption energy of oxygen on Pt_3X (111) surface with an overlayer.

-	a		
	Side view		Top view
1	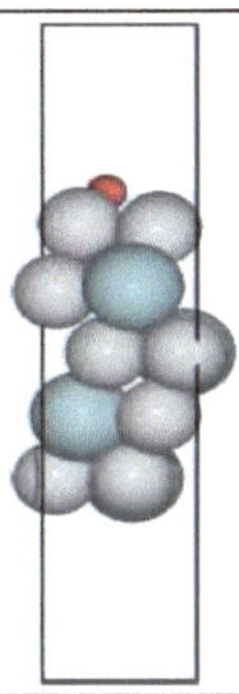	2	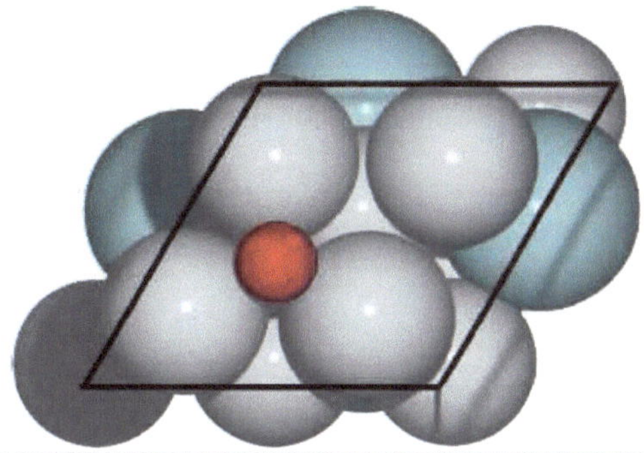

b		
Overlayer	Pt_3Nb	Pt_3Ti
Pt skin	0.2060	0.0884
X skin	-3.5071	-4.8677

The calculated oxygen adsorption energy on Pt_3X referenced to Pt(111) surface is presented below. The presence of Pt skin on the Pt_3X(111) surface leads to weakened bonding of oxygen on the surface. This effect is highly pronounced on the Pt_3Nb surface, with the E_{ads} of oxygen shifting towards the optimal value for ORR, as shown in the table below, compared to the Pt_3Ti surface. The presence of X (Ti or Nb) skin makes oxygen binding stronger to the surface. Overall, the X-skin surface configuration is not energetically favoured. Noteworthy, the presence of Pt-skin on Pt_3X significantly improved the adsorption energies.

Antisite and direct exchange: Tables **7** and **8** show oxygen's integrated and differential adsorption energy on Pt_3X(111) surfaces. The integrated E_{ads} of oxygen give the resultant adsorption of oxygen, whereas the differential E_{ads} show the individual signatures of the surface species.

Table **7** (see **2** & **3**) clearly shows the effect of different adsorption sites on the E_{ads}. The calculated **I** and **D** are almost similar in Table **7 (2)**, whereas they are different in Table **7** (**3**). This is because of different adsorption sites. As mentioned earlier, the calculated differential E_{ads} captures the individual effects of the different sites.

Table 7. Calculated integrated (I) and differential (D) adsorption energy of oxygen on Pt_3Nb (111) antisite surface configuration.

-	Antisite	I (eV)	D (eV)	-	Antisite	I (eV)	D (eV)
1	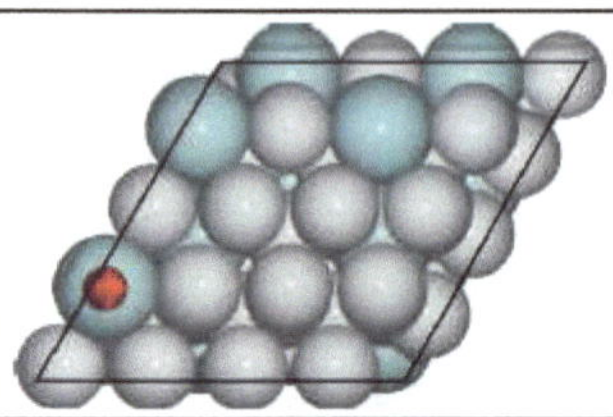	-3.4350	-3.4350	4	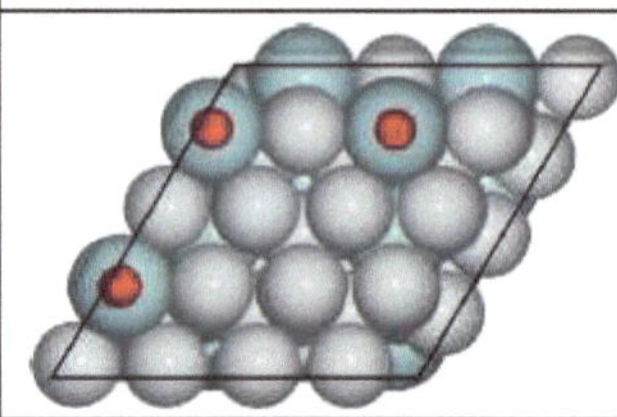	-3.4341	-3.3994
2	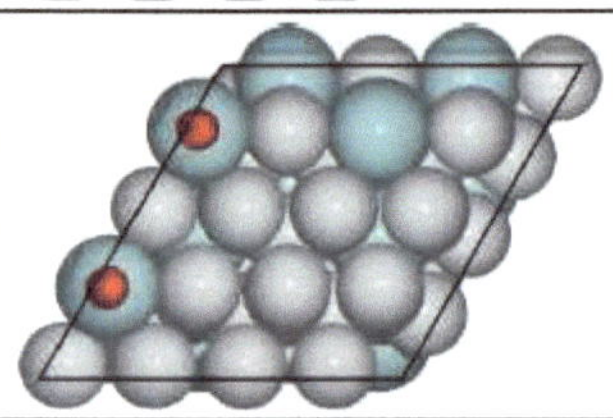	-3.4514	-3.4679	5	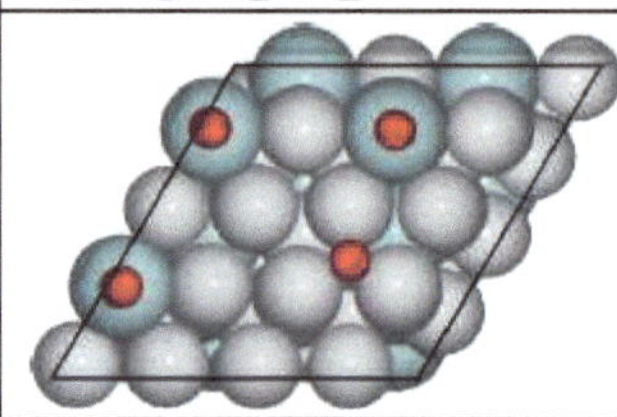	-2.7756	-0.8001
3	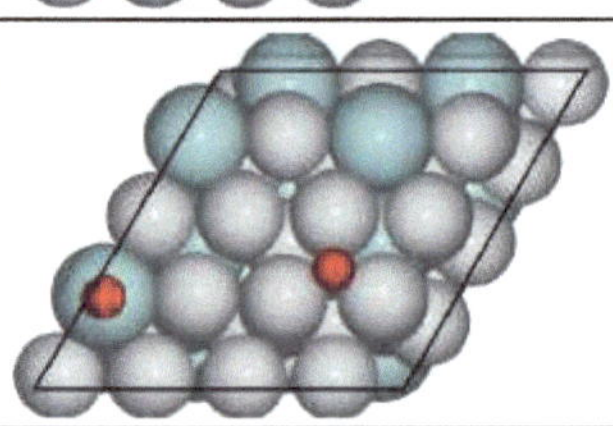	-2.1025	-0.7699	6	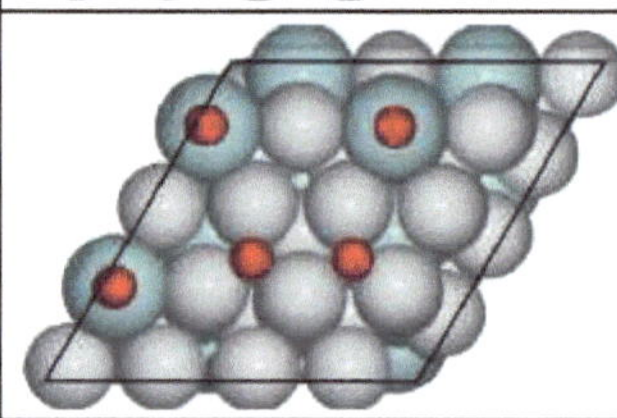	-2.3490	-0.6425

Table 8. Calculated integrated (I) and differential (D) adsorption energy of oxygen on Pt_3Ti (111) antisite surface configuration.

-	Antisite	I (eV)	D (eV)	-	Antisite	I (eV)	D (eV)
1	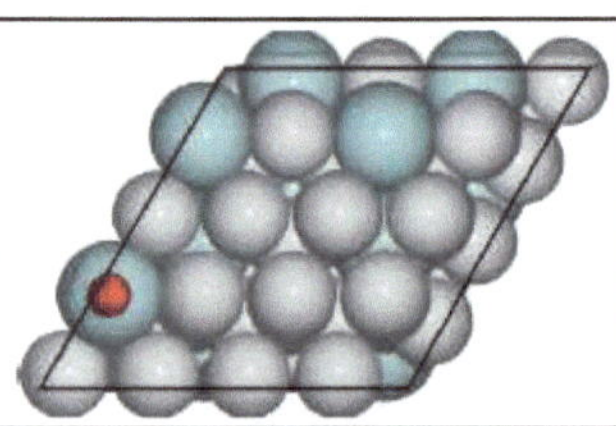	-2.5520	-2.5520	3	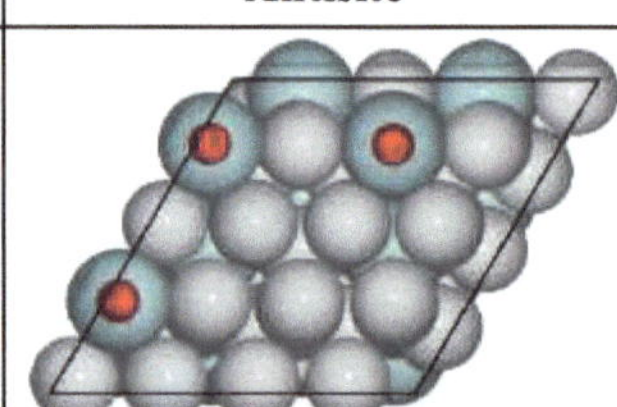	-2.5161	-2.5275
2	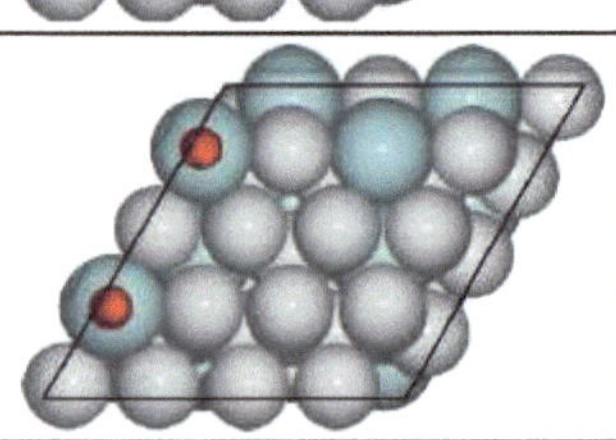	-2.5247	-	4	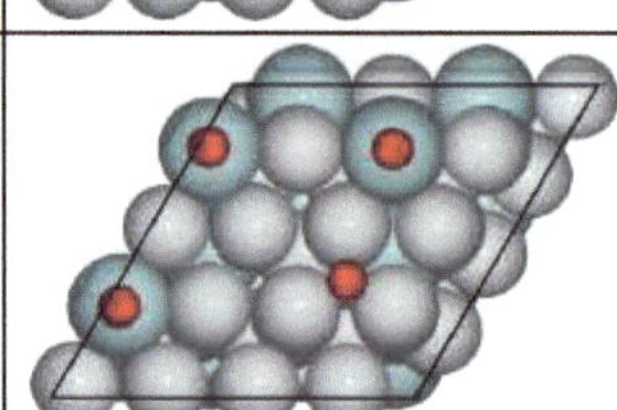	-2.1001	-0.8521

Table **9** provides a clear summary of Tables **7** and **8**. The calculated differential

Table **9** provides a clear summary of Tables **7** and **8**. The calculated differential E_{ads} for the surface with antisite migration referenced to Pt (111) surface shows that (i) the oxygen would initially adsorb on the X-atom with lower energy compared to the preferential FCC Pt site, (ii) the binding of an oxygen atom to the preferential FCC Pt site is slightly lower in energy on Pt_3Nb compared with pristine Pt(111) surface but similar for Pt_3Ti surface.

Table 9. Calculated differential (D) adsorption energy of oxygen on Pt_3X (111) antisite and direct surface configuration referenced to Pt (111) surface.

Antisite Migration				
Pt_3X	*Single*	*Double*	*Triple*	*Quad*
Pt_{Ti}	-1.689	-1.6944	-1.6657	-0.0190
Pt_{Nb}	-2.6018	-2.6347	-2.5662	0.0330
Direct Exchange				
Pt_3X	single	-	-	-
Pt_{Nb}	0.3432	-	-	-

Only the Pt_3Nb(111) surface is feasible, considering the direct exchange route. The calculated differential E_{ads} for this surface referenced to Pt(111) shows that the energy is about 0.34 eV higher than that of the Pt surface. The implication is that it is about 0.14 eV above the optimal value. This is relatively good compared to pristine Pt for ORR reaction.

CONCLUSION

This study presents the surface segregation in (111) surface of Pt and Pt_3X (where X = Nb, Ti) based alloys using density functional theory techniques. Our results suggest that surface segregation due to a direct exchange of X (where X=Nb or Ti) with Pt at the surface is energetically favourable for Pt_3Nb (-0.3833 eV) but not for the Pt_3Ti (0.516 eV) ordered alloy. In contrast, a direct exchange of Pt with either Nb or Ti at the surface is unfavourable for the Pt3X systems studied in this work. Moreover, the direct exchange of Pt in the top layer with Nb from the third layer and beyond of the Pt_3Nb five-layered slab model results in endothermic segregation energy in all cases. This showed that the formation of pristine bulk like Pt (111) surface from a Pt_3Nb surface *via* surface segregation is unlikely. Our investigation of surface segregation due to antisite migration effects showed that the migration of excess Pt atoms from the bulk towards the outermost layer of both Pt_3X surfaces is energetically favourable, while the reverse migration of X atoms from bulk to replace Pt atoms on the surface (X_{Pt}) is unfavourable, *i.e.*, endothermic in energy. This shows that antisite migration due to off-stoichiometric effects in the Pt_3X-based alloys will result in the formation of an overlayer or skin of Platinum atoms. In the second part of this work, we compared the adsorption energies of oxygen atoms on the Pristine Pt(111) surface to that of the Pt_3X (X = Nb or Ti) alloy surfaces. To model the effect of surface segregation due to direct exchange or antisite migrations on the catalytic activity of the Pt_3X alloys for ORR, we also computed the adsorption energy of the O-atom on the various surface segregated configurations. For the pristine Pt (111) surface, we found that a minute doping with Nb in the subsurface is favourable for modifying the energetics of O-atom adsorption on this surface to enhance its activity for ORR. In the case of Pt_3X alloys, the presence of Pt skin on the surface leads to a weakened oxygen bonding on their surfaces. This effect is more pronounced on the Pt3Nb surface. Thus, the E_{ads} of oxygen on the Pt_3Nb surface shifted towards the optimal value for ORR in the presence of a Pt overlayer.

In contrast, surface segregation that leads to the formation of an X skin results in stronger oxygen binding to the surface, and this will hinder the activity of Pt_3X alloys for ORR. Our results also show that the binding of an oxygen atom to the preferential fcc Pt site is lower in energy on the Pt_3Nb than the Pristine Pt (111) surface. However, the binding of an oxygen atom to the fcc Pt site in Pt_3Ti and pristine Pt (111) surface is comparable in energy. Overall, our results highlight the suitability of ordered Pt_3X alloys as a lower-cost and promising catalyst material for ORR. Our DFT calculations show the Pt_3Nb alloy to be a good example of such materials. Moreover, it is crucial to note that these Pt_3Nb alloys should be designed and fabricated in such a way that their surface segregation processes favour the migration of Pt to the alloy surface, thus leading to the formation of Pt 'skin.' This might improve their activity as a catalyst for the ORR and similar reactions.

ACKNOWLEDGEMENT

The authors thank the Centre for High-Performance Computing (CHPC), Cape Town, South Africa, for the computational resources used in this study and the HySA-Infrastructure Centre of Competence, Faculty of Engineering, North-West University, for financial support. KOO thanks the National Institute of Theoretical and Computational Sciences, Johannesburg, South Africa.

REFERENCES

[1] Guerrero Moreno, N.; Cisneros Molina, M.; Gervasio, D.; Pérez Robles, J.F. Approaches to polymer electrolyte membrane fuel cells (PEMFCs) and their cost. *Renew. Sustain. Energy Rev.,* **2015**, *52*, 897-906.
[http://dx.doi.org/10.1016/j.rser.2015.07.157]

[2] Appleby, A.J.; Foulkes, F.R. *Fuel Cell Handbook*; Van Nostrand Reinhold: New York, **1989**.

[3] Strasser, P.; Koh, S.; Anniyev, T.; Greeley, J.; More, K.; Yu, C.; Liu, Z.; Kaya, S.; Nordlund, D.; Ogasawara, H.; Toney, M.F.; Nilsson, A. Lattice-strain control of the activity in dealloyed core–shell

fuel cell catalysts. *Nat. Chem.,* **2010**, *2*(6), 454-460. [http://dx.doi.org/10.1038/nchem.623] [PMID: 20489713]

[4] Othman, R.; Dicks, A.L.; Zhu, Z. Non precious metal catalysts for the PEM fuel cell cathode. *Int. J. Hydrogen Energy,* **2012**, *37*(1), 357-372. [http://dx.doi.org/10.1016/j.ijhydene.2011.08.095]

[5] Kulkarni, A.; Siahrostami, S.; Patel, A.; Nørskov, J.K. Understanding catalytic activity trends in the oxygen reduction reaction. *Chem. Rev.,* **2018**, *118*(5), 2302-2312. [http://dx.doi.org/10.1021/acs.chemrev.7b00488] [PMID: 29405702]

[6] Gomez-Marin, A. U.; Feliu, A. M. Oxygen reduction on platinum single crystal electrodes. In: *Encyclopedia of Interfacial Chemistry*; , **2017**; pp. 820-830. [http://dx.doi.org/10.1016/B978-0-12-409547-2.13333-5]

[7] Hwang, S.J.; Kim, S.K.; Lee, J.G.; Lee, S.C.; Jang, J.H.; Kim, P.; Lim, T.H.; Sung, Y.E.; Yoo, S.J. Role of electronic perturbation in stability and activity of Pt-based alloy nanocatalysts for oxygen reduction. *J. Am. Chem. Soc.,* **2012**, *134*(48), 19508-19511. [http://dx.doi.org/10.1021/ja307951y] [PMID: 23131009]

[8] Jinnouchi, R.; Kodama, K.; Hatanaka, T.; Morimoto, Y. First principles based mean field model for oxygen reduction reaction. *Phys. Chem. Chem. Phys.,* **2011**, *13*(47), 21070-21083. [http://dx.doi.org/10.1039/c1cp21349k] [PMID: 22064886]

[9] Stamenković, V.; Schmidt, T.J.; Ross, P.N.; Marković, N.M. Surface composition effects in electrocatalysis: Kinetics of oxygen reduction on well-defined Pt_3 Ni and Pt_3 Co alloy surfaces. *J. Phys. Chem. B,* **2002**, *106*(46), 11970-11979. [http://dx.doi.org/10.1021/jp021182h]

[10] Sha, Y.; Yu, T.H.; Merinov, B.V.; Shirvanian, P.; Goddard, W.A., III Mechanism for oxygen reduction reaction on Pt_3 Ni alloy fuel cell cathode. *J. Phys. Chem. C,* **2012**, *116*(40), 21334-21342. [http://dx.doi.org/10.1021/jp303966u]

[11] Ma, Y.; Balbuena, P.B. Role of iridium in Pt-based alloy catalysts for the ORR: Surface adsorption and stabilization studies. *J. Electrochem. Soc.,* **2010**, *157*(6), B959-B963. [http://dx.doi.org/10.1149/1.3409559]

[12] Bonnet, N.; Otani, M.; Sugino, O. Self-poisoning dynamical effects in the oxygen reduction reaction on Pt(111) from a top-down kinetic analysis. *J. Phys. Chem. C,* **2014**, *118*(25), 13638-13643. [http://dx.doi.org/10.1021/jp502807z]

[13] Ma, Y.; Balbuena, P.B. Pt surface segregation in bimetallic Pt3M alloys: A density functional theory study. *Surf. Sci.,* **2008**, *602*(1), 107-113. [http://dx.doi.org/10.1016/j.susc.2007.09.052]

[14] Ma, Y.; Balbuena, P.B. Surface segregation in bimetallic Pt3M (M=Fe, Co, Ni) alloys with adsorbed oxygen. *Surf. Sci.,* **2009**, *603*(2), 349-353. [http://dx.doi.org/10.1016/j.susc.2008.11.036]

[15] Abrams, B.L.; Vesborg, P.C.K.; Bonde, J.L.; Jaramillo, T.F.; Chorkendorff, I. Dynamics of surface exchange reactions between Au and Pt for HER and HOR. *J. Electrochem. Soc.,* **2009**, *156*(2), B273-B282. [http://dx.doi.org/10.1149/1.3040509]

[16] Tréglia, G.; Legrand, B.; Ducastelle, F.; Saúl, A.; Gallis, C.; Meunier, I.; Mottet, C.; Senhaji, A. Alloy surfaces: Segregation, reconstruction and phase transitions. *Comput. Mater. Sci.,* **1999**, *15*(2), 196-235. [http://dx.doi.org/10.1016/S0927-0256(99)00004-X]

[17] Müller, S.; Stöhr, M.; Wieckhorst, O. Structure and stability of binary alloy surfaces: Segregation, relaxation, and ordering from first-principles calculations. *Appl. Phys., A Mater. Sci. Process.,* **2006**, *82*(3), 415-419. [http://dx.doi.org/10.1007/s00339-005-3362-6]

[18] Yuge, K.; Seko, A.; Kuwabara, A.; Oba, F.; Tanaka, I. First-principles study of bulk ordering and surface segregation in Pt-Rh binary alloys. *Phys. Rev. B Condens. Matter Mater. Phys.,* **2006**, *74*(17), 174202.
[http://dx.doi.org/10.1103/PhysRevB.74.174202]

[19] Pourovskii, L.V.; Ruban, A.V.; Abrikosov, I.A.; Vekilov, Y.K.; Johansson, B. Bulk ordering and surface segregation in Ni 50 Pt 50. *Phys. Rev. B Condens. Matter,* **2001**, *64*(3), 035421.
[http://dx.doi.org/10.1103/PhysRevB.64.035421]

[20] Goringe, C.M.; Bowler, D.R.; Hernández, E. Tight-binding modelling of materials. *Rep. Prog. Phys.,* **1997**, *60*(12), 1447-1512.
[http://dx.doi.org/10.1088/0034-4885/60/12/001]

[21] Kohn, W.; Becke, A.D.; Parr, R.G. Density functional theory of electronic structure. *J. Phys. Chem.,* **1996**, *100*(31), 12974-12980.
[http://dx.doi.org/10.1021/jp960669l]

[22] Daw, M.S.; Foiles, S.M.; Baskes, M.I. The embedded-atom method: A review of theory and applications. *Mater. Sci. Rep.,* **1993**, *9*(7-8), 251-310.
[http://dx.doi.org/10.1016/0920-2307(93)90001-U]

[23] Lomocso, T.L.; Baranova, E.A. Electrochemical oxidation of ammonia on carbon-supported bi-metallic PtM (M=Ir, Pd, SnOx) nanoparticles. *Electrochim. Acta,* **2011**, *56*(24), 8551-8558.
[http://dx.doi.org/10.1016/j.electacta.2011.07.041]

[24] Tayal, J.; Rawat, B.; Basu, S. Bi-metallic and tri-metallic Pt–Sn/C, Pt–Ir/C, Pt–Ir–Sn/C catalysts for electro-oxidation of ethanol in direct ethanol fuel cell. *Int. J. Hydrogen Energy,* **2011**, *36*(22), 14884-14897.
[http://dx.doi.org/10.1016/j.ijhydene.2011.03.035]

[25] Wu, H.; La Parola, V.; Pantaleo, G.; Puleo, F.; Venezia, A.; Liotta, L. Ni-based catalysts for low temperature methane steam reforming: Recent results on Ni-Au and comparison with other Bi-metallic systems. *Catalysts,* **2013**, *3*(2), 563-583.
[http://dx.doi.org/10.3390/catal3020563]

[26] Mohanapriya, S.; Suganthi, S.; Raj, V. Mesoporous Pt–Ni catalyst and their electro catalytic activity towards methanol oxidation. *J. Porous Mater.,* **2017**, *24*(2), 355-365.
[http://dx.doi.org/10.1007/s10934-016-0268-y]

[27] Sahoo, P.K.; Panigrahy, B.; Bahadur, D. Facile synthesis of reduced graphene oxide/Pt–Ni nanocatalysts: Their magnetic and catalytic properties. *RSC Advances,* **2014**, *4*(89), 48563-48571.
[http://dx.doi.org/10.1039/C4RA07686A]

[28] Tanksale, A.; Zhou, C.H.; Beltramini, J.N.; Lu, G.Q. Hydrogen production by aqueous phase reforming of sorbitol using bimetallic Ni–Pt catalysts: Metal support interaction. *J. Incl. Phenom. Macrocycl. Chem.,* **2009**, *65*(1-2), 83-88.
[http://dx.doi.org/10.1007/s10847-009-9618-6]

[29] Lin, R.; Che, L.; Shen, D.; Cai, X. High durability of Pt-Ni-Ir/C ternary catalyst of PEMFC by stepwise reduction synthesis. *Electrochim. Acta,* **2020**, *330*, 135251.
[http://dx.doi.org/10.1016/j.electacta.2019.135251]

[30] Zhang, P.; Dai, X.; Zhang, X.; Chen, Z.; Yang, Y.; Sun, H.; Wang, X.; Wang, H.; Wang, M.; Su, H.; Li, D.; Li, X.; Qin, Y. One-pot synthesis of ternary Pt–Ni–Cu nanocrystals with high catalytic performance. *Chem. Mater.,* **2015**, *27*(18), 6402-6410.
[http://dx.doi.org/10.1021/acs.chemmater.5b02575]

[31] Mao, L.; Ba, Q.; Liu, S.; Jia, X.; Liu, H.; Chen, W.; Li, X. Pt–Ni_x alloy nanoparticles: A new strategy for cocatalyst design on a CdS surface for photo-catalytic hydrogen generation. *RSC Advances,* **2018**, *8*(55), 31529-31537.
[http://dx.doi.org/10.1039/C8RA06581K] [PMID: 35548231]

[32] Stamenkovic, V.R.; Fowler, B.; Mun, B.S.; Wang, G.; Ross, P.N.; Lucas, C.A.; Marković, N.M. Improved oxygen reduction activity on Pt3Ni(111) *via* increased surface site availability. *Science,* **2007**, *315*(5811), 493-497. [http://dx.doi.org/10.1126/science.1135941] [PMID: 17218494]

[33] Hu, J.; Kuttiyiel, K.; Sasaki, K.; Su, D.; Yang, T.H.; Park, G.G.; Zhang, C.; Chen, G.; Adzic, R. Pt monolayer shell on nitrided alloy core—A path to highly stable oxygen reduction catalyst. *Catalysts,* **2015**, *5*(3), 1321-1332. [http://dx.doi.org/10.3390/catal5031321]

[34] Brankovic, S.R.; Wang, J.X.; Adžić, R.R. Metal monolayer deposition by replacement of metal adlayers on electrode surfaces. *Surf. Sci.,* **2001**, *474*(1-3), L173-L179. [http://dx.doi.org/10.1016/S0039-6028(00)01103-1]

[35] Ding, E.; More, K.L.; He, T. Preparation and characterization of carbon-supported PtTi alloy electrocatalysts. *J. Power Sources,* **2008**, *175*(2), 794-799. [http://dx.doi.org/10.1016/j.jpowsour.2007.10.010]

[36] Paßens, M.; Caciuc, V.; Atodiresei, N.; Moors, M.; Blügel, S.; Waser, R.; Karthäuser, S. Tuning the surface electronic structure of a $Pt_3Ti(111)$ electro catalyst. *Nanoscale,* **2016**, *8*(29), 13924-13933. [http://dx.doi.org/10.1039/C5NR08420B] [PMID: 26865393]

[37] Mun, B.S.; Watanabe, M.; Rossi, M.; Stamenkovic, V.; Markovic, N.M.; Ross, P.N., Jr A study of electronic structures of Pt3M (M=Ti,V,Cr,Fe,Co,Ni) polycrystalline alloys with valence-band photoemission spectroscopy. *J. Chem. Phys.,* **2005**, *123*(20), 204717. [http://dx.doi.org/10.1063/1.2126662] [PMID: 16351303]

[38] Duan, Z.; Zhong, J.; Wang, G. Modeling surface segregation phenomena in the (111) surface of ordered Pt3Ti crystal. *J. Chem. Phys.,* **2010**, *133*(11), 114701. [http://dx.doi.org/10.1063/1.3490792] [PMID: 20866148]

[39] Wang, X.; Li, Z.; Qu, Y.; Yuan, T.; Wang, W.; Wu, Y.; Li, Y. Review of metal catalysts for oxygen reduction reaction: From nanoscale engineering to atomic design. *Chem,* **2019**, *5*(6), 1486-1511. [http://dx.doi.org/10.1016/j.chempr.2019.03.002]

[40] Sarkar, S.; Peter, S.C. An overview on Pt_3X electrocatalysts for oxygen reduction reaction. *Chem. Asian J.,* **2021**, *16*(10), 1184-1197. [http://dx.doi.org/10.1002/asia.202100166] [PMID: 33749999]

[41] Biz, C.; Fianchini, M.; Polo, V.; Gracia, J. Magnetism and heterogeneous catalysis: In depth on the quantum spin-exchange interactions in Pt3M (M = V, Cr, Mn, Fe, Co, Ni, and Y)(111) Alloys. *ACS Appl. Mater. Interfaces,* **2020**, *12*(45), 50484-50494.

[42] Mortensen, J.J.; Hansen, L.B.; Jacobsen, K.W. Real-space grid implementation of the projector augmented wave method. *Phys. Rev. B Condens. Matter Mater. Phys.,* **2005**, *71*(3), 035109. [http://dx.doi.org/10.1103/PhysRevB.71.035109]

[43] Hjorth Larsen, A.; Jørgen Mortensen, J.; Blomqvist, J.; Castelli, I.E.; Christensen, R.; Dułak, M.; Friis, J.; Groves, M.N.; Hammer, B.; Hargus, C.; Hermes, E.D.; Jennings, P.C.; Bjerre Jensen, P.; Kermode, J.; Kitchin, J.R.; Leonhard Kolsbjerg, E.; Kubal, J.; Kaasbjerg, K.; Lysgaard, S.; Bergmann Maronsson, J.; Maxson, T.; Olsen, T.; Pastewka, L.; Peterson, A.; Rostgaard, C.; Schiøtz, J.; Schütt, O.; Strange, M.; Thygesen, K.S.; Vegge, T.; Vilhelmsen, L.; Walter, M.; Zeng, Z.; Jacobsen, K.W. The atomic simulation environment—a Python library for working with atoms. *J. Phys. Condens. Matter,* **2017**, *29*(27), 273002. [http://dx.doi.org/10.1088/1361-648X/aa680e] [PMID: 28323250]

[44] Perdew, J.P.; Burke, K.; Ernzerhof, M. Generalized gradient approximation made simple. *Phys. Rev. Lett.,* **1996**, *77*(18), 3865-3868. [http://dx.doi.org/10.1103/PhysRevLett.77.3865] [PMID: 10062328]

[45] Lehtola, S.; Steigemann, C.; Oliveira, M.J.T.; Marques, M.A.L. Recent developments in libxc — A

comprehensive library of functionals for density functional theory. *SoftwareX,* **2018**, *7*, 1-5. [http://dx.doi.org/10.1016/j.softx.2017.11.002]

[46] Monkhorst, H.J.; Pack, J.D. Special points for Brillouin-zone integrations. *Phys. Rev., B, Solid State,* **1976**, *13*(12), 5188-5192. [http://dx.doi.org/10.1103/PhysRevB.13.5188]

[47] Methfessel, M.; Paxton, A.T. High-precision sampling for Brillouin-zone integration in metals. *Phys. Rev. B Condens. Matter,* **1989**, *40*(6), 3616-3621. [http://dx.doi.org/10.1103/PhysRevB.40.3616] [PMID: 9992329]

[48] Kresse, G.; Joubert, D. From ultrasoft pseudopotentials to the projector augmented-wave method. *Phys. Rev. B Condens. Matter,* **1999**, *59*(3), 1758-1775. [http://dx.doi.org/10.1103/PhysRevB.59.1758]

CHAPTER 8

Nanoparticles and Environmental Health

Aparna Roy[1], **Kumaresh Mandal**[1], **Shishir Tamang**[1,2], **Soni Subba**[1], **Saptaparni De**[1], **Divya Rai**[1], **Biswajit Roy**[3] and **Rakesh Tamang**[1,*]

[1] *Department of Zoology, University of Calcutta, Kolkata, West Bengal, India*

[2] *Department of Zoology, Darjeeling Government College, Darjeeling, West Bengal, India*

[3] *CSIR-Centre for Cellular and Molecular Biology, Hyderabad, Telangana, India*

Abstract: The size range of nanoparticles between 1-100nm is unique because of their extremely small structure with a very high surface area to volume ratio. Besides naturally produced nanoparticles, there is a huge worldwide demand for synthetic nanoparticles. These synthetic nanoparticles are modified to some extent according to the specific need. These manipulations at the nano-scale paved the way for a popular branch of science called nanotechnology. However, with the massive use of nanoparticle-based industrial products in our day-to-day lives, we knowingly or unknowingly ignore their impact on the environment. The air, water, and soil quality determines environmental health, which is reflected by a healthy ecosystem and its biodiversity. The existing intricate interaction between humans and their surrounding environment is important for maintaining a fine balance in the ecosystem. Any change in this interaction may lead to adverse consequences. The nanoparticles released in the environment cause a varying degree of effects on the ecosystem based on the type, surface coating, and degree of its environmental transformation. Some nanoparticles are harmful to the environment and some are beneficial. Some of the nanoparticles in the environment get bioaccumulated in plants and animals, disturbing their growth and productivity. Remediation by nanoparticles has been effective in removing some toxic compounds from the environment, thereby providing a way to minimize pollution efficiently. Thus, in this review, we have tried to present an overview of the sources, fate, and effects of nanoparticles available in air, water, and soil. We strongly advocate for the long-term assessment of nanoparticles, and the formulation of strict guidelines for their usage by the concerned industries for better environmental health, and in turn a healthy ecosystem.

Keywords: Environment, Nanoparticle, Nanotechnology, Pollutant, Remediation.

[*] **Corresponding author Rakesh Tamang:** Department of Zoology, University of Calcutta, Kolkata, West Bengal, India; Tel: +919674129387; E-mail: rtzoo@caluniv.ac.in

Dibya Prakash Rai, Kingsley O. Obodo, & Jitendra Pal Singh (Eds.)

INTRODUCTION

The definition of nanoparticles (NPs) or nanomaterials put forward by the European Commission says the nanoparticles are natural, incidental, or manufactured material containing particles in the unbound or aggregated form where 50% or more of the particles are in the size range of 1nm-100nm in at least one dimension [1]. Since the discovery of nanotechnology in 1959 by physicist Richard Feynman, its application in different industries like electronics, healthcare, energy storage and production, construction, cosmetics and personal care, pest control, agriculture, water treatment, *etc*. has increased tremendously. The NPs are found in all environmental compartments: soil, water, and air, with the ocean holding the largest amount of NPs on earth [2]. The largest market for nanomaterials is North America, and its fastest-growing market is in the Asia Pacific. In the Asia Pacific region, the leading countries in terms of nanomaterial production are China, India, Japan, and South Korea [3]. Mainly, the surge in demand for nanomaterials is noted in the fields of energy storage and conversion, treatment of diseases, agricultural productivity, personal care products, water treatment, and remediation. This increased use of nanoparticle-based products has resulted in growing concern over its effect on environmental health. This may be due to the scattered research on its risk assessment and unaccounted disposal in the environment. Thus, this review intends to summarize the available research articles in nanoparticles pertaining to environmental health, and put forward the future perspective on its sustainable use.

NATURAL NPs

The NPs can be produced naturally as well as synthetically. Natural NPs are either organic or inorganic. Organic natural NPs are produced by various natural processes, like photochemical reactions, weathering of rocks and minerals, wind erosions, occasional forest fires, volcanic eruptions, organismal activities, and extra-terrestrial processes. Examples of natural organic NPs are Oceanic spray, volcanic ash, mineral NPs, naturally produced carbon-based NPs, aerosols, forest fire smoke, *etc*. Some typical representatives of naturally occurring inorganic are FeS_2, Ag, Au, SiO_2, Fe_2O_3, Al_2O_3, and MnO_2 [4]. Natural NPs play an important role in the biogeochemical processes on Earth. The natural nano-sized geomaterials perform key roles in the distribution of different minerals on Earth by complex physiochemical interactions with biotic and abiotic factors, different transformation procedures, and weathering [2]. Due to their high surface area to volume ratio and high reactivity, NPs contain a significantly higher amount of minerals than bulk soil mass, which ultimately leads to increased mineral availability to plants, microorganisms, and higher taxa [5]. The nano-sized mineral dust, particulate matter generated by forest fires, and volcanic ashes can cause serious environmental and human health issues [6]. Moreover naturally formed metal NPs are highly stable, and thus can be transported to large distances without transformation to any other forms. The coating of organic matter and ROS on these metal NPs significantly affect their toxicity [4].

ENGINEERED NPs

Nanoparticles that are developed with synthetically definite chemical composition, size, and surface coatings, either simple or conjugated with other bioactive molecules to enhance some specific properties are called Engineered nanoparticles (ENPs) [7]. Designed for specificity, controlled reactivity, increased sensitivity, and cost-effectiveness, The ENPs have become an integral part of several industries worldwide, such as electronics, medical diagnosis, personal care, textile, food packaging, construction, *etc.* The different types of ENPs are produced using metal, metal oxide, carbon, nano-clay, quantum dots, *etc.* The ENPs get released into the environment during their production, use, and disposal. Currently, its fate in the environment is mostly unaccountable or unmanaged. Thus, it is essential to study the fate of ENPs in the environment and their mechanisms of interactions with aquatic and terrestrial ecosystems. There is a high possibility of ENPsentry into the environment during production. Therefore, production volumes are a good indicator of ENPs released in the environment. Of the total ENPs produced, 0.1-2% are released into the environment. Of these released ENPs, 20-80% go directly into the environment, from which 10-40% go to the air and 10-40% to the water bodies. The direct release of ENPs during product use is dependent on the type of product produced. A significant amount is released if the product is in liquid, gas, or powder form [8]. The indirect emission of ENPs in the environment occurs *via* effluents of wastewater treatment plants and the disposition of untreated biosolids produced as by-products of NPs into soil landfills. During the whole life cycle from production to disposal, the global estimation of ENPs released into the environment is almost 63-91% into the landfills, 8-28% in soil, 7% in the aquatic environment, and 1.5% in the air [9].

Among engineered NPs, Quantum Dots (QDs) are also used frequently. QDs are nanocrystalline structures having semi-conductivity and unique fluorescent properties. The characteristic fluorescent property described as "quantum yield" of QDs is based on the ability of QDs to emit light (quantum) after absorption, and it is directly proportional to the QDs efficacy in different fields like imaging, display industry (liquid crystal display, light emitting diodes), medical imaging, solar cells. QDs can be engineered to be made of only one element, like carbon or silicone, or they can be combined with other compounds based on the properties of the desired material. The release of QDs in the environment most likely occurs during the synthesis step and end-product formation step. After release in the environment, the QDs can be transformed by the surrounding environmental factors that determine the fate of QDs. In freshwater aquatic systems, QDs dissolve faster and tend to settle down in sediments. Thus, high concentrations of QDs can show potentially toxic effects in benthic aquatic organisms.

THE FATE OF NPs IN THE ENVIRONMENT

Mostly, the NPs released in the environment undergo various physical, chemical, and biological transformations which either increase or decrease their toxicity and bioavailability. So, transformation plays a major role in the fate of ENMs in the environment. The surrounding environmental conditions like media chemistry, amount of organic matter, temperature, and pH play a major role in the transformation process. NPs are degraded partially or completely depending on the structural properties and composition of NPs. In some cases, uncontrolled degradation may even lead to the formation of more toxic products. Therefore, proper risk assessment of NPs should consider the transformation of NPs with time and all the toxic by-products produced by their interactions with other components in the immediate environment [8].

To know the effect of NPs in the environment, it is important to understand how they aggregate. The physiochemical transformation of NPs due to their surrounding environmental factors determines this aggregation pattern. The study of this aggregation pattern of NPs becomes theoretically pretty much understandable based on how colloid particles aggregate since the size range of colloidal particles is also <100nm, like NPs. However, the colloidal theory cannot completely describe the aggregation pattern of NPs because of its unique properties [10]. When two particles come in contact, they can either attract or repel each other. The main attractive force, the van der walls force, acts based on fluctuating electromagnetic fields generated by the polarization effect between the particles and depends on the properties like size and distance between the interacting particles. The charged surface of the particles, along with counter-ions in its vicinity, forms an electrical double layer. When two particles come in close contact such that their respective electrical double layers overlap, repulsive force is generated [11]. Among the different theories that explain the aggregation of particles in the nanometer range, the Derjaguin-Landau-Verwey-Overbeak (DLVO) theory is the most widely accepted theory that explains aggregation. DLVO theory predicts the resultant force acting between the particles by simply combining van der walls attractive and electrical double-layer repulsive force. The thermodynamic interaction in this short range that allows the particles to collide can lead to aggregate formation once it exceeds the repulsive force acting as a barrier between the particles [11]. The small size and different structural shapes of NPs, apart from only being spherical, chemical composition, and surface coatings provide unique properties to the NP aggregates, which cannot be described by classical DLVO theory has been shown that the heavy metal adsorption ability of ferric iron oxyhydroxide NPs is controlled by its aggregation state, which is governed by the available surface area and porosity of the aggregates. The tendency of ferric iron oxyhydroxide NPs to form aggregates reduces the long-term adsorption ability of heavy metal ions from the aquatic environment due to the reduction of the surface area of the aggregates for adsorption and fewer pores available to sequester the heavy metal ions into the interior of a compact aggregate structure [12]. Aggregates formed can be homo-aggregates or hetero-aggregates. Naturally occurring NPs tend to form homo-aggregates, whereas engineered NPs generally form hetero-aggregates with natural NPs since natural NPs are exceedingly abundant in nature compared to engineered NPs [10]. Though the physical properties of these aggregates control the function, toxicity, and transport of the NPs in the environment, the unpredictable nature of the NP aggregate formation does not allow us to fully understand the fate of NPs in the environment [10]. Additionally, uncontrolled aggregate formation shows adverse effects on the environment, whereas controlled aggregation of the NPs is more beneficial [11].

Physiochemical Transformation

The physical transformation of ENPs includes aggregation and agglomeration. The ENPs can aggregate among themselves because of their high surface energy and form into either homoaggregates or hetero-aggregates [13]. The NPs are involved in a reversible process of adhesion called agglomeration with other particles and precipitates by phase separation, forming a large colloidal mass [8]. The aggregates of metal-based NPs in an aqueous environment influence the overall distribution pattern. TiO_2 NP can agglomeratein water. It is shown that a small aggregate formation by TiO_2 NP of size 17nm at *p*H= 2, can influence its toxicity [14]. In the aquatic system, agglomeration increases the velocity of sedimentation and thus results in increased uptake by lower invertebrates [15].

The chemical transformation of ENPsis is dependent on their size, shape, surface coating, and also on the surrounding environmental parameters like temperature, pH, salinity, and presence of organic and inorganic matter [9]. When the metal NPs used in the textile industry, paints, and cosmetics come in contact with water bodies show toxicity due to metal ions formation [8]. The ZnO-NP is modified in the environment depending on the chemistry of the surrounding solution. IHP (Myo-inositol 25 hexakisphosphate), which is a common organic phosphate, enhances the dissolution of ZnO-NP to a greater extent. IHP can immobilize Zn^{2+} ions, transforming them into Zn-IHP precipitate [16]. The redox reactions of NPs with the environmental compounds lead to the formation of oxides and sulfides.

Their interactions with other NPs may lead to the formation of new NPs.The fate of AgNP in water is influenced by both physical and chemical transformations. Mg^{2+} and Ca^{2+} ions in surface water enhance AgNP aggregate formation. The dissolved organic matter (DOM) inhibits the aggregate formation and photo transformation of AgNP. The AgNP in water is oxidized to Ag^+ by dissolved oxygen in the presence of sunlight, which is again reduced to Ag by DOM. The high concentration of natural organic matter stabilizes AgNP in water [17]. Covellite CuS-NP in humid conditions shows surface transformation. Molecular oxygen present in water gets adsorbed preferentially to the CuS-NP resulting in increased oxidation of CuS to $CuSO_4$ at high relative humidity (87%) [18].

Biological Transformation of ENPs

The NPs enter the organisms from the surrounding environment and get transformed within the organism. The process of transformation of NPs within the organisms determines their ability of bioaccumulation and biotransformation which in turn regulates the fate of the NPs.An in-depth study of Cerium oxide (CeO_2)NP by P. Zhang et al. 2012, provided evidence of biotransformation in the cucumber plant. This study states that when roots absorb CO_2-NP, a portion of it is transformed to Ce(III), which then reacts with PO_4^{3-} present in the roots to form insoluble $CePO_4$ [19]. They found the roots to contain 66% CeO_2 and 34% $CePO_4$. Whereas, CeO_2 concentration was higher in stem (86.4%) and leaves(78.5%). This indicates biotransformation of CeO_2-NP is in roots [19]. Another study suggested that the Selenium-NP (SeNP) absorbed by plants is converted to eight different organic forms of Selenium [20]. The Roots absorbSeNP by passive diffusion and are converted mainly to Se(IV) and SE(VI). A small amount of SeNP is transported from roots to shoots and is rapidly assimilated to Se(IV)and other forms in shoots [20].

NPs IN SOIL

Soil maintains a healthy ecosystem around us. It provides shelter and nutrients to all the living organisms residing or dependent on soil. Soil has an important role in nutrient cycling, atmospheric temperature regulation, and water filtration, which altogether have a direct effect on the growth and survival of living organisms, and maintaining biodiversity. The NPs in soil have both beneficial and detrimental effects on soil microbiota, plants, and other terrestrial organisms. Therefore, the maintenance of healthy soil is necessary for maintaining a healthy biota.

Source of Soil NPs

Soil harbours a wide range of organic and inorganic nanoparticles naturally. The most abundant of them are clay minerals, followed by metal oxide and hydroxides, humic substances, allophane, and imogolite [21]. Additionally, anthropogenic NPs are also found in soil (Fig. **1**). The NPs enter the soil either directly or indirectly through sewage sludge from WWTP, which is used as biosolids in the agricultural field. The nano-based fertilizers and anti-microbial agents applied to the crop field may also act as a source for introducing nanoparticles into the environment. The biosolid-laden soil contains several toxic metals and nanoparticles (Cu, Ag, TiO_2) [22].

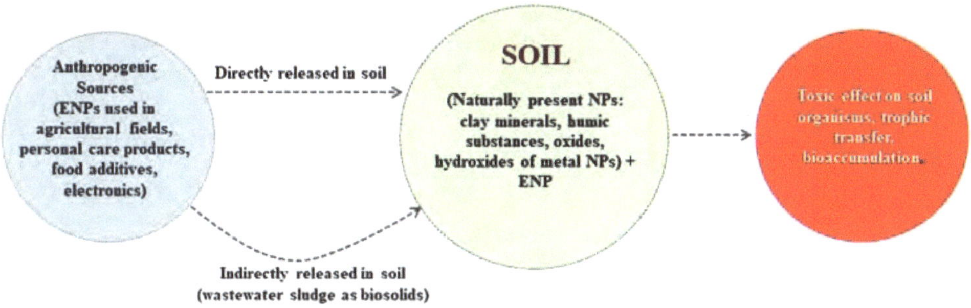

Fig. (1). Source and effect of NPs in soil.

Effect on Soil

The NPs in the soil system are modified to various forms by physicochemical transformations based on soil type, pH, and chemical interactions with organic and inorganic molecules in the soil. These transformations, in turn, change the bioavailability of the nanoparticles and bioaccumulation by the terrestrial food chain and may lead to biomagnifications in higher trophic levels (Fig. **1** and Table **1**). Thus, along with the high usage of the NP-based products, their effects on the soil community and health must be evaluated carefully.

Table 1. List of NPs in the environment and their beneficial and harmful effects. These information are gathered from the published literature ([26, 31, 37, 43, 46, 47, 50, 51, 53, 55, 68, 69, 75 - 100]).

NPs	Use	Beneficial Role on the Environment	The Toxic Effect on the Environment
TiO_2	The photocatalytic effect, Sunscreen, food, and personal care products	The photocatalytic ability of TiO_2 was successful in completely removing toxic chemical constituents of cigarette smoke	TiO_2-NP in natural water was found to increase bioaccumulation of copper in *Daphnia magna*, hence Cu-toxicity in freshwater increases

(Table 1) cont.....

NPs	Use	Beneficial Role on the Environment	The Toxic Effect on the Environment
Ag	Antibacterial agents, sensors, optical probes	Green synthesized Ag-NP can be used for remediation of a phenyl pyrazole insecticide, Fipronil-contaminated water.	Ag-NPs are uptaken mainly by roots in plants and have a detrimental effect on plant growth and nutrient content
Au	Sensors, therapeutic agent delivery, probes	Treatment of green synthesized Au-NP in DDT (a harmful organochlorine pesticide) contaminated soil, showed a high degradation rate of DDT, indicating the efficiency of Au-NP in environmental remediation.	Au-NPs were taken by the primary producer of tobacco plants and found to be present throughout the plant tissue. Primary consumer hornworms consumed the leaf materials, and Au-NPs accumulated in hornworms, though the concentration was much lesser, so it signifies that Au-NP can bioaccumulate in higher trophic levels
Cu, CuO	Anti-bacterial, the anti-fungal agent added to plastics, coatings, textiles, ceramic resistors, superconducting material, and semiconductors.	Highly effective adsorbent CuO-NP removed arsenic almost completely from groundwater.	Cu-NP harmed the seedling growth on both test plants *P.radiatus* and *T.aestivum,* Cu-NP was bioavailable to the plants, and the rate of bioaccumulation in the plants was concentration-dependent.
Zn, ZnO, ZnS	Inhibit UV radiation, breast cancer treatment, antimicrobial, photocatalytic effect, cosmetics, and rubber industry.	(ZnS) NP aerogels enabled very effective removal of Pb^{2+} and Hg^{2+} metal ions from aqueous solution, suggesting ZnS NP gels can be used in real-life applications for water remediation	Zn-NP, ZnO-NP can cause toxic effects on zooplanktons- *D.magna*, *A.salina*, by increasing intracellular oxidative stress
Fe_2O_3, nZVI	Application in MRI for tumour detection, drug delivery, coatings	Fe-based NPs are important for water pollutant remediation purposes. Nano Fe^o was highly effective in removing uranium (98% removal within 2 hours of reaction) in a carbonate-rich complex aquatic environment	Fe_2O_3 NP present in the aquatic environment has a detrimental physiological effect on fish *L.rohita*.A significant decrease in WBC count, changes in ion regulation, and gill Na+/K+ ATPase activity was observed indicating a negative impact on the aquatic system.
SiO_2	Additive for the manufacture of rubber and plastics, drug delivery	SiO_2 -NP was most effective for the remediation of all three heavy metals: Zn(zinc), Cd(cadmium), and Ni(nickel) in calcareous soil	Nano TiO_2, SiO_2 inhibits the growth of Daphnia by producing ROS in the presence of light, in dark the inhibition decreases

(Table 1) cont.....

NPs	Use	Beneficial Role on the Environment	The Toxic Effect on the Environment
CeO_2	Corrosion protection, solar cell, chemical/mechanical polishing	The photocatalytic ability of CeO_2-NP is useful for treating different pollutants present in industrial wastewater. Being an n-type semiconductor with a wide bandgap, CeO_2-NP can be used to photoelectrocatalytically degrade pollutants in either single or composite forms doped with other metals and non-metals.	CeO_2-NP induced 52.9% loss in total chlorophyll in water silk (*Spirogyra borealis*) and 41.3% loss in water lettuce (*Pista stratiotes*) after 90days treatment
Carbon-based NPs -Carbon nanotubes (CNTs), fullerenes, graphene	Biomedicine, diagnosis, drug delivery gene therapy	Carbon nanostructures like CNTs, fullerenes, and graphene have high efficacy as nanoscale adsorbents to treat greenhouse gases. CNTs target NO_x, CO_2, and CO; fullerene target CO_2. N_2O; graphene targets CO_2, NH_3, SO_2, H_2S gases	Carbon-based nanoparticles can bioaccumulate in different aquatic organisms, protozoa, and fishes. After 21 days of treatment with nC_{60}, there was a delay in the moulting of Daphnia with a significant decrease in reproduction. potential.
Al_2O_3	Biosensor, Pharmaceuticals, Mechanical industry	Green synthesized α-Al_2O_3-NP from Pepsi can waste show antifungal activity against some mycotoxin, aflatoxin-producing fungi	Excessive bioaccumulation of Al_2O_3-NP in organisms and plants causes increased mortality in organisms and growth retardation in plants. The toxic effect of the NP is directly proportional to a smaller size, dose and exposure time.
NiO	Gas-sensors, electronic devices, catalysts, fuel cells	The non-toxic, chemically stable, semiconductor nature of NiO nanoparticles allowed it to be a very efficient photo-catalyst, which can degrade harmful azo dyes in an aqueous environment in the presence of UV light.	The toxic effect of NiO-NP on freshwater alga *Pseudokirchneriella subcapitata* is severe, shows reduced metabolic activity, disrupts plasma membrane integrity, and accumulates intracellular reactive oxygen species causing bleaching of photosynthetic pigment chlorophyll, ultimately affecting photosynthesis.

(Table 1) cont.....

NPs	Use	Beneficial Role on the Environment	The Toxic Effect on the Environment
CdO	Electronic devices, Biomedical sensors, optical sensors	The eco-friendly CdO-NP synthesized from *Corchorus olitorus* leaves has enhanced biocatalytic activity helpful for the successful degradation of toxic direct violet dye by an advanced oxidative method	The non-biodegradable nature of CdO-NP, makes it more harmful to the environment, improperly affecting the entire ecosystem. The variety of toxic effects of Cd-containing NPs is diverse-interfering with microfaunal diversity causes growth retardation and reduced photosynthetic activity in plants, bioaccumulated in different organisms, showing carcinogenic effect, thus affecting multiple systems in organisms.
SnO_2	Photovoltaic cells, sensors, photocatalysts, solar heat collector	The green synthesis of SnO_2 NP from methanolic extract of *Cyphomandra betacea* plant was successfully used for photocatalytic degradation of methylene blue, providing an environment-friendly option for the degradation of toxic waste	In a study, the effect of SnO_2 NP on two environmental bacteria Bacillus subtilis (gram-positive) and Escherichia coli (gram-negative) showed that specific interaction between the negative surface charges on SnO_2 NP and the positive charges on the surface of *B. subtilis* bacteria causes lethal effect by damaging the cell membrane and cellular function disruption, whereas E. coli remain unaffected.
ZrO_2	Ceramic industry, electronic devices, sensor devices, drug industry	Green synthesized, stable and biocompatible ZrO_2-NP is useful for bioremediation purposes and shows antibacterial. Antifungal, photocatalytic, anticancer property. The diversity in the surface coatings during the synthesis of these NPs from different botanical and microbial extracts forms a stable capping around the NPs, which has a greater implication in the environmental remediation scenario	The unique properties attributed to this NP consist of its biocompatibility, rigidity, toughness, anti-corrosive nature, and chemical stability. Despite very little available data regarding its effect on the environment, this NP has a much less harmful effect on the environment than others. The available research articles suggest ZrO_2 NP has a potentially toxic effect in plants due to high phyto-availability, and bio-accumulation in organisms and it is capable of producing oxidative stress.

(Table 1) cont.....

NPs	Use	Beneficial Role on the Environment	The Toxic Effect on the Environment
MoO_3	Electrical capacitors, coatings, textile industry, photocatalyst	The photocatalytic activity of MoO_3 was enhanced by creating α-MoO_3 nanosheet structure, which offers a new membrane technology for the removal of organic pollutants like methylene blue and rhodamine B by photodegradation	In a study on a soybean plant, the toxic effects caused by molybdenum-based NPs affect both the physiological and biochemical properties of the plant. The molybdenum-based NPs, when enter the plant, accumulate in the plant tissues and induce oxidative stress in the plant cells. The NP inhibits the growth and development of plants, affects the microstructure of root cells, and interferes with the symbiotic relationship of soybean with rhizobium, The impaired nitrogen fixation capacity of soybean plants in term affects the nutrient content of plants.
WO_3	Electronic devices, sensors, photocatalyst	The photocatalytic activity of WO_3 NP is used for the degradation of organic dyes in aquatic systems, water dissociation providing a cost-effective solution for environmental pollution	The toxicity of WO_3-NP is enhanced in the presence of UV radiation against bacterial species and mammalian cells. The enhanced photocatalytic activity under UV irradiation causes membrane damage and alters cellular metabolic activity, leading to cell death.
Co	Lithium-ion batteries, gas sensors, capacitors	Cobalt oxide NP synthesized from Punica granatum plant peel extract enhanced the photocatalytic activity under sunlight, it can degrade 78.45% of toxic Remazol Brilliant Orange 3R (RBO 3R) dye within 50min. The sufficient amount of bioactive compound present in the peel extract was able to maintain the eco-friendly cobalt-oxide NP activity.	The physical properties of Co-NP determine the toxicity of this NP in the ecosystem. In an experiment, the effect of particle size of Co-NP was assessed on different bacterial activities, algal growth, root-shoot ratio and germination in plants. The smaller size of Co-NP has a more inhibitory effect on bacterial activities, algal growth and germination of plants.

(Table 1) cont.....

NPs	Use	Beneficial Role on the Environment	The Toxic Effect on the Environment
Pt	Imaging, drug delivery, biomedical application	The Pt-NPs are synthesized by the green synthesis method from different types of parts of plant extracts. The organic compounds present in the plant extracts like flavonoids, amides, alcohols, terpenes, cannabinoids, sugars, and amines can form a protective organic coating over Pt-NP thus stabilising the NP and minimising its side effects. eco-friendly Pt-NP show antifungal, antibacterial, anticancer, and antidiabetic activity.	The Pt-NP causes oxidative stress by producing abiotic reactive oxygen species in the algal species, Pseudokirchneriella subcapitata and *Chlamydomonas reinhardtii*, leading to membrane damage and inhibition in the growth of micro algal cells.
Quantum Dots	Display systems, biomedical imaging, solar cells	Quantum dots synthesized by the green synthesis method are used for environmental remediation purposes, they can rapidly detect contaminants like heavy metals, pesticides, and antibiotics from food. Carbon quantum dots are more reliable in terms of environmental safety.	QDs can transfer in the food chain of aquatic ecosystems, from protozoa, Astasia longa to the second consumer in the food chain zooplankton, *Moina macrocopa*, and finally to fish, *Danio rerio*.

Nanoscale zero-valent iron (nZVI) mostly used for environmental remediation purposes, has positive effects on plants. Then nZVI exposed *Arabidopsis thaliana* was found to have a 38% higher dry weight of rosette, elevating the CO_2 assimilation rate, and the enhanced stomatal opening. These altogether indicated a significant increase in photosynthetic activity [23].

A study on a land snail, *Helix aspersa,* indicated that iron oxide NPs (IONPs) and iron oxide NP incorporated in a zeolite increase oxidative stress, DNA damage, and lipid peroxidation [24]. The size and surface chemistry of IONPs allow it to be degraded into iron ion NPs after entering the lysosomes, causing lysosomal membrane damage and resulting in ROS production. Then ROS produces a multitude of oxidative stress inside cells like DNA damage, protein ubiquitylation, and cell death by apoptosis [24]. Cu^+ ions from Copperoxide NPs (CuO-NPs) can inhibit denitrification by a reduction in nitrate reductase activity [25]. The CuO NP released in soil forms Cu ions, after which it can form different exchangeable forms, and alters the functional genes for denitrification, the effect of which is directly proportional to the concentration of Cu ions [25].

The Silver NPs (Ag-NPs) are taken up mainly by roots in plants and have a detrimental effect on plant growth and nutrient content [26]. The accumulation of Ag-NP in wheat grains occurs in a dose-dependent manner, thereby possessing a potential risk of entering the food chain. Also, Ag-NP reduces the nutrient content of plants. Micro-elements like Fe, Zn, and Cu markedly reduced in the presence of Ag-NP, causing negative effects in chlorophyll synthesis, auxin biogenesis, vital enzyme synthesis, and electron transfer of light reaction in addition to inducing oxidative stress in plants. Thus it is evident that wheat plants show significant growth retardation, declining crop quality and yield as an outcome of the toxic effect of Ag-NPs [26]. In a microcosm study, an experimental setup mimicking the natural environment, Titanium dioxide NPs (TiO_2-NPs) was found to inhibit plant growth and reduce shoot biomass [27]. Interestingly, the high loam soil is found to be more capable of inhibiting Gold NP (Au-NPs) uptake than sandy soil. The bioavailability of Au-NP is higher in agricultural soil, and thus, it shows a possibility of Au-NPstrophic transfer [28]. It has been shown that the bioavailability of Au-NP depends primarily on the soil characteristics. The surface coatings of Au-NP and the physical properties of soil affect the uptake of Au-NP by plants, majorly in agricultural fields, because of the extensive use of pesticides and fertilizers containing ENPs. The Au-NP up taken by plants also shows toxic effects at a cellular level, which is evident from the significant fold change of stress response genes in tomato plants in the presence of Au-NP compared to the control [28].

The influence of nanoparticles on the enzyme of soil organisms was found to be dependent on the type and concentration of nanoparticles. The CeO_2-NPs stimulate phosphatase activity and inhibitβ-glucosidase and urease activities [29]. In an experiment on the trophic transfer of CeO_2-NP, it was shown that CeO_2-NP can be accumulated in a significant amount in lettuce plant leaves, which in turn is consumed by the herbivore, snails. The CeO_2 concentration was higher in visceral organs than in the foot and shell, which are the edible parts. Therefore, the possibility of CeO_2-NP transfer to humans is comparatively less [30]. However, a study on Au-NPs showed that it is taken up by the tobacco plant(primary producer) and its traces are found throughout the plant tissue. When the hornworms, the primary consumer, consume the tobacco leaf, results in the trophic transfer of Au-NPs. However, the concentration of accumulated Au-NPs was much lesser. This signifies that the Au-NPs can bioaccumulate at higher trophic levels, but the possibility of biomagnifications is lesser [31]. Another study on the trophic transfer of Ag-NPs and TiO_2-NPs from plants to the primary consumers (PCs) showed their differential biomagnification ability and their varied adverse effects on the PCs [32]. They found that the PCs excreted almost 40% of Ag-NP and 10% TiO_2-NP through feces, retaining a significant amount in the digestive tract. The combination of Ag-NP and TiO_2-NP results in a cumulative adverse effect on the growth and activity of snails, possibly because TiO_2-NP can act as a carrier that facilitates the uptake of Ag-NP. The long-term feeding of Ag-NP, TiO_2-NP treated lettuce leaves to the PC (snails) causes the distribution of these NPs in different tissues in snails. The NPs in particulate form can not cross the epithelium membrane of soft tissues and thus remain mostly in the digestive tract or end up being in the intestine for excretion, whereas the ionic form of the metals can easily assimilate, eventually translocating into foot, mantle, shell, even into the brain of the snails. The higher retention of the NPs in the digestive gland affects their excretion through feces, thereby disrupting its function, producing higher oxidative stress and strongly affecting their metabolic activity. The long term feeding on contaminated leaves by PCs affects the activity and behaviour of PCs, thereby causing potential threat in the food chain in terrestrial ecosystem [32]. Graphene nanoparticles are known for their incredible mechanical strength and super-capacitor properties. Graphene oxide nanoparticles undergo physiochemical transformation in the environment and result in a variety of eco-toxicological effects like growth retardation, reduction in photosynthesis, nutrient loss in plants, membrane disruption, and oxidative stress in living organisms [33].

Soil Remediation with NPs

The NPs sometimes help to eliminate various toxic metals, pollutants, and contaminants from the soil (Table **1**). One of the major NPs used for remediation purposes is nZVI. ThenZVI produced from neem and mint leaves by the green synthesis method was found to be more efficient in removing Pb and Ni than chemically synthesized nZVI [34]. Also, green synthesized iron oxide nanoparticles from the leaf extract of *Euphorbia cochinchinensis* were able to reduce arsenic availability significantly in soil [35]. The ZnO-NPs containing nano fertilizers were found to restore the deficiency symptoms in rice plants grown in Zn-deficient soil [36]. High surface area to volume ratio and charge density are the most important parameters of NPs underlying their effectiveness for the immobilization of heavy metals in soil. Al_2O_3, SiO_2, and TiO_2 NPs were found to reduce heavy metals like Zn, Cd, and Ni from the soil depending on soil type. The SiO_2-NPs were most effective for the remediation of all three heavy metals in calcareous soil [37]. A study in China showed the beneficial effect of modified carbon black nanoparticles(MNCB) on heavy metal-contaminated soil. MNCB was found to be efficient for the immobilization of Cd and Ni reducing their bioavailability, thereby increasing the biomass of the plants which otherwise show decreased biomass [38].

NPs IN THE AQUATIC ENVIRONMENT

Water is essential for maintaining life on earth. Both ocean and freshwater ecosystems have an important role in nutrient recycling, climate regulation, flood control, providing food and habitat for a wide variety of organisms and groundwater regulation. Therefore, a disturbance in an aquatic environment can affect the entire aquatic ecosystem (Fig. **2** and Table **1**).

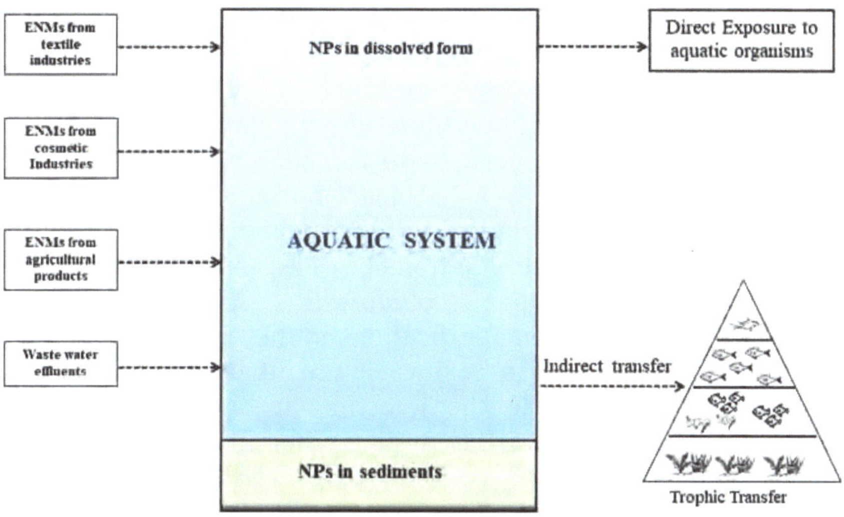

Fig. (2). Source and effects of NPs in an aquatic environment.

Source

The NPs are present in the aquatic system naturally as colloids formed from humic acid and in mineral form originating from weathering rocks. In addition, man-made NPs enter the aquatic system from different sources (Fig. **2**). Most of these settle in the sediments, and few are in the dissolved state. The aquatic environment plays a major role in the transformation of NPs, indirectly affecting their bioavailability. The trophic transfer and bioaccumulation of the NPs in various freshwater and marine organisms and subsequent biomagnification to higher trophic levels is a serious threat. The NPs can enter the human body from an aquatic food source or through underground drinking water. A large number of nanoparticles, mainly ZnO, TiO_2, Ag-NM, Cu-NP, and carbon black, enter the aquatic bodies through untreated water from different industries. If the contaminated water is not treated, this may lead to serious health hazards [8].

Effect of NPs in Water

It was found that the TiO_2-NP accumulates in the gills and intestine of goldfish (*Carassius auratus*), and was responsible for reducing 19.7% of body size at a high dose(100mg/L) of the NP [39]. In an experiment on freshwater teleost, *P. promelas* when exposed to an environmentally available concentration of TiO_2-NP, the NP accumulation occurred in the kidney and spleen, affecting hematopoiesis and the immune system. The neutrophil population present in the kidney encounters the TiO_2-NP because the major sink of TiO_2-NP bioaccumulation is the kidney. The raised concentration of TiO_2-NP in the kidney leads to affected neutrophils, and inefficient antibacterial response, thereby increasing the mortality rate in the presence of various pathogens. The morphological changes of the glomerulus, endothelial lining disruption, and cardiomyocyte structure damage lead to disrupted renal filtering and cardiac function due to the toxic effect of TiO_2-NP. Thus, during pathogenic infection, the effect of TiO_2-NP on the fish population will be detrimental to the survival of the organisms [40]. A study on the trophic transfer of Ag-NPsrevealed that Ag-NPs transferred from green algae (*Chlorella sp.)* to water flea (*Moinamacrocopa*)showed no biomagnifications in the subsequent higher trophic levels [41]. The CeO_2-NP present in freshwater can attach to the gills of fish, and when ingested, it accumulates in the intestine at high concentrations. The bioaccumulation of CeO_2-NP is different for different species of fish, which affects the subsequent food chain accordingly. CeO_2-NP can form aggregates in water-sediment interfaces in aquatic environments that can be bioaccumulated in organisms, or eliminated by sedimentation. The higher bioavailability of CeO_2-NP in aquatic environments can be a potential threat to the entire aquatic ecosystem [42]. Another study on CeO_2-NP in a freshwater ecosystem provided in-depth insight into the bioaccumulation and effect of long-term exposure (10 months) to the nano-particle. The abnormalities were observed only in hydrophytes in the form of chlorophyll loss, causing an imbalance in productivity in the water ecosystem [43]. It was observed that the CdSe/ZnS quantum dots (QDs) in algal food show a higher mortality rate in amphipods than in the normal dissolved form of Cd. This indicates a higher risk of bioaccumulation of nanoforms of heavy metals [44]. A study conducted in Old Danube Lake of Vienna, Austria, suggested that the TiO_2-NP in sunscreen is released into the water which aggregates in sediments rather than remaining suspended in water [45]. The TiO_2-NP concentration measured from the lake water was found significantly higher during summer than in other seasons, which led to a better understanding of the effect of high concentrations of NPs in the environment. The combined effect of natural colloids, physiochemical properties of the aquatic environment, and radiation from the sun, cause the removal of the protective organic coating around NPs in the sunscreen and exposes the TiO_2-NPs in the surrounding environment. Then, fast aggregation of these NPs allows them to sediment and may have effects on the aquatic environment. So, further evaluation of the aquatic ecosystem in this scenario will be useful to study the adverse effect of the NPs in the lake seasonally [45]. TiO_2-NP in natural water was found to increase the bioaccumulation of copper in *Daphnia magna*, which in turn increases Cu-toxicity [46]. In a comparative study, it was observed that nanosized TiO_2 and SiO_2 inhibit the growth of Daphnia by producing increased ROS in the presence of light. The inhibition decreasesin the dark [47]. A study on zebrafish embryo provided the in detail mechanism of Fe_2O_3-NP toxicity in the various developmental stages. As a common mechanism for major NPs in aquatic environment, Fe_2O_3-NP also show tendency to remain in sediments after aggregation. The benthic aquatic organisms that remain in close contact of sediments are the ones directly encountering the high concentration of NPs. As a model organism to study the effect of the NPs in benthic environment, zebrafish embryos were studied. The process by which Fe_2O_3-NPs aggregate, transform, and interact with other NPs can affect the rate of its toxicity. The developmental toxicity mechanism of Fe_2O_3-NP occurs primarily by adhering to the surface of this larva, probably induce a hypoxic state and cause increased mortality, and decrease in hatching of zebrafish larva [48].

Remediation of Contaminants in Water

Iron-based NPs are important for water pollutant remediation. Nano Fe^o was found to be highly effective in removing uranium (98% removal within 2 hours of reaction) in a carbonate-rich aquatic environment in the Lisava Valley, Romania, whereas Fe_3O_4 was less effective. The difference is mainly attributed to the comparatively better crystallinity of Fe^0-NP than Fe_3O_4 NP [50]. The efficiency of NPs in heavy metal remediation of drinking water is of major significance. The highly porous and large surface area of metal sulfide (e.g. ZnS) NP aerogels is very effective in removing Pb^{2+} and Hg^{2+} metal ions from aqueous solution by cation-exchange reaction [51]. In another study, the TiO_2 NP doped with 1% HfO_2 was found to degrade almost 90% of methylene blue (one of the non-biodegradable dyes) within 10 minutes [52]. It is well known that the groundwater in many places in West Bengal, India, has a high arsenic content. The treatment of this contaminated water with the CuO-NP showed a promising result. The highly effective adsorbent CuO-NP removed arsenic almost completely from groundwater [53]. Azo dyes coming from textile industries present in wastewater are harmful as they can form toxic aromatic amines. The non-toxic, chemically stable, semiconductor nature of NiO nanoparticles make it an efficient photo-catalyst, which can degrade harmful azo dyes in an aqueous environment in the presence of UV light. This photo-catalytic nature of NiO nanoparticles plays an important role in its anti-microbial activity [54]. Among the different polymorphs of Molybdenum oxides, α-MoO_3 is the most studied. The cost-effective, non-toxic, and semi-conducting structure of MoO_3 has a vital role in its extensive use in different fields. The photocatalytic activity of MoO_3 was enhanced by creating α-MoO_3 nanosheet structure, which offers a new membrane technology for the removal of organic pollutants like methylene blue and rhodamineB by photodegradation [55]. Carbon-based nanoparticles, like cylindrical fullerene derivative or carbon nanotubes (CNTs), graphene oxide, and multi-walled CNTs, have been widely used in recent years for environmental pollutant treatment because of their physiochemical properties, such as large surface areas, thermostable nature, and high acid stability. The adsorption efficiency and photocatalytic properties of carbon-based NPs are used for heavy metal and organic pollutant remediation from the environment, which primarily depends upon the interaction between the organic and metal pollutants with oxygen-containing groups present on the surface of carbon NPs [56].

A healthy aquatic system provides valuable information about the entire region. The excessive use of nanoparticles and their long-term effect on the aquatic system must be evaluated to decrease the risk of bioaccumulation and biomagnifications.

NPs IN AIR

The quality of air that we breathe in is an indicator of different health problems worldwide. Serious conditions like pulmonary diseases, cardiac disease, and many cancers are the long-term effects of poor air quality. The nanosized particles present in the air (<100nm) originate from different natural and anthropogenic sources. They are formed in the atmosphere from natural vegetation, ocean spray, forest fires, volcanic eruptions, *etc*. Anthropogenic sources of nanosized particles include the combustion of fuels and biomass and industrial products and their byproducts entering directly into the air [57]. Nanoparticles of anthropogenic sources are higher in urban areas than in rural areas, which is why urban people suffer most from air pollution-related health problems. Aerosols can scatter and absorb sunlight and form both heating and cooling effects in the atmosphere [58].

Source

The amount of metal NPs in the air is increasing day by day due to its growing demand in human life (Fig. **3**). It has turned into one of the major concerns to deal with. A large amount of gas and ashes containing silicates, and metal NPs (Fe, Cd, Cu, Pb, *etc*.) from volcanic eruptions, dust storms, and cosmic dust are the major sources of naturally occurring metal NPs [59]. In a study in Latin America, dust from the roads of Manizales was found to contain different toxic nanoparticles like As, Cd, Hg, As, Pb, *etc*., from volcanic emissions of Nevado del Ruiz volcano [60]. Metallurgical industries and chemical industries generate ultrafine metal particles. The extent of human exposure to these nanoparticles is massive for the people working in those industries [59]. Vehicular exhaust is one of the major contributors to aerosol in urban areas. Fe-NPs in their different oxides forms, are emitted from vehicles and natural sources, wood burning in a household, *etc*. [61]. A study in Delhi, India, shows that nanoparticles entering the atmosphere from vehicles are the major contributor to an increase in deaths from air pollution [62]. Diesel contains chemically bound sulfur, which, upon combustion, gets oxidized to form sulfates. The sulfates and black carbon released maximally from heavy-duty diesel vehicles contribute to ultrafine particles in urban air [52, 53]. Black carbon and metallic NPs like Fe, Cu, and Pd are emitted from an air-conditioning machine that ultimately gets distributed to the indoor airdust, which acts as a sink for these NPs [58].

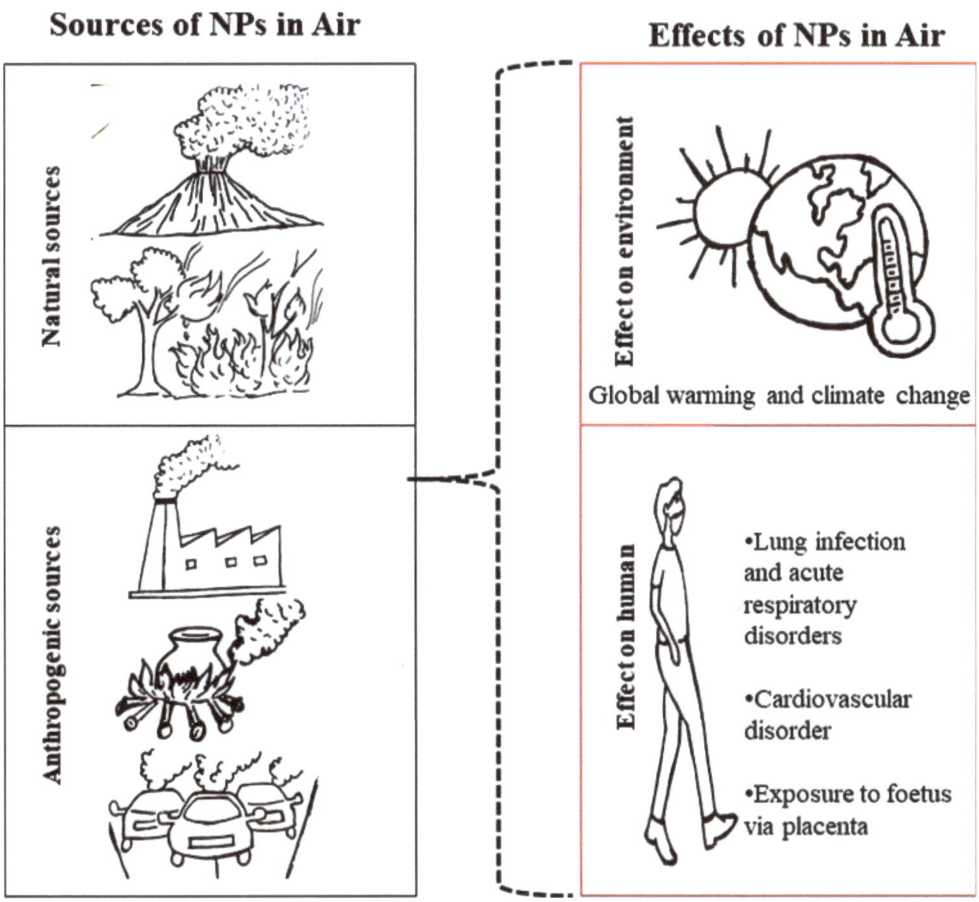

Fig. (3). Source and effect of NPs in air.

Nanoparticles in air undergo aggregation, sedimentation, coagulation, and dissolution. The ultrafine particles remain in the atmosphere, whereas the aggregated forms sediment and end up being distributed in soil and water. Chemical transformations like photocatalytic degradation, redox reactions, and reactions with other nanoparticles affect the properties of nanoparticles in an atmosphere that is dependent upon the environmental conditions [59].

Effect of NPs in the Air

Nanoparticles Suspended in the Air Enter the Respiratory Tract and Cause Lung Infections and Acute Respiratory Syndromes (Fig. **3**). Respiratory disorders are common worldwide. Cardiovascular disorders are also inevitable. The particulate matter in the air can increase susceptibility to ischemic heart disease, thereby providing a link between air pollution with mortality risk from cardiovascular disease [65]. The nanoparticles in the air, which are of different sizes and different chemistry, cause air toxicity by various mechanisms. The particulate matter present in the air causes a decrease in oxygen supply to the myocardium, thereby causing ischemic heart disease [65]. Extreme loads of nanoparticles in urban air inhaled by pregnant ladies can translocate to their placenta. The macrophage-enriched placental cells from 15 placentas provided evidence for the presence of carbonaceous and metal NPs, even titanium, and cerium NPs. Thus fetus gets exposed to these nanosized particles and their health may be affected [66]. The presence of exogenous carbon and metal-containing NPs in Macrophage-enriched placental cells is indicative of the inhaled nano-sized particulate matter present in the urban air, which mainly comes from heavy traffic roads, and the combustion of fossil fuels. This is of serious concern because the effect of these NPs in healthy placental tissues can have long-term effects on the fetus. The translocated NPs are taken up by tissue-specific phagocytes in distal organs. The effect is not only limited to the person inhaling these NPs from air, the fetus is also at a potential risk [66]. Nanoparticles released from diesel vehicles may cause visibility impairment, and poor air quality is a major cause of climate change [57]. The primary target for the ultrafine particles present in the air is the lung followed by the cardiovascular system. It can enter the circulation, thereby showing toxic effects on vascular endothelium, and blood coagulation, and alter cardiac functions by triggering the autonomic nervous system. The interactions between the disease gene and ultrafine particle determine the outcome of the toxic effect in lungs and cardiac tissues. The effect of the NPs at the gene level needs to be studied further as the DNA damage response owing to increased DNA fragmentation due to the NPs, can help answer the serious impact of these NPs at the molecular level [67].

Remediation of Pollutants in the Air by NPs

The NPs have a significant role in controlling air pollution because of their very small size and high surface area to volume ratio. They can react with toxic material and degrade it into a non-toxic product. Interestingly, the use of NPs in coatings was found to reduce the pollutant load. Carbon nanostructures like CNTs, fullerenes, and graphene have high efficacy as nanoscale adsorbents to treat greenhouse gases (Table **1**). CNTs target NO_x, CO_2, and CO. Fullerene targets CO_2 and N_2O. Graphene targets CO_2, NH_3, SO_2, and H_2S gases [68]. The photocatalytic ability of TiO_2 is significant because of its very strong oxidizing power. For this reason, TiO_2-impregnated titanium mesh filters (TMiP) have been developed for air purification purposes. This novel strategy of utilizing the photocatalytic ability of TiO_2 was successful in completely removing toxic chemical constituents of cigarette smoke. The favoured crystalline structure and high surface area to volume ratio of TiO_2 NP enable TMiP to decompose the chemical compounds of cigarette smoke efficiently [69]. Apart from the photo-oxidative properties of photo-catalytically active material, their ability to affect the surface area and surface energy of material has a significant role in controlling air pollution. Nitrogen oxides are one of the important air pollutant gases to have negative impacts on human health and the environment. The photo-catalytically active coatings of TiO_2 NP and a mixture of SiO_2-TiO_2sol applied to a concrete surface are highly efficient for the treatment of NO_x pollution [70]. Thus air pollutant remediation using NPs can be achieved by pollutant degradation by nanocatalysts, sensing and detecting pollutant load, separation, and filtration of toxic pollutants by nanofilters [68]. MoO_3-based NPs show a superior quality of pollutant resistance thus they are also used for air pollutant remediation [71].

LIMITATIONS AND FUTURE PERSPECTIVES

The promising aspect of NPs in diverse fields has proven to be beneficial to modern human society. We are becoming increasingly dependent upon these nano-sized materials without completely understanding their story. The lifecycle of NPs from industrial manufacturing to their release upon use and following delivery into the environment, leading to its complex biochemical and biophysical modifications by a wide array of environmental parameters, are often ignored due to a dearth of available scientific investigations as well as a lack of awareness amongst the masses. Though scattered information is available that delved into the harmful effects of these promising new age tools with the potential of revolutionizing many industries, including environmental remediation, we are still far from decrypting its significance and subsequent regulations that would have allowed optimal utilization of these technologies. Therefore, it is not tough to predict that there are still many challenges to overcome before nanoparticles can be widely used in the environment.

The toxic effects of NPs are greatly reduced with the green synthesis method. The NPs synthesized from the different plant part extracts, algae, bacteria, and fungi are environment-friendly and cost-effective [72, 73]. The non-toxic method of NP synthesis has been proven to be useful for environmental remediation purposes. As has been already described in this chapter, the utility of green synthesis of different NPs is immense, and research should be continued in this field to develop a sustainable environment. The properties like small size, large surface area, high reactivity, and unique properties of different nanoparticles can be used for specific pollutant removal from the environment. A novel type of nanopolymer called Molecularly-imprinted polymers (MIPs) is developed, in which an artificial receptor site is created that can recognize a specific substrate. Thus, the MIPs are useful for selectively removing environmental pollutants without being harsh to the environment [74]. Similar to the concept of enzyme-substrate recognition, the polymerization technique developed in such a way that several functional monomers of MIPs self-assemble around the target molecules. The specificity of interaction between the functional group of the monomers with the functional group of the target substrate molecule enables selective removal of the heavy metal and toxic dyes from the environment. Adsorption of heavy metal ions like mercury, chromium, copper, lead, arsenic, nickel *etc.* and some common toxic dyes like reactive red, basic red dye, methylene blue, and rhodamine B are done by specifically designed MIPs [74]. The adsorption-desorption of the MIPs with the template causes template removal by either solvent extraction, template cleavage or other specific methods. The selective absorptivity, along with the reusability of MIPs, will be useful for future use for environmental safety.

CONCLUSION

The application of NP is increasing gradually in different industries, including molecular diagnostics, agricultural, and environmental remediation. Nanotechnology has provided a broad field to manipulate NPs according to the requirement. The use of NPs and their routes to the environment may create long-term harmful effects on the environment. Therefore a proper risk management system for NPs use must be evaluated before introducing them in the market. The life cycle assessment of NPs is also important to know the possible exposure routes, release rates, quantification, transformation procedures affecting their properties, and interaction with a biological system. This information may help us to delineate the environmental risk assessment (EPA) of the NPs precisely. Despite having some negative impacts, NPs are widely used as sensors for pollutants and pollutant remediation. Hence, NPs should be used judiciously. The global collaboration on the creation of a database for all the NPs, including its chemistry, application, interaction, and involved risks, is highly necessary.

ACKNOWLEDGMENTS

The authors are thankful to the Department of Zoology, University of Calcutta, India, for the necessary help. RT acknowledges the financial support from DST-SERB grants CRG/2018/001727 and EEQ/2019/000750.

REFERENCES

[1] Eu, T. *Definition of a nanomaterial*, **2011**, pp. 21-23.

[2] H. J., M. F. Nanogeoscience : From origins to cutting-edge applications. *Elements,* **2008**, 373-380. [http://dx.doi.org/10.2113/gselements.4.6.373]

[3] Market, T.N.; Type, P.; Nanomaterials, N.O.; Nanomaterials, P.; Industry, E. *Nanomaterials Market - Growth, Trends, Covid-19 Impact*; And Forecasts, **2022**, pp. 1-8.

[4] Sharma, V.K.; Filip, J.; Zboril, R.; Varma, R.S. Natural inorganic nanoparticles – formation, fate, and toxicity in the environment. *Chem. Soc. Rev.,* **2015**, *44*(23), 8410-8423. [http://dx.doi.org/10.1039/C5CS00236B] [PMID: 26435358]

[5] Taylor, P. Critical reviews in environmental science and natural nanoparticles. In: *Implications for Environment and Human Health*; , **2014**; pp. 37-41. [http://dx.doi.org/10.1080/10643389.2014.921975]

[6] Buzea, C.; Pacheco, I.I.; Robbie, K.; Buzea, C. Nanomaterials and nanoparticles: Sources and toxicity. *Biointerphases,* **2016**, *2*(4), MR17-MR71. [http://dx.doi.org/10.1116/1.2815690]

[7] Lespes, G.; Faucher, S.; Slaveykova, V.I. Natural nanoparticles, anthropogenic nanoparticles, where is the frontier? *Front. Environ. Sci.,* **2020**, *8*(May), 71. [http://dx.doi.org/10.3389/fenvs.2020.00071]

[8] Mortimer, M.; Holden, P.A. *Fate of engineered nanomaterials in natural environments and impacts on ecosystems*; Chapter 3, Elsevier Inc., **2019**. [http://dx.doi.org/10.1016/B978-0-12-814835-8.00003-0]

[9] Bundschuh, M.; Filser, J.; Lüderwald, S.; McKee, M.S.; Metreveli, G.; Schaumann, G.E.; Schulz, R.; Wagner, S. Nanoparticles in the environment: Where do we come from, where do we go to? *Environ. Sci. Eur.,* **2018**, *30*(1), 6. [http://dx.doi.org/10.1186/s12302-018-0132-6] [PMID: 29456907]

[10] Hotze, E.M.; Phenrat, T.; Lowry, G.V. Nanoparticle aggregation: Challenges to understanding transport and reactivity in the environment. *J. Environ. Qual.,* **2010**, *39*(6), 1909-1924. [http://dx.doi.org/10.2134/jeq2009.0462] [PMID: 21284288]

[11] Shrestha, S.; Wang, B.; Dutta, P. Nanoparticle processing: Understanding and controlling aggregation. *Adv. Colloid Interface Sci.,* **2020**, *279*, 102162. [http://dx.doi.org/10.1016/j.cis.2020.102162] [PMID: 32334131]

[12] Gilbert, B.; Ono, R.K.; Ching, K.A.; Kim, C.S. The effects of nanoparticle aggregation processes on aggregate structure and metal uptake. *J. Colloid Interface Sci.,* **2009**, *339*(2), 285-295. [http://dx.doi.org/10.1016/j.jcis.2009.07.058] [PMID: 19709669]

[13] Peng, C.; Zhang, W.; Gao, H.; Li, Y.; Tong, X.; Li, K. Behavior and potential impacts of metal-based engineered nanoparticles in aquatic environments. *Nanomaterials,* **2017**, *7*(1), 21. [http://dx.doi.org/10.3390/nano7010021]

[14] Murugadoss, S. Agglomeration of titanium dioxide nanoparticles increases toxicological responses *in vitro* and *in vivo*. *Part Fibre Toxicol.,* **2020**, *7*, 1-14.

[15] Wang, J.; Bruinink, A.; Wang, J.; Wick, P. Effect of particle agglomeration in nanotoxicology. *Arch Toxicol.,* **2015**, *89*(5), 659-675.
[http://dx.doi.org/10.1007/s00204-015-1460-6]

[16] Feng, X. Enhanced dissolution and transformation of ZnO nanoparticles : The role of inositol hexakisphosphate enhanced dissolution and transformation of ZnO nanoparticles. *The Role of Inositol Hexakisphosphate,* **2016**.
[http://dx.doi.org/10.1021/acs.est.6b00268]

[17] Yin, Y. *Water chemistry controlled aggregation and photo-transformation of silver nanoparticles in environmental waters*; JES, **2015**, pp. 1-10.
[http://dx.doi.org/10.1016/j.jes.2015.04.005]

[18] Manuscript, A. A high impact, peer reviewed journal publishing experimental and theoretical work across the breadth of nanoscience and nanotechnology. *Nanoscale,* **2020**.
[http://dx.doi.org/10.1039/D0NR05934J]

[19] Zhang, P. Biotransformation of ceria nanoparticles in cucumber plants. *ACS Nano.,* **2012**, (11), 9943-9950.

[20] Hu, T.; Li, H.; Li, J.; Zhao, G.; Wu, W.; Liu, L. Absorption and bio-transformation of selenium nanoparticles by wheat seedlings (*Triticum aestivum* L .) *Front Plant Sci,* **2018**, *9*, 597.
[http://dx.doi.org/10.3389/fpls.2018.00597]

[21] Theng, B.K.G.; Yuan, G. Nanoparticles in the soil environment. *Elements,* **2008**, *4*(6), 395-400.
[http://dx.doi.org/10.2113/gselements.4.6.395]

[22] Yang, Y.; Wang, Y.; Westerhoff, P.; Hristovski, K.; Jin, V.L.; Johnson, M.V.V.; Arnold, J.G. Metal and nanoparticle occurrence in biosolid-amended soils. *Sci. Total Environ.,* **2014**, *485-486*, 441-449.
[http://dx.doi.org/10.1016/j.scitotenv.2014.03.122] [PMID: 24742554]

[23] Yoon, H.; Kang, Y.; Chang, Y. Effects of zerovalent iron nanoparticles on photosynthesis and biochemical adaptation of soil-grown arabidopsis thaliana. *Nanomaterials,* **2019**, *9*(11), 1543.

[24] Sidiropoulou, E.; Feidantsis, K.; Kalogiannis, S.; Gallios, G.P.; Kastrinaki, G.; Papaioannou, E.; Václavíková, M.; Kaloyianni, M. Insights into the toxicity of iron oxides nanoparticles in land snails. *Comp. Biochem. Physiol. C Toxicol. Pharmacol.,* **2018**, *206-207*(February), 1-10.
[http://dx.doi.org/10.1016/j.cbpc.2018.02.001] [PMID: 29408432]

[25] Zhao, S.; Su, X.; Wang, Y.; Yang, X.; Bi, M.; He, Q. Copper oxide nanoparticles inhibited denitrifying enzymes and electron transport system activities to influence soil denitrification and N_2O emission. *ECSN,* **2019**, 125394.
[http://dx.doi.org/10.1016/j.chemosphere.2019.125394]

[26] Yang, J. Alteration of crop yield and quality of wheat upon exposure to silver nanoparticles in a life cycle study. *J. Agric. Food Chem.,* **2018**, *66*(11), 2589-2597.
[http://dx.doi.org/10.1021/acs.jafc.7b04904]

[27] Vijayaraj, V. Transfer and ecotoxicity of titanium dioxide nanoparticles in terrestrial and aquatic ecosystems: A microcosm study. *Environ Sci Technol.,* **2018**, *52*(21), 12757-12764.
[http://dx.doi.org/10.1021/acs.est.8b02970]

[28] Agtuca, B. J. Digital commons @ ESF gold nanoparticles in the environment : Studying the genetic toxicity and bioavailability in soils and hydroponic exposures with lycopersicon esculentum (Tomato 'Brandywine') *Beverly J. Agtuca,* **2014**.

[29] Li, B.; Chen, Y.; Liang, W.; Mu, L.; Bridges, W.C.; Jacobson, A.R.; Darnault, C.J.G. Influence of cerium oxide nanoparticles on the soil enzyme activities in a soil-grass microcosm system. *Geoderma,* **2017**, *299*, 54-62.

[http://dx.doi.org/10.1016/j.geoderma.2017.03.027]

[30] Ma, Y. Ecotoxicology and human environmental health trophic transfer and transformation of CeO_2 nanoparticles along a terrestrial food chain : Influence of exposure routes trophic transfer and transformation of CeO_2 nanoparticles along a terrestrial food chain. *Environ. Sci. Technol.,* **2018**, *52*(14), 7921-7927. [http://dx.doi.org/10.1021/acs.est.8b00596]

[31] Judy, J.D.; Unrine, J.M.; Bertsch, P.M. Evidence for biomagnification of gold nanoparticles within a terrestrial food chain. *Environ Sci Technol.,* **2011**, *45*(2), 776-781. [http://dx.doi.org/10.1021/es103031a]

[32] Wu, J.; Bosker, T.; Vijver, M.G.; Peijnenburg, W.J.G.M. Trophic transfer and toxicity of (Mixtures of) Ag and TiO_2 nanoparticles in the lettuce–terrestrial snail food chain. *Environ. Sci. Technol.,* **2021**, *55*(24), 16563-16572. [http://dx.doi.org/10.1021/acs.est.1c05006] [PMID: 34841858]

[33] Ding, X.; Pu, Y.; Tang, M.; Zhang, T. Environmental and health effects of graphene-family nanomaterials: Potential release pathways, transformation, environmental fate and health risks. *Nano Today,* **2022**, *42*, 101379. [http://dx.doi.org/10.1016/j.nantod.2022.101379]

[34] Francy, N.; Shanthakumar, S.; Chiampo, F. Using nanoscale zero-valent iron (nZVI) particles synthesized using green leaves: First results. *Processes,* **2020**, *8*, 1453.

[35] Su, B.; Lin, J.; Owens, G.; Chen, Z. Impact of green synthesized iron oxide nanoparticles on the distribution and transformation of As species in contaminated soil. *Environ. Pollut.,* **2019**. [http://dx.doi.org/10.1016/j.envpol.2019.113668] [PMID: 31796319]

[36] Bala, R.; Kalia, A.; Dhaliwal, S. S. Evaluation of efficacy of zno nanoparticles as remedial zinc nanofertilizer for rice. *J Soil Sci Plant Nutr,* **2019**, *19*, 379-389. [http://dx.doi.org/10.1007/s42729-019-00040-z]

[37] Jalali, E.N.P.M. Application of three nanoparticles (Al_2O_3, SiO_2 and TiO_2) for metal - contaminated soil remediation (measuring and modeling). *Int. J. Environ. Sci. Technol.,* **2018**, *16*, 7207-7220. [http://dx.doi.org/10.1007/s13762-018-2134-8]

[38] Cheng, J.; Sun, Z.; Yu, Y.; Li, X.; Li, T. Effects of modified carbon black nanoparticles on plant-microbe remediation of petroleum and heavy metal co-contaminated soils. *Int. J. Phytoremediation,* **2019**, *21*(7), 634-642. [http://dx.doi.org/10.1080/15226514.2018.1556581] [PMID: 30656965]

[39] Ates, M.; Demir, V.; Adiguzel, R.; Arslan, Z. Distribution of engineered titanium dioxide nanoparticles in goldfish (*Carassius auratus*). *J Nanomater.,* **2013**, 460518.

[40] Jovanović, B.; Whitley, E.M.; Kimura, K.; Crumpton, A.; Palić, D. Titanium dioxide nanoparticles enhance mortality of fish exposed to bacterial pathogens. *Environ. Pollut.,* **2015**, *203*, 153-164. [http://dx.doi.org/10.1016/j.envpol.2015.04.003] [PMID: 25884347]

[41] Taylor, P.; Street, M.; Wt, L. Water flea (*Moina macrocopa*), blood worm Toxicity, bioaccumulation and biomagnification of silver nanoparticles in green algae (*Chlorella* sp.), water flea (*Moina macrocopa*), blood worm (*Chironomus* spp.) and silver barb (*Barbonymus goniono*). In: *Chem. Speciat. Bioavailab*; , **2014**; 26, pp. (4)37-41. [http://dx.doi.org/10.3184/095422914X14144332205573]

[42] Lu, K.; Zhang, Z.; He, X.; Ma, Y.; Zhou, K.; Zhang, H.; Bai, W.; Ding, Y.; Wu, Z.; Zhao, Y.; Chai, Z. Bioavailability and distribution and of ceria nanoparticles in simulated aquatic ecosystems, quantification with a radiotracer technique. *J. Nanosci. Nanotechnol.,* **2010**, *10*(12), 8658-8662. [http://dx.doi.org/10.1166/jnn.2010.2494] [PMID: 21121379]

[43] Zhao, X. Distribution, bioaccumulation, trophic transfer, and in fl uences of CeO_2 nanoparticles in a constructed aquatic food web. *Environ Sci Technol.,* **2017**, *51*(9), 5205-5214.

[http://dx.doi.org/10.1021/acs.est.6b05875]

[44] Jackson, B. P.; Bugge, D.; Ranville, J. F.; Chen, C. Y. Bioavailability, toxicity, and bioaccumulation of quantum dot nanoparticles to the amphipod leptocheirus plumulosus *Environ. Sci. Technol.,* **2012**, *46*(10), 5550-5556.
[http://dx.doi.org/10.1021/es202864r]

[45] Gondikas, A. P.; Von Der Kammer, F.; Reed, R. B.; Wagner, S.; Ranville, J. F.; Hofmann, T. Release of TiO_2 nanoparticles from sunscreens into surface waters: A one-year survey at the old danube recreational lake. *Environ. Sci. Technol.,* **2014**, *48*(10), 5415-5422.

[46] Fan, W.; Cui, M.; Liu, H.; Wang, C.; Shi, Z.; Tan, C.; Yang, X. Nano-TiO_2 enhances the toxicity of copper in natural water to *Daphnia magna. Environ. Pollut.,* **2011**, *159*(3), 729-734.
[http://dx.doi.org/10.1016/j.envpol.2010.11.030] [PMID: 21177008]

[47] Adams, L.K.; Lyon, D.Y.; Mcintosh, A.; Alvarez, P.J.J. Comparative toxicity of nano-scale TiO_2, SiO_2 and ZnO water suspensions. In: *Water Sci Technol*; , **2006**; 54, pp. (11-12)327-334.
[http://dx.doi.org/10.2166/wst.2006.891]

[48] Zhu, X.; Tian, S.; Cai, Z. Toxicity assessment of iron oxide nanoparticles in zebrafish (*Danio rerio*) early life stages. *PLoS One,* **2012**, *7*(9), e46286.
[http://dx.doi.org/10.1371/journal.pone.0046286] [PMID: 23029464]

[49] Giroux, M.S.; Zahra, Z.; Salawu, O.A.; Burgess, R.M.; Ho, K.T.; Adeleye, A.S. Assessing the environmental effects related to quantum dot structure, function, synthesis and exposure. *Environ. Sci. Nano,* **2022**, *9*(3), 867-910.
[http://dx.doi.org/10.1039/D1EN00712B] [PMID: 35401985]

[50] Crane, R.A.; Dickinson, M.; Popescu, I.C.; Scott, T.B. Magnetite and zero-valent iron nanoparticles for the remediation of uranium contaminated environmental water. *Water Res.,* **2011**, *45*(9), 2931-2942.
[http://dx.doi.org/10.1016/j.watres.2011.03.012] [PMID: 21470652]

[51] Pala, I. R.; Brock, S. L. ZnS nanoparticle gels for remediation of Pb^{2+} and Hg^{2+} polluted water. *ACS Appl Mater Interfaces.,* **2012**, *4*(4), 2160-2167.

[52] Hfo, H. Heterostructured HfO_2/TiO_2 spherical nanoparticles for visible photocatalytic water remediation. *Mater. Lett.,* **2018**, *231*, 225-228.
[http://dx.doi.org/10.1016/j.matlet.2018.08.053]

[53] Mcdonald, K. Removal of arsenic from groundwater in West Bengal, India using CuO nanoparticle adsorbent removal of arsenic from groundwater in West Bengal, India using CuO nanoparticle adsorbent. *Environ Earth Sci,* **2014**, *73*, 3593-3601.
[http://dx.doi.org/10.1007/s12665-014-3645-3]

[54] Akbari, A.; Sabouri, Z.; Hosseini, H.A.; Hashemzadeh, A.; Khatami, M.; Darroudi, M. Effect of nickel oxide nanoparticles as a photocatalyst in dyes degradation and evaluation of effective parameters in their removal from aqueous environments. *Inorg. Chem. Commun.,* **2020**, *115*(January), 107867.
[http://dx.doi.org/10.1016/j.inoche.2020.107867]

[55] Etman, A.S.; Abdelhamid, H.N.; Yuan, Y.; Wang, L.; Zou, X.; Sun, J. Facile water-based strategy for synthesizing MoO $_{3-x}$ nanosheets: Efficient visible light photocatalysts for dye degradation. *ACS Omega,* **2018**, *3*(2), 2193-2201.
[http://dx.doi.org/10.1021/acsomega.8b00012] [PMID: 31458524]

[56] Liu, Z.; Ling, Q.; Cai, Y.; Xu, L.; Su, J.; Yu, K.; Wu, X.; Xu, J.; Hu, B.; Wang, X. Synthesis of carbon-based nanomaterials and their application in pollution management. *Nanoscale Adv.,* **2022**, *4*(5), 1246-1262.
[http://dx.doi.org/10.1039/D1NA00843A] [PMID: 36133685]

[57] Access, O. We are intechopen, the world's leading publisher of Open Access books Built by scientists, for scientists TOP 1 %.

[58] Buseck, P.R.; Adachi, K.; Qc, N.R. Molecular Sciences, School of (SMS)Geographical Sciences and Urban Planning, School of (SGSUP)Earth and Space Exploration, School of (SESE) In: *Nanoparticles in the atmosphere*; , **2008**; pp. 389-394.
[http://dx.doi.org/10.2113/gselements.4.6.389]

[59] Manuscript, A. Metal nanoparticles in the air: state of the art and future perspectives. *Environ. Sci.,* (11)**2020**,
[http://dx.doi.org/10.1039/D0EN00536C]

[60] Trejos, E.M. Geoscience frontiers volcanic emissions and atmospheric pollution: A study of nanoparticles. *Geosci. Front.,* **2021**, *12*, 746-755.
[http://dx.doi.org/10.1016/j.gsf.2020.08.013]

[61] Gonet, T.; Maher, B.A.; Accepted, J. Airborne, vehicle-derived Fe-bearing nanoparticles in the urban environment – A review. *Environ. Sci. Technol.,* **2019**, *53*(17), 9970-9991.
[http://dx.doi.org/10.1021/acs.est.9b01505] [PMID: 31381310]

[62] Kumar, P.; Gurjar, B.R.; Nagpure, A.S.; Harrison, R.M. Preliminary estimates of nanoparticle number emissions from road vehicles in megacity delhi and associated health impacts. In: *Environ Sci Technol*; , **2011**; 45, pp. (13)5514-5521.

[63] Morawska, L.; Ristovski, Z.; Jayaratne, E.R.; Keogh, D.U.; Ling, X. Ambient nano and ultrafine particles from motor vehicle emissions: Characteristics, ambient processing and implications on human exposure. *Atmos. Environ.,* **2008**, *42*(35), 8113-8138.
[http://dx.doi.org/10.1016/j.atmosenv.2008.07.050]

[64] Kirchstetter, T.W.; Harley, R.A.; Kreisberg, N.M.; Stolzenburg, M.R.; Hering, S.V. On-road measurement of fine particle and nitrogen oxide emissions from light- and heavy-duty motor vehicles. *Atmos. Environ.,* **1999**, *33*(18), 2955-2968.

[65] Pekkanen, J. Particulate air pollution and risk of st-segment depression during repeated submaximal exercise tests among subjects with coronary heart disease the exposure and risk assessment for fine and ultrafine particles in Ambient Air (ULTRA) study. *Circulation,* **2002**, *106*(8), 933-938.
[http://dx.doi.org/10.1161/01.CIR.0000027561.41736.3C]

[66] Liu, N.M.; Miyashita, L.; Maher, B.A.; McPhail, G.; Jones, C.J.P.; Barratt, B.; Thangaratinam, S.; Karloukovski, V.; Ahmed, I.A.; Aslam, Z.; Grigg, J. Evidence for the presence of air pollution nanoparticles in placental tissue cells. *Sci. Total Environ.,* **2021**, *751*, 142235.
[http://dx.doi.org/10.1016/j.scitotenv.2020.142235] [PMID: 33181987]

[67] Terzano, C.; Di Stefano, F.; Conti, V.; Graziani, E.; Petroianni, A. Air pollution ultrafine particles: Toxicity beyond the lung. *Eur. Rev. Med. Pharmacol. Sci.,* **2010**, *14*(10), 809-821.
[PMID: 21222367]

[68] Mohamed, E.F. Nanotechnology. *Environmental Management and Sustainable Development,* **2017**, *6*(2), 429-454.
[http://dx.doi.org/10.5296/emsd.v6i2.12047]

[69] Slimen, H. Photocatalytic decomposition of cigarette smoke using a TiO_2 -impregnated titanium mesh filter. In: *Build Environ*; , **2012**; pp. 587-590.

[70] Faraldos, M.; Kropp, R.; Anderson, M.A.; Sobolev, K. Photocatalytic hydrophobic concrete coatings to combat air pollution. *Catal. Today,* **2015**.
[http://dx.doi.org/10.1016/j.cattod.2015.07.025]

[71] Zhu, Y.; Yao, Y.; Luo, Z.; Pan, C.; Yang, J.; Fang, Y.; Deng, H.; Liu, C.; Tan, Q.; Liu, F.; Guo, Y. Nanostructured MoO_3 for efficient energy and environmental catalysis. *Molecules,* **2019**, *25*(1), 18.
[http://dx.doi.org/10.3390/molecules25010018] [PMID: 31861563]

[72] Hano, C.; Abbasi, B.H. Plant-based green synthesis of nanoparticles: Production, characterization and applications. *Biomolecules,* **2021**, *12*(1), 31.
[http://dx.doi.org/10.3390/biom12010031] [PMID: 35053179]

[73] Narayanan, K.B.; Sakthivel, N. Green synthesis of biogenic metal nanoparticles by terrestrial and aquatic phototrophic and heterotrophic eukaryotes and biocompatible agents. *Adv. Colloid Interface Sci.,* **2011**, *169*(2), 59-79. [http://dx.doi.org/10.1016/j.cis.2011.08.004] [PMID: 21981929]

[74] Sharma, G.; Kandasubramanian, B. Molecularly imprinted polymers for selective recognition and extraction of heavy metal ions and toxic dyes. *J. Chem. Eng. Data,* **2020**, *65*(2), 396-418. [http://dx.doi.org/10.1021/acs.jced.9b00953]

[75] Ali, A.; Romeh, A. Green silver nanoparticles for enhancing the phytoremediation of soil and water contaminated by fipronil and degradation products. *Water Air Soil Pollut.,* **2018**, *229*, 147.

[76] Abd El-Aziz, A.R.; Al-Othman, M.R.; Mahmoud, M.A. Degradation of DDT by gold nanoparticles synthesised using *Lawsonia inermis* for environmental safety. *Biotechnol. Biotechnol. Equip.,* **2018**, *32*(5), 1174-1182. [http://dx.doi.org/10.1080/13102818.2018.1502051]

[77] Aestivum, T. Toxicity and bioavailability of copper nanoparticles to the terrestrial plants mung bean (*Phaseolus radiatus*) and wheat (*Triticum aestivum*): Plant agar test for water-insoluble nanoparticles. In: *Environ Toxicol Chem*; , **2008**; 27, pp. (9)1915-1921.

[78] Ates, M.; Danabas, D.; Ertit Tastan, B.; Unal, I.; Cicek Cimen, I.C.; Aksu, O.; Kutlu, B.; Arslan, Z. Assessment of oxidative stress on artemia salina and daphnia magna after exposure to Zn and ZnO nanoparticles. *Bull. Environ. Contam. Toxicol.,* **2020**, *104*(2), 206-214. [http://dx.doi.org/10.1007/s00128-019-02751-6] [PMID: 31748865]

[79] Sadanandan, A.; Ramesh, M.; Saravanan, M.; Krishnan, R.; Bharathi, S.; Nataraj, D. Iron oxide nanoparticles to an Indian major carp, Labeo rohita : Impacts on hematology, iono regulation and gill Na + / K + ATPase activity. *J. King Saud Univ. Sci.,* **2014**. [http://dx.doi.org/10.1016/j.jksus.2014.11.002]

[80] Kusmierek, E. A CeO_2 semiconductor as a photocatalytic and photoelectrocatalytic material for the remediation of pollutants in industrial wastewater. *RE:view,* **2020**, 1-54.

[81] Oberdo, E.; Zhu, S.; Blickley, T.M. Ecotoxicology of carbon-based engineered nanoparticles : Effects of fullerene (C_{60}) on aquatic organisms. *Carbon,* **2006**, *44*(6), 1112-1120. [http://dx.doi.org/10.1016/j.carbon.2005.11.008]

[82] Freixa, A.; Acuña, V.; Sanchís, J.; Farré, M.; Barceló, D.; Sabater, S. Ecotoxicological effects of carbon based nanomaterials in aquatic organisms. *Sci. Total Environ.,* **2018**, *619-620*, 328-337. [http://dx.doi.org/10.1016/j.scitotenv.2017.11.095] [PMID: 29154051]

[83] Samy, A. *Synthesis of α -Al_2O_3 Nanoparticles From Pepsi Cans Wastes and Its Fungicidal Effect on Some Mycotoxins Producing Fungal Isolates,* **2021**.

[84] Ogolo, N.A.; Onyekonwu, M.O. Effect of aluminum oxide nanoparticles on biotic factors of the environment-A Review İD İD. *Arab. J. Chem. Environ. Res.,* **2021**, *08*(2), 247-258.

[85] Sousa, C.A.; Soares, H.M.V.M.; Soares, E.V. Toxic effects of nickel oxide (NiO) nanoparticles on the freshwater alga *Pseudokirchneriella subcapitata. Aquat. Toxicol.,* **2018**, *204*, 80-90. [http://dx.doi.org/10.1016/j.aquatox.2018.08.022] [PMID: 30205248]

[86] Aljedaani, R.O.; Kosa, S.A.; Abdel Salam, M. Ecofriendly green synthesis of copper (II) oxide nanoparticles using *corchorus olitorus* leaves (molokhaia) extract and their application for the environmental remediation of direct violet dye *via* advanced oxidation process. *Molecules,* **2022**, *28*(1), 16. [http://dx.doi.org/10.3390/molecules28010016] [PMID: 36615210]

[87] Suhani, I.; Sahab, S.; Srivastava, V.; Singh, R.P. Impact of cadmium pollution on food safety and human health. *Curr. Opin. Toxicol.,* **2021**, *27*, 1-7. [http://dx.doi.org/10.1016/j.cotox.2021.04.004]

[88] Chávez-Calderón, A.; Paraguay-Delgado, F.; Orrantia-Borunda, E.; Luna-Velasco, A. Size effect of SnO_2 nanoparticles on bacteria toxicity and their membrane damage. *Chemosphere,* **2016**, *165*, 33-40. [http://dx.doi.org/10.1016/j.chemosphere.2016.09.003] [PMID: 27639075]

[89] Elango, G.; Roopan, S.M. Efficacy of SnO_2 nanoparticles toward photocatalytic degradation of methylene blue dye. *J. Photochem. Photobiol. B,* **2016**, *155*, 34-38. [http://dx.doi.org/10.1016/j.jphotobiol.2015.12.010] [PMID: 26724726]

[90] Tran, T.V.; Nguyen, D.T.C.; Kumar, P.S.; Din, A.T.M.; Jalil, A.A.; Vo, D.V.N. Green synthesis of ZrO_2 nanoparticles and nanocomposites for biomedical and environmental applications: A review. *Environ. Chem. Lett.,* **2022**, *20*(2), 1309-1331. [http://dx.doi.org/10.1007/s10311-021-01367-9] [PMID: 35035338]

[91] Hichem, N.; Hadjer, Z.; Fateh, S.; Feriel, L.; Wang, Z. The potential exposure and hazards of zirconia nanoparticles: A review. *Ecotoxicol. Environ. Contam.,* **2022**, *17*(1), 1-21. [http://dx.doi.org/10.5132/eec.2022.01.01]

[92] Yang, J.; Song, Z.; Ma, J.; Han, H.; Han, H. Toxicity of molybdenum-based nanomaterials on the soybean–rhizobia symbiotic system: Implications for nutrition. *ACS Appl. Nano Mater.,* **2020**, *3*(6), 5773-5782. [http://dx.doi.org/10.1021/acsanm.0c00943]

[93] Ni, Z.; Wang, Q.; Guo, Y.; Liu, H.; Zhang, Q. Research progress of tungsten oxide-based catalysts in photocatalytic reactions. *Catalysts,* **2023**, *13*(3), 579. [http://dx.doi.org/10.3390/catal13030579]

[94] Popov, A. L. Photo-induced toxicity of tungsten oxide photochromic nanoparticles. *J. Photochem. Photobiol. B Biol.,* **2018**, *178*, 395-403. [http://dx.doi.org/10.1016/j.jphotobiol.2017.11.021]

[95] Kong, I.C.; Ko, K.S.; Koh, D.C.; Chon, C.M. Comparative effects of particle sizes of cobalt nanoparticles to nine biological activities. *Int. J. Mol. Sci.,* **2020**, *21*(18), 6767. [http://dx.doi.org/10.3390/ijms21186767] [PMID: 32942696]

[96] Bibi, I.; Nazar, N.; Iqbal, M.; Kamal, S.; Nawaz, H.; Nouren, S.; Safa, Y.; Jilani, K.; Sultan, M.; Ata, S.; Rehman, F.; Abbas, M. Green and eco-friendly synthesis of cobalt-oxide nanoparticle: Characterization and photo-catalytic activity. *Adv. Powder Technol.,* **2017**, *28*(9), 2035-2043. [http://dx.doi.org/10.1016/j.apt.2017.05.008]

[97] Sørensen, S.N.; Engelbrekt, C.; Lützhøft, H.C.H.; Jiménez-Lamana, J.; Noori, J.S.; Alatraktchi, F.A.; Delgado, C.G.; Slaveykova, V.I.; Baun, A. A multimethod approach for investigating algal toxicity of platinum nanoparticles. *Environ. Sci. Technol.,* **2016**, *50*(19), 10635-10643. [http://dx.doi.org/10.1021/acs.est.6b01072] [PMID: 27577171]

[98] Naseer, A.; Ali, A.; Ali, S.; Mahmood, A.; Kusuma, H.S.; Nazir, A.; Yaseen, M.; Khan, M.I.; Ghaffar, A.; Abbas, M.; Iqbal, M. Biogenic and eco-benign synthesis of platinum nanoparticles (Pt NPs) using plants aqueous extracts and biological derivatives: Environmental, biological and catalytic applications. *J. Mater. Res. Technol.,* **2020**, *9*(4), 9093-9107. [http://dx.doi.org/10.1016/j.jmrt.2020.06.013]

[99] Lee, W.M.; An, Y.J. Evidence of three-level trophic transfer of quantum dots in an aquatic food chain by using bioimaging. *Nanotoxicology,* **2015**, *9*(4), 407-412. [http://dx.doi.org/10.3109/17435390.2014.948517] [PMID: 25119416]

[100] Niu, C.; Yao, Z.; Jiang, S. Synthesis and application of quantum dots in detection of environmental contaminants in food: A comprehensive review. *Sci. Total Environ.,* **2023**, *882*(January), 163565. [http://dx.doi.org/10.1016/j.scitotenv.2023.163565] [PMID: 37080319]

CHAPTER 9

Investigation for Optimum Site for Adsorption and Population Effect of Lithium on Silicene Monolayer

Lalhumhima[1,2], Lalmuanchhana[1,2], Shivraj Gurung[1], Lalrinthara Pachuau[1], Lalmuanpuia Vanchhawng[1], Zodinmawia[2] and **Lalhriat Zuala[1,*]**

[1] *Department of Physics, Physical Sciences Research Centre (PSRC), Pachhunga University College, Aizawl, Mizoram, India*

[2] *Department of Physics, Mizoram University, Aizawl, Mizoram, India*

Abstract: This investigation holds the search for optimum sites for Lithium on Silicene Monolayer, by studying the adsorption energy on putting the Lithium atom on different possible sites. The center of the hexagonal structure in silicene was found to be the most favourable site. Using transition state search (TSS) the optimized stable state is confirmed. A low Diffusion Energy Barrier (DEB) of 0.348 eV indicates that the adsorption of Li over silicene can occur easily. Multiple adsorptions are considered, up to 4 Li atoms upon silicene substrate. The metallic property of pristine silicene is maintained throughout Li atom adsorption. Large adsorption energy in each of the adsorption suggests that silicene may be promising for Li-based battery material.

Keywords: Diffusion energy barrier, Lithium, Lithium-ion battery , Silicene monolayer.

INTRODUCTION

Over the years, electronics have been growing, and the search for newer technology comes with various challenges to make various components compact which involves an even more compact power source. Since the discovery of Lithium-ion Battery (LIB), its application has been widely explored for many electronics and automobiles [1 - 3], and it has been used conventionally in almost all electronics. For a long time, graphite has been used as an anode (the negative terminal of the battery); however, due to various limitations such as thickness, low capacity and safety concerns, newer materials such as SiO_2, alloy anodes, and conversion-type transition-metal compounds have been proposed by various researchers [4 - 6].

* **Corresponding author Lalhriat Zuala:** Department of Physics, Physical Sciences Research Centre (PSRC), Pachhunga University College, Aizawl, Mizoram, India; E-mail: hriata@pucollege.edu.in

Dibya Prakash Rai, Kingsley O. Obodo, & Jitendra Pal Singh (Eds.)

2D materials have found a wide area of applicability due to their large surface-to-volume ratio. Among them, silicene is one of the most considered materials. Silicene is a low-dimensional buckled monolayer of silicon. Lately, experimentally-fabricated materials with buckling structures have been developed. In fact, in recent experimental works, silicene monolayers have been synthesized by various methods such as Molecular-beam-epitaxy (MBE), by condensation of Silicon atoms on silver crystals in a vacuum chamber, Wet-chemical exfoliation method, and various methods of synthesis have also been presented by Balendhran. S. *et al.*, [7 - 10].

In general, the so-called silicene, a silicon (Si) equivalent of graphene, also receives a great deal of scientific attention. Similar to graphene, silicene is composed of hexagonal rings. However, they are not planar due to symmetric buckling caused by a pseudo-Jahn Teller effect forbidding them from being flattened; the instability of the high buckled and the planar silicene has also been proven by Roome. N.J. *et al.,* [11]. Interestingly, due to its buckled honeycomb structure compared to graphene, silicene demonstrates a substantially higher level of chemical reactivity, as well as significantly stronger adsorption of atoms and molecules [12 - 18] than graphene with promising new silicene-based electronic device applications [19, 20], Li-ion storage batteries [14, 21 - 23], hydrogen storage [24 - 26], catalysts [18, 27], thin-film solar cell absorbers [28, 29], and hydrogen separation membranes [30, 31], gas sensor and detection [32 - 34]. While sp^2 hybridization is favoured for 2D graphene, Si strongly favours sp^3 [35]. Li-ion secondary batteries are a prospective form of energy storage that can be used in both mobile and grid applications [36, 37]. Their energy density is significantly influenced by the individual charge capacities of the constituent electrodes [38]. Silicene has long been proposed as a substitute for conventional graphite electrodes in LIB. Low specific capacity is one of the few drawbacks of a graphite electrode (372 mAh g^{-1}) [39]. It is worth noting that the clustering of lithium atoms reduces the chargeability of the LIB and is also the primary cause of deterioration of battery health. A high theoretical capacity of silicon (4200 mAh g^{-1}) and a suitable operating voltage make it one of the most promising anode materials for a LIB [40]. The interaction between silicene and Li is stronger than that of graphene [41]. These characteristics have indicated the promising capabilities of silicene as an anode in lithium-ion batteries.

Specific optimum site adsorption of fully occupied lithium on silicene monolayers has not been observed in our literature survey. In this work, we have studied the adsorption for a single lithium atom to find the most optimum site considering four different sites. A transition state search is performed, taking energy footprints along the transition. We have interpreted its electronic properties and determined its character for an anode in Li-ion battery by implementing the optimum site for specific site adsorption and checking the effects of lithium population on silicene. Although this is not a large scale, this study provides a rough description of how a lithium atom reacts with a silicene monolayer monolayer (Fig. **1**).

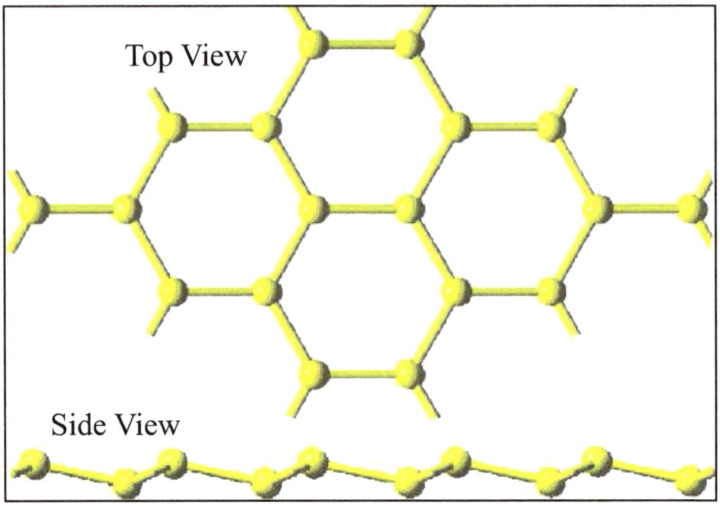

Fig. (1). Top view and side view of silicene.

COMPUTATIONAL DETAILS

In this work, all the calculations were based on density functional theory and the Generalized Gradient Approximation made simple by Perdew, Burke and Ernzerhof [42] exchange-correlation function was incorporated to describe the interactions which are integrated with VASP (Vienna Ab-initio Simulation Package). Van der Waals interactions are taken into consideration using DFT-D3 dispersion correction. Spin-polarized magnetic calculations were carried out (in case the compound is magnetic) using 'normal' precision and a plane-wave cutoff energy of 400 eV. Electronic iterations convergence of 10^{-5} eV is employed using the Fast (Davidson and RMM-DIIS) algorithm and reciprocal space projection operators. A k-mesh of 4x4x1 was employed, and Tetrahedron with Bloechl corrections [43] was chosen for the smearing method.

The study of adsorption energy helps in finding the most favourable site for adsorption. For this, we considered 8 Si (P1) 2×2×1 lattice. Geometric optimization is carried out, from which VASP energy is considered for the energy of the systems. Equation 1 (given below) is incorporated for determining adsorption energy (E_{ads}).

$$E_{ads} = \frac{(E_{sil} + N_{Li}E_{Li} - E_{Total})}{N_{Li}} \quad (1)$$

For further studying the effects of multiple lithium adsorption, a bigger 3×3×1 structure, which consists of 18 Si atoms, has been considered. This system could fit 4 Li atoms in each optimum site. Geometric optimization is carried out over again, from which the VASP energies are taken, applying the values for the same formula as above, and the adsorption energies are calculated using those energies. Bandstructure and density of states (DOS) have been calculated for the electronic properties for each inclusion of 1 to 4 lithium atoms.

RESULTS AND DISCUSSIONS

To understand the optimal binding configuration of Li atoms on the pristine silicene the adsorption studies were performed. Upon optimizing the structure, the height difference between the upper and the lower silicon atoms (the buckling parameter) was 0.407 Å, and the distance between each Si-Si bond was 2.274 Å, which agrees when compared to other reported data where the buckling parameter is reported as 0.44 Å and the bond distance have been reported as 2.2-2.4 Å [19]. Four adsorption sites are considered for Li molecule upon the relaxed Silicene monolayer. The center of the silicene hexagon (the hollow site) is named Site 1, the top of the upper Si atom is Site 2, the top of the lower Si atom is Site 3, and above the center of the Si-Si bond is Site 4 (Fig. **2**).

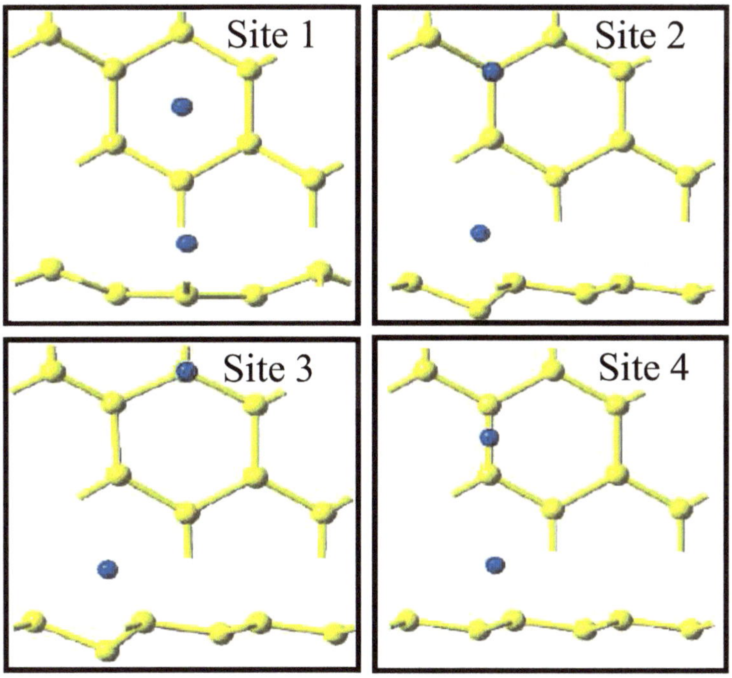

Fig. (2). Sites adopted for single lithium adsorption.

For the adsorbed molecules on the silicene monolayer, the adsorption energy is calculated by the relation given in Eqn. 1. Our calculation shows that the adsorption energy for Li molecules in the silicene ML is lowest in Site 1, *i.e.*, the center of the Silicene hexagon, provided in Table **1**, with a value of -2.2467 eV, when lithium is subjected to this configuration three of the nearest silicon atoms form a bond. For the proposed site 1, the distance of the Li-Si bond appears to be approximately 2.8 Å. We consider this site the most optimal site for lithium adsorption on silicene which was similarly observed by Xu, S. *et al.,* and Wang, X. *et al.,* [44, 45] and the majority of adsorption studies on 2D materials have revealed that the hollow site is the most optimum site or the defect site if available. Although Site 2 and Site 4 were considered, during structural optimization, the atoms shift down above the lower Si atom, forming a bond with a distance of approximately 2.65 Å between the Li-Si atoms, much like the considered Site 3. The calculated adsorption energies were -2.2165 eV, -2.2171 eV and -2.2170 eV for Site 2, Site 3 and Site 4, respectively. Almost equal adsorption energy for these sites can be explained by the selective movement of the atom on optimization.

Table 1. Observed adsorption energy for different sites.

Adsorption Site	Adsorption Energy (Eads)
Site 1 (Hollow site)	-2.2467 eV
Site 2 (Above upper Si)	-2.2165 eV
Site 3 (Above lower Si)	-2.2171 eV
Site 4 (Above the bond)	-2.2170 eV

Transition state search has been implemented for Diffusion Energy Search, which calculates the energy gradient along the transition of lithium from one optimum site to the other by going over the upper silicon atom and to the Si-Si bridge and the lower silicon atom and the other optimum site. This study is important to describe the mobility of lithium-ion on the silicene since charging characteristics rely on it. We have observed the highest diffusion energy above the bridge which is the transition state in Fig. (**3**). The total energy gradient along the diffusion pathway is provided. The beginning (initial state) and the end of the graph (final state) confirm the optimum site for adsorption as the hexagonal hollow site.The transition from one optimal site to the other is much easier, as indicated by the negative diffusion energy in the final state. Low Diffusion Energy of 0.348 eV shows that binding could occur without much effort.

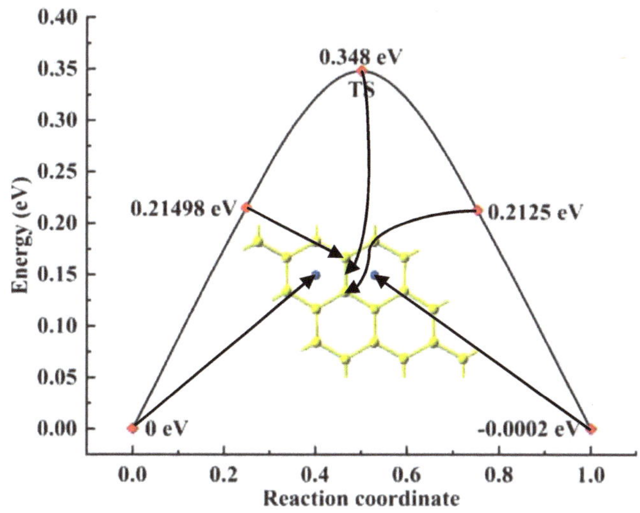

Fig. (3). Diffusion energy barrier as a function of diffusion coordinate and diffusion pathway of Li atom on Silicene.

To understand how the number of lithium atoms affects the adsorption energy, the optimum absorption site and its energy for lithium in site 1 are studied by creating a 3×3 supercell that consists of 18 silicon atoms, as shown in Fig. (**4**). On simulating the adsorption from 1 to 4 Li atoms, the adsorption energy is as given in Table **2**. The adsorption energy for one lithium atom is -2.393 eV, and it increases when two and three lithium atoms are adsorbed. However, in the case of four lithium atoms, the energy strangely decreases again. This could be explained by the ionic repulsion between the lithium atoms, which makes the lithium atoms' adsorption more helpful [46].

Bandstructure for the above structure on lithium adsorption has been calculated to identify the electronic behaviour of multiple lithium adsorption. Γ-B-K-Γ points were considered for the high symmetry points of the first Brillouin zone. Although pristine silicene is reported to exhibit a Dirac fermion-like nature with ð-ð* band, which appears to vanish near the fermi line in Density of States (DOS) [47] like those in the graphene sheet, due to the concentration of lithium in this configuration, the electronic properties show a metallic behaviour. The metallic behaviour has also been proven by Xu, S. *et. al.,* and Wang X., *et. al.*, [44, 45] for different concentrations of lithium and the semiconducting nature is only exhibited particularly when lithium and silicon composition is one-to-one. It is acceptable to state that lithium atoms impart metallicity to what was previously a semiconducting compound. On increasing the concentration of Li atoms, silicene maintains its metallicity throughout the multiple lithium adsorption. Although spin-polarised calculations were carried out, the band structures show only negligible differences, which is visible in the DOS. Therefore, band structures are only shown for spin-up calculation.

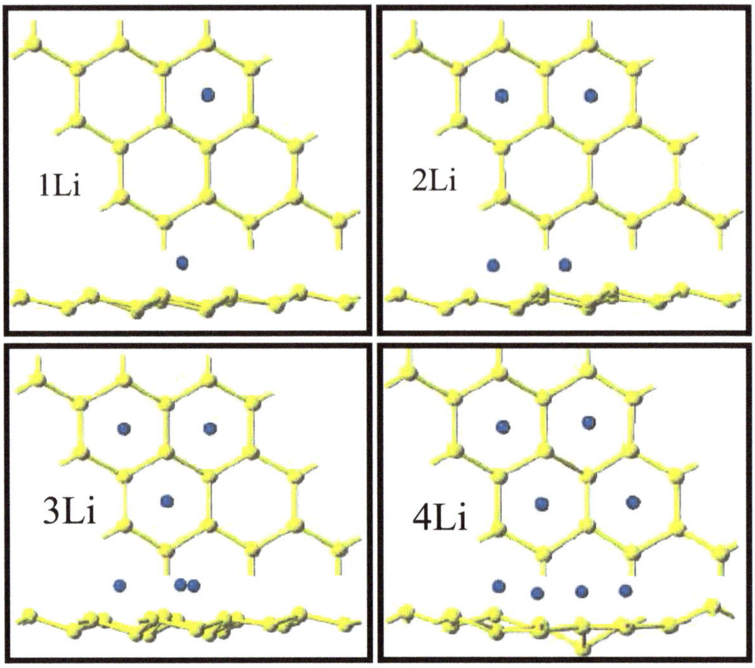

Fig. (4). Adsorption by increasing no. of Li atoms.

Table 2. Observation by increasing the no. of Li on Si (18) with 4 optimum adsorption sites.

No. of Li Atoms	Adsorption Energy (Eads)
1	-2.393 eV
2	-2.348 eV
3	-2.334 eV
4	-2.377 eV

Lastly, DOS calculation was carried out for the same structure for which the band structures were calculated. DOS for spin-polarised calculations are shown in Fig. (5) on the right side of each corresponding band structure. The contributions from Si-p orbitals dominate near the fermi levels in every case, while minority contributions are attributed to Li-s orbitals. The strong hybridization between the Si-p and the Li-s orbital signifies typical chemical adsorption. Since the spin states cancel at almost every level of energy this compound is suggested as a non-magnetic metal.

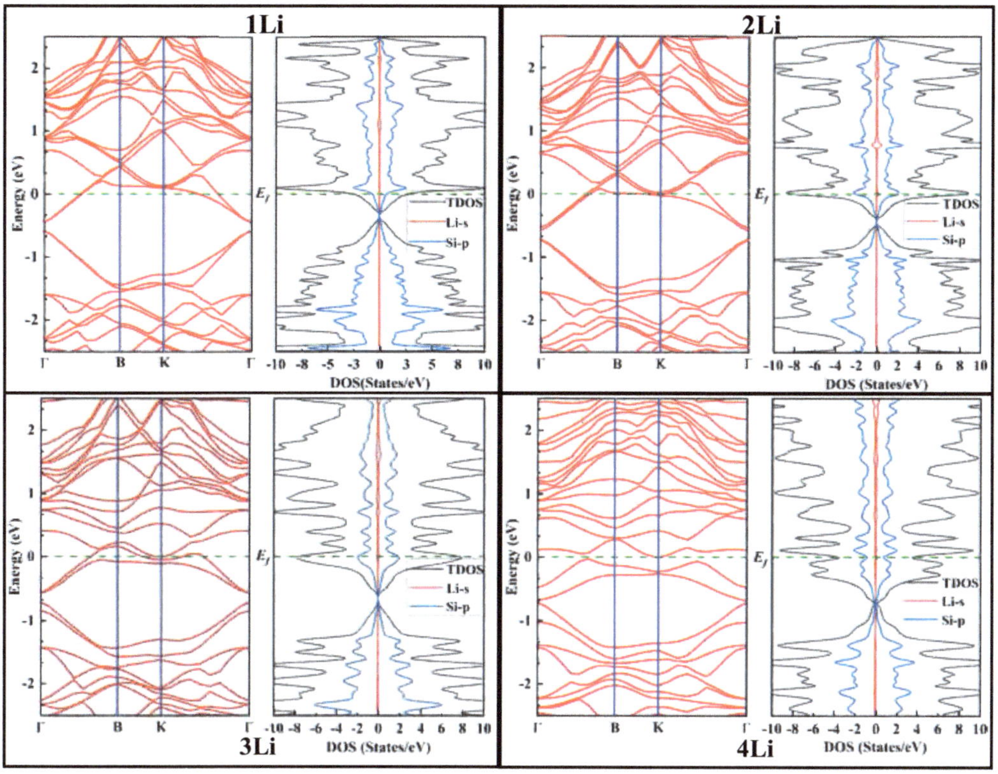

Fig. (5). Bandstructure and corresponding DOS for 1 Li to 4 Li atoms adsorption on silicene.

CONCLUSION

The adsorption on the hollow hexagonal site was the most favourable setting, as shown by the results of the adsorption investigation. This shows that the lithium atom is more likely to head towards the hollow region when introduced to the silicene monolayer. Based on the electronic properties and adsorption studies of the larger system, the system's properties remain unchanged even after lithium atoms have filled all of the optimal sites for adsorption. With said adsorption energies in the effective range for a decent anode for Li-ion battery and maintaining its property throughout the adsorption, silicene monolayer can indeed be an excellent anode.

REFERENCES

[1] Chen, W.; Liang, J.; Yang, Z.; Li, G. A review of lithium-ion battery for electric vehicle applications and beyond. *Energy Procedia,* **2019**, *158*, 4363-4368. [http://dx.doi.org/10.1016/j.egypro.2019.01.783]

[2] Masias, A.; Marcicki, J.; Paxton, W.A. Opportunities and challenges of lithium ion batteries in automotive applications. *ACS Energy Lett.,* **2021**, *6*(2), 621-630.

[http://dx.doi.org/10.1021/acsenergylett.0c02584]

[3] Megahed, S.; Ebner, W. Lithium-ion battery for electronic applications. *J. Power Sources,* **1995**, *54*(1), 155-162. [http://dx.doi.org/10.1016/0378-7753(94)02059-C]

[4] Chang, W.S.; Park, C-M.; Kim, J-H.; Kim, Y-U.; Jeong, G.; Sohn, H-J. Quartz (SiO_2): A new energy storage anode material for Li-ion batteries. *Energy Environ. Sci.,* **2012**, *5*(5), 6895-6899. [http://dx.doi.org/10.1039/c2ee00003b]

[5] Barbosa, J.C.; Gonçalves, R.; Costa, C.M.; Lanceros-Mendez, S. Recent advances on materials for lithium-ion batteries. *Energies,* **2021**, *14*(11), 3145. [http://dx.doi.org/10.3390/en14113145]

[6] Nzereogu, P.U.; Omah, A.D.; Ezema, F.I.; Iwuoha, E.I.; Nwanya, A.C. Anode materials for lithium-ion batteries: A review. *Appl. Surf. Sci. Adv.,* **2022**, *9*, 100233. [http://dx.doi.org/10.1016/j.apsadv.2022.100233]

[7] Zhuang, J.; Xu, X.; Peleckis, G.; Hao, W.; Dou, S.X.; Du, Y. Silicene: A promising anode for lithium-ion batteries. *Adv. Mater.,* **2017**, *29*(48), 1606716. [http://dx.doi.org/10.1002/adma.201606716] [PMID: 28328167]

[8] Ornes, S. Beyond graphene. *Proc. Natl. Acad. Sci. USA,* **2015**, *112*(43), 13128-13130. [http://dx.doi.org/10.1073/pnas.1515378112] [PMID: 26508206]

[9] Lin, H.; Qiu, W.; Liu, J.; Yu, L.; Gao, S.; Yao, H.; Chen, Y.; Shi, J. Silicene: Wet-chemical exfoliation synthesis and biodegradable tumor nanomedicine. *Adv. Mater.,* **2019**, *31*(37), 1903013. [http://dx.doi.org/10.1002/adma.201903013] [PMID: 31347215]

[10] Balendhran, S.; Walia, S.; Nili, H.; Sriram, S.; Bhaskaran, M. Elemental analogues of graphene: Silicene, germanene, stanene, and phosphorene. *Small,* **2015**, *11*(6), 640-652. [http://dx.doi.org/10.1002/smll.201402041] [PMID: 25380184]

[11] Roome, N.J.; Carey, J.D. Beyond graphene: Stable elemental monolayers of silicene and germanene. *ACS Appl. Mater. Interfaces,* **2014**, *6*(10), 7743-7750. [http://dx.doi.org/10.1021/am501022x] [PMID: 24724967]

[12] Kang, J.; Wu, F.; Li, J. Symmetry-dependent transport properties and magnetoresistance in zigzag silicene nanoribbons. *Appl. Phys. Lett.,* **2012**, *100*(23), 233122. [http://dx.doi.org/10.1063/1.4726276]

[13] Jose, D.; Datta, A. Structures and chemical properties of silicene: Unlike graphene. *Acc. Chem. Res.,* **2014**, *47*(2), 593-602. [http://dx.doi.org/10.1021/ar400180e] [PMID: 24215179]

[14] Kara, A.; Enriquez, H.; Seitsonen, A.P.; Lew Yan Voon, L.C.; Vizzini, S.; Aufray, B.; Oughaddou, H. A review on silicene — New candidate for electronics. *Surf. Sci. Rep.,* **2012**, *67*(1), 1-18. [http://dx.doi.org/10.1016/j.surfrep.2011.10.001]

[15] Lin, X.; Ni, J. Much stronger binding of metal adatoms to silicene than to graphene: A first-principles study. *Phys. Rev. B Condens. Matter Mater. Phys.,* **2012**, *86*(7), 075440. [http://dx.doi.org/10.1103/PhysRevB.86.075440]

[16] Sivek, J.; Sahin, H.; Partoens, B.; Peeters, F.M. Adsorption and absorption of boron, nitrogen, aluminum, and phosphorus on silicene: Stability and electronic and phonon properties. *Phys. Rev. B Condens. Matter Mater. Phys.,* **2013**, *87*(8), 085444. [http://dx.doi.org/10.1103/PhysRevB.87.085444]

[17] Tritsaris, G.A.; Kaxiras, E.; Meng, S.; Wang, E. Adsorption and diffusion of lithium on layered silicon for Li-ion storage. *Nano Lett.,* **2013**, *13*(5), 2258-2263. [http://dx.doi.org/10.1021/nl400830u] [PMID: 23611247]

[18] Li, C.; Yang, S.; Li, S-S.; Xia, J-B.; Li, J. Au-decorated silicene: Design of a high-activity catalyst

toward co oxidation. *J. Phys. Chem. C,* **2013**, *117*(1), 483-488. [http://dx.doi.org/10.1021/jp310746m]

[19] Kharadi, M.A.; Malik, G.F.A.; Khanday, F.A.; Shah, K.A.; Mittal, S.; Kaushik, B.K. Review—silicene: From material to device applications. *ECS J. Solid State Sci. Technol.,* **2020**, *9*(11), 115031. [http://dx.doi.org/10.1149/2162-8777/abd09a]

[20] Molle, A.; Grazianetti, C.; Tao, L.; Taneja, D.; Alam, M.H.; Akinwande, D. Silicene, silicene derivatives, and their device applications. *Chem. Soc. Rev.,* **2018**, *47*(16), 6370-6387. [http://dx.doi.org/10.1039/C8CS00338F] [PMID: 30065980]

[21] Galashev, A.Y.; Ivanichkina, K.A. Silicene anodes for lithium-ion batteries on metal substrates. *J. Electrochem. Soc.,* **2020**, *167*(5), 050510. [http://dx.doi.org/10.1149/1945-7111/ab717a]

[22] Nematzadeh, M.; Massoudi, A.; Nangir, M. Optimization of silicene oxidation as lithium-ion battery anode. *Mater. Today Proc.,* **2021**, *42*, 1588-1591. [http://dx.doi.org/10.1016/j.matpr.2020.03.210]

[23] Vargas, D.D.; Cardoso, G.L.; Piquini, P.C.; Ahuja, R.; Baierle, R.J. 2D dumbbell silicene as a high storage capacity and fast ion diffusion anode for Li-ion batteries. *ACS Appl. Mater. Interfaces,* **2022**, *14*(41), 47262-47271. [http://dx.doi.org/10.1021/acsami.2c13535] [PMID: 36205921]

[24] Zhang, Y.; Liu, P.; Zhu, X. Li decorated penta-silicene as a high capacity hydrogen storage material: A density functional theory study. *Int. J. Hydrogen Energy,* **2021**, *46*(5), 4188-4200. [http://dx.doi.org/10.1016/j.ijhydene.2020.10.193]

[25] Hussain, T.; Chakraborty, S.; Ahuja, R. Metal-functionalized silicene for efficient hydrogen storage. *ChemPhysChem,* **2013**, *14*(15), 3463-3466. [http://dx.doi.org/10.1002/cphc.201300548] [PMID: 24009141]

[26] Kalwar, B.A.; Fangzong, W.; Ahmed, I.; Saeed, M.H. Ti atom doped single vacancy silicene for hydrogen energy storage: DFT study. *J. Chin. Chem. Soc.,* **2021**, *68*(12), 2243-2253. [http://dx.doi.org/10.1002/jccs.202100369]

[27] Zhou, S.; Pei, W.; Zhao, J.; Du, A. Silicene catalysts for CO_2 hydrogenation: The number of layers controls selectivity. *Nanoscale,* **2019**, *11*(16), 7734-7743. [http://dx.doi.org/10.1039/C9NR01336A] [PMID: 30949654]

[28] Huang, B.; Xiang, H.J.; Wei, S.H. Chemical functionalization of silicene: Spontaneous structural transition and exotic electronic properties. *Phys. Rev. Lett.,* **2013**, *111*(14), 145502. [http://dx.doi.org/10.1103/PhysRevLett.111.145502] [PMID: 24138253]

[29] Lv, J.; Xu, M.; Lin, S.; Shao, X.; Zhang, X.; Liu, Y.; Wang, Y.; Chen, Z.; Ma, Y. Direct-gap semiconducting tri-layer silicene with 29% photovoltaic efficiency. *Nano Energy,* **2018**, *51*, 489-495. [http://dx.doi.org/10.1016/j.nanoen.2018.06.079]

[30] Hu, W.; Wu, X.; Li, Z.; Yang, J. Porous silicene as a hydrogen purification membrane. *Phys. Chem. Chem. Phys.,* **2013**, *15*(16), 5753-5757. [http://dx.doi.org/10.1039/c3cp00066d] [PMID: 23486830]

[31] Hu, W.; Li, Z.; Yang, J. Water on silicene: A hydrogen bond-autocatalyzed physisorption-chemisorption-dissociation transition. *Nano Res.,* **2017**, *10*(7), 2223-2233. [http://dx.doi.org/10.1007/s12274-016-1411-4]

[32] Nguyen, D.K.; Hoang, D.Q.; Hoat, D.M. Exploring a silicene monolayer as a promising sensor platform to detect and capture NO and CO gas. *RSC Advances,* **2022**, *12*(16), 9828-9835. [http://dx.doi.org/10.1039/D2RA00442A] [PMID: 35424916]

[33] Prasongkit, J.; Amorim, R.G.; Chakraborty, S.; Ahuja, R.; Scheicher, R.H.; Amornkitbamrung, V. Highly sensitive and selective gas detection based on silicene. *J. Phys. Chem. C,* **2015**, *119*(29),

16934-16940.
[http://dx.doi.org/10.1021/acs.jpcc.5b03635]

[34] Walia, G.K.; Randhawa, D.K.K.; Malhi, K.S. Rise of silicene and its applications in gas sensing. *J. Mol. Model.*, **2021**, *27*(10), 277.
[http://dx.doi.org/10.1007/s00894-021-04892-0] [PMID: 34482432]

[35] Wang, H.; Wu, M.; Tian, Z.; Xu, B.; Ouyang, C. First principles study of penta-siligraphene as high-performance anode material for li-ion batteries. *Nanoscale Res. Lett.*, **2019**, *14*(1), 260.
[http://dx.doi.org/10.1186/s11671-019-3097-5] [PMID: 31363867]

[36] Choi, D.; Shamim, N.; Crawford, A.; Huang, Q.; Vartanian, C.K.; Viswanathan, V.V.; Paiss, M.D.; Alam, M.J.E.; Reed, D.M.; Sprenkle, V.L. Li-ion battery technology for grid application. *J. Power Sources,* **2021**, *511*, 230419.
[http://dx.doi.org/10.1016/j.jpowsour.2021.230419]

[37] Chen, T.; Jin, Y.; Lv, H.; Yang, A.; Liu, M.; Chen, B.; Xie, Y.; Chen, Q. Applications of lithium-ion batteries in grid-scale energy storage systems. *Trans. Tianjin Univ.*, **2020**, *26*(3), 208-217.
[http://dx.doi.org/10.1007/s12209-020-00236-w]

[38] Liu, Y.; Jiang, S.P.; Shao, Z. Intercalation pseudocapacitance in electrochemical energy storage: Recent advances in fundamental understanding and materials development. *Mater. Today Adv.*, **2020**, *7*, 100072.
[http://dx.doi.org/10.1016/j.mtadv.2020.100072]

[39] Kali, R.; Padya, B.; Rao, T.N.; Jain, P.K. Solid waste-derived carbon as anode for high performance lithium-ion batteries. *Diam. Relat. Mater.*, **2019**, *98*, 107517.
[http://dx.doi.org/10.1016/j.diamond.2019.107517]

[40] Gu, M.; He, Y.; Zheng, J.; Wang, C. Nanoscale silicon as anode for Li-ion batteries: The fundamentals, promises, and challenges. *Nano Energy,* **2015**, *17*, 366-383.
[http://dx.doi.org/10.1016/j.nanoen.2015.08.025]

[41] Wang, H.; Wu, M.; Lei, X.; Tian, Z.; Xu, B.; Huang, K.; Ouyang, C. Siligraphene as a promising anode material for lithium-ion batteries predicted from first-principles calculations. *Nano Energy,* **2018**, *49*, 67-76.
[http://dx.doi.org/10.1016/j.nanoen.2018.04.038]

[42] Perdew, J.P.; Burke, K.; Ernzerhof, M. Generalized gradient approximation made simple. *Phys. Rev. Lett.*, **1996**, *77*(18), 3865-3868.
[http://dx.doi.org/10.1103/PhysRevLett.77.3865] [PMID: 10062328]

[43] Blöchl, P.E.; Jepsen, O.; Andersen, O.K. Improved tetrahedron method for Brillouin-zone integrations. *Phys. Rev. B Condens. Matter,* **1994**, *49*(23), 16223-16233.
[http://dx.doi.org/10.1103/PhysRevB.49.16223] [PMID: 10010769]

[44] Xu, S.; Fan, X.; Liu, J.; Singh, D.J.; Jiang, Q.; Zheng, W. Adsorption of Li on single-layer silicene for anodes of Li-ion batteries. *Phys. Chem. Chem. Phys.*, **2018**, *20*(13), 8887-8896.
[http://dx.doi.org/10.1039/C7CP08036K] [PMID: 29547213]

[45] Wang, X.; Liu, H.; Huttula, M.; Luo, Y.; Zhang, M.; Cao, W. First-principles studies of lithium adsorption and diffusion on silicene with grain boundaries. *Int. J. Quantum Chem.*, **2019**, *119*(13), e25913.
[http://dx.doi.org/10.1002/qua.25913]

[46] Pattarapongdilok, N.; Parasuk, V. Adsorptions of lithium ion/atom and packing of Li ions on graphene quantum dots: Application for Li-ion battery. *Comput. Theor. Chem.*, **2020**, *1177*, 112779.
[http://dx.doi.org/10.1016/j.comptc.2020.112779]

[47] Qin, R.; Wang, C-H.; Zhu, W.; Zhang, Y. First-principles calculations of mechanical and electronic properties of silicene under strain. *AIP Adv.*, **2012**, *2*(2), 022159.
[http://dx.doi.org/10.1063/1.4732134]

CHAPTER 10

Strategies for Synthesizing Metal Oxide Nanoparticles and the Challenges

Harish Bishwakarma[1], **Mukul Anand**[1] and **Alok Kumar Das**[1,*]

[1] *Indian Institute of Technology, Indian School of Mines, Dhanbad, India*

Abstract: The development of nanoscience and nanotechnology has improved our quality of life. The new class of materials known as nanoparticles (NPs) contributes to the development of nanotechnology. For the NPs, at least one dimension of particles should be 1 to 100 nm. The synthesis approaches can modify NPs structure and size, which is crucial in molecular biology, physics, chemistry, medicine, and material science. The high surface area of NPs can be achieved *via* synthesis approaches, providing increased value and imperative parameters like surface reactivity. Several approaches to synthesizing NPs can be used, mainly categorized into two parts: bottom-up and top-down. These two categories are classified based on the starting materials used to synthesize the NPs. This review discussed the brief of synthesis approaches and their utilization in the field of nanotechnology and nanoscience. The novel approach to the synthesis of NPs *i.e.*, the electrochemical discharge process, is discussed in detail. The materials synthesis like ZnO, carbon, graphene, and other metal oxide and their composite are discussed in tabular form. Finally, the challenges, advantages, disadvantages, conclusions and NPs synthesis are discussed.

Keywords: Bottom-up, Nanoscience, Nanotechnology, Nanoparticle, Top-down.

INTRODUCTION

The NPs are a class of materials that have high demand in molecular biology, physics, chemistry, medicine, and material science [1]. The study, structure, formation, synthesis, manipulation, and utilization of materials, devices, and functional systems through nanoscale control are all included in nanoscience and nanotechnology (NST). This also consists of exploiting phenomena and properties of the material at the nanoscale. The word "nano" comes from the Greek word "nanos" or the Latin word "nanus," both of which mean "dwarf". or "extremely little" and represent one-millionth of a meter (10^{-9} nm) [5]. The dimension of nano can also be understood through the best example, the arrangement of five silicon

* **Corresponding author Alok Kumar Das:** Indian Institute of Technology, Indian School of Mines, Dhanbad, India; E-mail: alokmech@iitism.ac.in

Dibya Prakash Rai, Kingsley O. Obodo, & Jitendra Pal Singh (Eds.)

atoms in a line, or the arrangement of ten hydrogen atoms in a line is considered as one nano-meter [6]. The daily life object, which we can see from an open eye or not in size scale (atom to meter), is shown in Fig. (**1**).

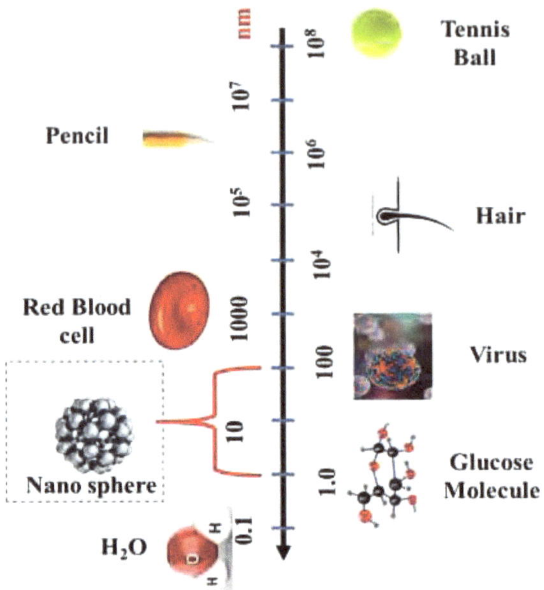

Fig. (1). Size distribution of NPs with different element.

NANOSCIENCE AND NANOTECHNOLOGY

At the time of Democritus and the Greeks in the fifth century B.C., scientists debated whether the matter is continuous and thus infinitely divisible into smaller pieces or whether it is made up of tiny, indivisible, and unbreakable particles or atoms. This debate is the result of people starting to think about nanoscience. It can observe, measure, assemble, control and manufacture the materials at the nanoscale. Nanotechnology can be stated as the development, synthesis, characterization and application of materials and devices at the nanoscale. The nanoscience and nanotechnology are co-related, under which knowledge of physics, materials science, chemistry, solid-state physics, and quantum mechanics is required. As a result, profound knowledge in one field will not be sufficient; a combination of physics, chemistry, material science, solid state, and biosciences is required. The nanoscience provides knowledge of atom arrangements and their fundamental properties at the nanoscale. Nanotechnology controls matter at the atomic level to create novel nanomaterials with various properties [6]. Nanotechnology has engulfed almost the entire engineering field. However, most people are unaware of its applications despite the fact that it covers almost every branch of science. Even after progress over time, there is still space for improvement in the era of nanotechnology and nanoscience for developing new materials and their application for humanity.

NANOTECHNOLOGY IN THE 20TH CENTURY

The term nano was first introduced by scientist Richard P. Feynman (1959-1960) in his lecture that. "Feynman posed the question, "Why can't we write the entirety of the 24 volumes of the Encyclopedia Britannica on the head of a pin?" in this lecture. "and presented a vision of building everything from larger machines down to molecules using machines". They introduce materials that can be broken into minor portions, *i.e.* nano and movement can be done from atom to atom to form nano-size [7]. Further future discovered the scanning electron microscope by Binning and Rhorer in 1983. Recent progress has made it possible for scientists to create unique nanomaterials and miniature technologies.

NANOMATERIALS CLASSIFICATION

NPs are the building block of nanotechnology, which is one billion meters. All the physics laws applied in our daily life use materials, but in the case of nano, it differs. It is observed that physics low is not applicable in NPs as the gold colour is nice yellowish, but 100 atoms of gold in the cube are red. It has a high surface-to-volume ratio, making it different physicochemical properties than bulk. Unexpectedly, the shape and size of the nanomaterials are altered at the nanoscale level, giving them a distinct personality with new powers. NPs can be synthesized in different forms, including rods, sheets, and flowers; each shape has unique features due to its dimensionality. Nanorods and nanotubes fall within the one-dimensional category of nanostructures, while films and layers-type particles are two-dimensional nanomaterials. Their individual functions typically classify them. The attractiveness of nanomaterials lies in the fact that their physical properties can be changed by combining them with other NPs. The following categories can be applied to materials based on their dimensions (< 100nm) [8].

1. Zero-dimension particles (0-D) – Nanomaterials with all three dimensions are at the nanoscale referred to be zero-dimensional NPs. The graphene quantum dot is the best example of a zero-dimensional nanoparticle because all three dimensions are on the nanoscale.

2. One-dimensional nanoparticle (1-D)- In this category, nanomaterials having one of the dimensions need to be nanoscale; the other two can be any size. Materials in this category comprise nanotubes, nanowires, and nanorods.

3. Two- Two-dimensional NPs (2-D)- Among this nanomaterials group, only two dimensions need to be of nanoscale; the third can lie outside this scale. Nano films, nanolayers, and nano coatings are examples.

4. Three-dimensional NPs (3-D)- These NPs are significantly larger than nanoscale in every dimension. Therefore, their scale in three arbitrary dimensions is more significant than 100 nm. Nanowire and nanotube bundles are examples of NPs under this category [9].

DIFFERENT METHODS OF GENERATION OF NPs

NPs are synthesized and stabilized *via* a variety of physical and chemical procedures, such as heat treatment, chemical reduction, and electrochemical treatments. During NPs synthesis, the kinematics of metal ions interact with the reducing agent, and the adsorption of the stabilizing agent results in the phase change of NPs.Thus, NPs and various experimental techniques affect their morphology (structure and size), stability, and physicochemical properties [10 - 12]. These techniques are divided into two main categories: (1) Bottom-up approach and (2) Top-down approaches (Fig. **2**) [13 - 15]. The main difference between these two techniques is the raw materials used to produce NPs. While atoms or molecules are the starting material in bottom-up approaches, top-down methods start with bulk material and use various physical, chemical, and mechanical processes to reduce the particle size and tune the morphology of NPs [16, 17].

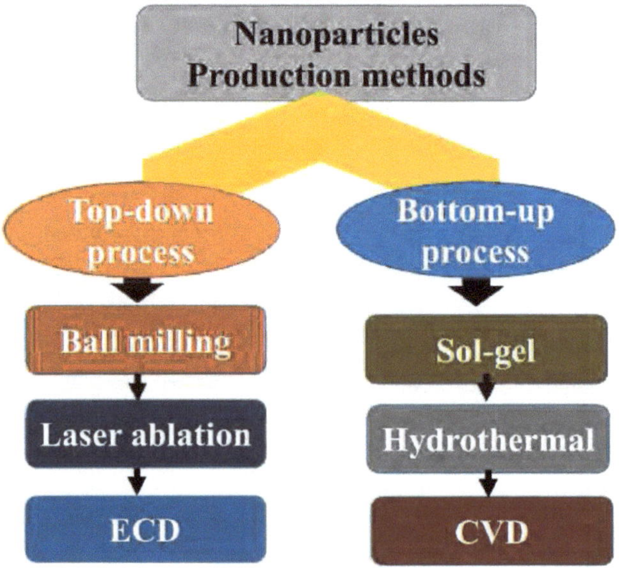

Fig. (2). Shows various approaches to the synthesis of NPs.

BOTTOM-UP APPROACHES

It is the most popular and successful strategy for the synthesis of NPs. Here, atoms self-assemble to form a new nucleus that grows into nano-sized and develops NPs. The chemical, hydrothermal and chemical vapour deposition methods are under this category.

Sol-gel Method

The sol-gel process is a chemical method for synthesizing nanostructures, especially metal oxide NPs. It is also known as the "wet chemical method". In this method, the molecular precursor, which is usually a metal alkoxide, is dissolved in water or alcohol and then heated and stirred to turn it into a gel through the process called hydrolysis or alcoholysis. Since the gel from the hydrolysis/alcoholysis process is wet, it needs to be dried to obtain desired NPs properties. Fig. (**3**) depicts the numerous steps and the transformation of the precursor into the finished product. Binary or tertiary hybrid systems are synthesized using different chemical precursors [18, 19]. Each sol-gel precursor has a different reaction rate [20, 21]. This reaction rate is affected by temperature, solvent type, solvent concentration, and pH. Table **1** shows the synthesis materials through sol-gel methods and their shape, size and challenges.

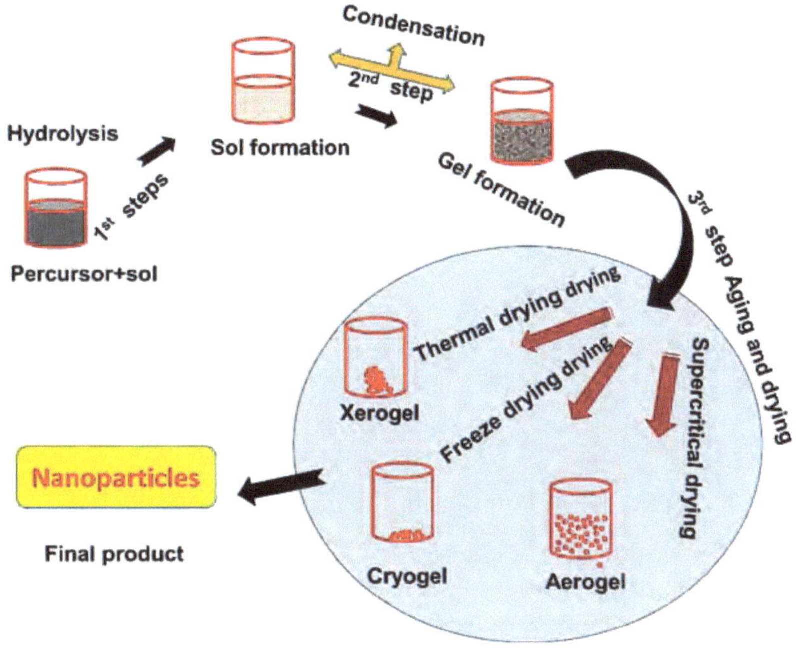

Fig. (3). Different steps and drying techniques of sol-gel methods.

Table 1. Synthesis materials through sol-gel methods.

NPs/year	Chemical	Oven Temperature	Time(h)	Synthesis Conditions	NPs Size, Shape (nm)	Drawback
MgO, CaO, SiO_2, Al_2O_3 (2022) [22]	$C_6H_8O_7$, metal precursor	500°C-600°C	10	Pichni type sol-gel	8.4, 20.2 Amorphous	Low density, foam formation
Cu-SiO_2, ZnO-SiO_2 (2021) [23]	$SiC_8H_{20}O_4$,$ZnC_4H_6O_4$, PVA	160 °C - 2 h	12	Gel, 3g Cu acetate, mix	Spherical shape	3-steps process, TC
Ligand-Decorated ZnO (2022) [24]	ZnO NCs,	-	24	Room temp, ligand exchange, 24 h	-	TC
Ag/GO/ZnO (2021) [25]	$Zn(CH_3COO)_2 \cdot 2H_2O$,GO,HCl, NaOH	400 °C and 500 °C	30	ultrasonication, aging at dark room	5.57, Hexagonal wutzite	More than 26h, Dark room required
Silver doped zinc oxide (2022) [26]	NaCMC, R1= amino acid, $(Zn(NO3)2.4H_2O)$, $AgNO_3$	700 °C	6	Zn^{2+} and Ag^{2+} gelatin liquid required.	Spherical shape. Size 27-36nm	the gelatin stabilizer sol-gel required

Advantage

- Simple process.
- NPs can be produced in different shapes.
- Economical Synthesis of NPs.
- The size and shape of particles can be controlled by monitoring the reaction parameters.

Disadvantage

- Due to the process in the solution phase, the precursors are chemically reactive.
- Processing time is relatively longer.
- Some pre and post-treatment required.
- Unwanted impurities may occur due to various chemical solution.

Hydrothermal Method

Hydrothermal is the wet chemical method in which an aqueous solution is used as a reaction chamber. The specified closed environment consists of a high temperature, high-pressure or vapour-pressure for synthesizing metal oxide NPs. By this, the solution precursor breaks down and transforms into a desired form, which does not dissolve or transform under normal conditions [27 - 29]. Fig. (**4**) shows the steps involved in the hydrothermal process. Table **2** shows the parameters and synthesis conditions for the number of NPs and their composite using the hydrothermal technique.

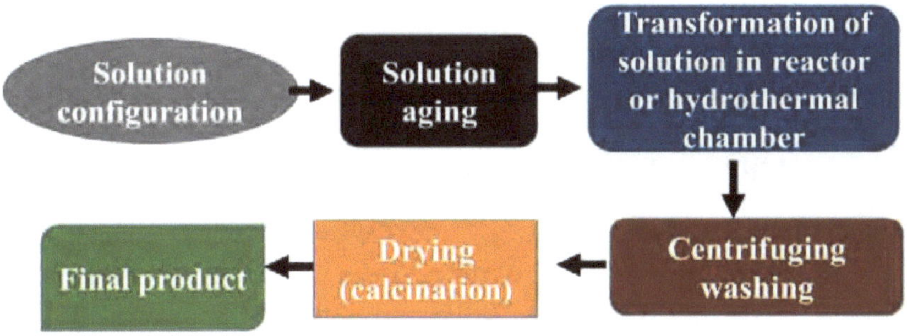

Fig. (4). Steps of the hydrothermal process to synthesize NPs.

Table 2. Syntheses of NPs using hydrothermal process.

NPs/Year	Chemical/ Precursor	Autoclave Conditions	Synthesis Conditions	Size- nm, Shape	Challenges
TiO_2/rGO (2017) [30]	Graphite, TiO_2, H_2SO_4, HCl, NaOH	160°C-10 h	TiO_2-GO composite 180 °C - 9 h	35,9 TiO_2, 0.74- GO	High processing time, harmful gases is used
Flower shape-ZnO nanorod (2020) [31]	$H_{12}N_2O_{12}Zn$, $(CH_2)_6N_4$, NaOH	90°C-24 h	ZnO-90°C 2hr, GO-ZnO 200°C 4h.	128-810, Flower	Chemical waste, dropcasting coated bond is weak
GO-ZnO-GO Sandwich (2020) [32]	ZnO Precursor	-	60°C- GO 90°C-GO-ZnO-GO	61-ZnO, 50- GO-Zn- -GO ZnO -Rod	3-steps process, TC, agglomeration
CuO/ZnO, rGO-CuO/ZnO (2020) [33]	$ZnSO_4$,7H_2O, $Cu(NO_3)_2$,3H_2 O,H2SO_4	180 °C -8 h.	60 °C for 2 h under vacuum condition	-	2-step process, harmful gases used, TC

Advantages

- The size and morphology of NPs are controllable.
- The hydrothermal method is used to produce oxides and oxygenated salts.
- Synthesized high-crystalline NPs.

Disadvantage

- High-pressure and temperature working environments require.
- Process control is difficult.
- Hard to monitor the in situ process.

Chemical Vapour Deposition Method

The CVD method is one of the most common ways to synthesize high-quality electronic nanomaterials. In this method, coating is done using a sample material over the substrate. A thin layer of gaseous reactants is laid on a substrate. A chemical reaction occurs when gas touches a heated substrate in a reaction chamber [34, 35]. This thin film is cut out and used as the final product. First, the reactant gases are sent to the reactor, which goes through homogeneous reaction stages in the gas phase to make intermediate reactants and by effects of gases, or they diffuse directly onto the heated substrate *via* a boundary layer. In both cases, intermediate reactants and reactant gases adsorb and diffuse onto the heated substrate surface. Evaporation of growing species to move them through a carrier gas to a surface where they can gather and grow. The heterogeneous reaction between a gas layer and solid substrates results in a thin film of products due to coalescence, growth reactions and nucleation. At some point, unwanted gas and species that haven't dissolved or desorbed from the surface of the reaction zone are taken away [36]. Gas-phase processes occur when the temperature is high enough or when more energy is added (like in the form of plasma). A heterogeneous reaction is needed if the deposition reaction depends on catalysis at the surface of the substrate, like with specific metal surface catalytic growth [37]. The schematic set-up shown in Fig. (**5**), has the varying controlling configuration. Table **3** depicts the NPs and their composite produced through CVD methods.

- Controlled delivery of gas-phase reactants.
- Enclosed reaction chamber (Gas absorb and discharge facilities).
- Control pressure reaction (delivery of energy source) for chemical reactions.

• Cleaning of exhaust gases to reach safe and non-toxic levels.

• An automated system is required to control the process and stabilize the deposition film.

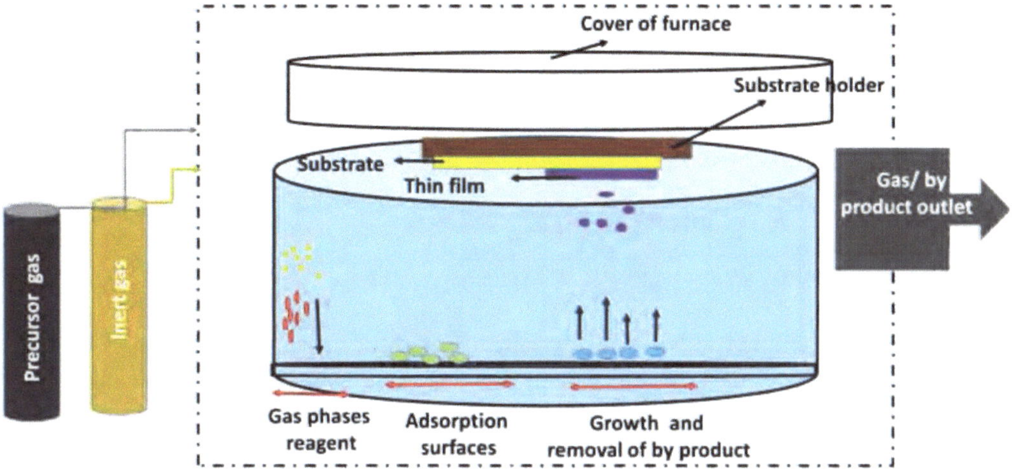

Fig. (5). Schematic diagram of CVD technique.

Table 3. Synthesis of NPs *via* CVD methods.

NPs/Year	Target/Substrate	Precursor	Temp	Time	Size, Shape	Challenges
Gr -coated on Cu (2020) [38]	Graphene/ polycrystalline copper	Ethanol	850°C	5-6 min		1.3 kPa pressure, heater controller
ZnO (2020) [39]	Ceramic, glass	$C_4H_{12}O_6Zn$	400°C	-	17 nm	60 L h^{-1} – gas flow & temp control
F-SnO_2 thin films (2017) [40]	fluorine-doped tin oxide glass,	$C_4H_9Cl_3Sn$, methanol, NH4F, BuSnCl3	550 ◦C	5min		Film thickness control on repetition is difficult.
Cu-doped ZnO (2012) [41]	ZnO:Cu, Si	Terpineol, ethyl cellulose,	650◦C 10◦C/min	30-120 min	Hollow spherical	Control of gas pressure mixture Ar, O_2

Advantages

- The most effective way to produce two-dimensional materials.
- It is relatively simple to synthesize thin films.
- High crystalline NPs are produced.

Disadvantages

- The selection of catalysts is tricky.
- A complex set-up is required.

TOP-DOWN APPROACHES TO THE SYNTHESIS OF NPs

In a top-down method, the bulk materials are reduced or broken down into small nano-sized particles that will compound with other materials to produce novel NPs [42]. The advantages of this strategy include machine simplicity, cost-effectiveness, and independence (unlike the bottom-up approach) [43]. NPs stability, shape, and size can be precisely regulated by carefully adjusting the temperature, pH, reduction environment, metal-electrolyte concentration, and reduction length. This method allows for more control over the process parameters [44].

Ball Milling Method

Ball milling represents a mechanical process involving alloying, nanosizing, and ultrafine grinding. It operates by transferring kinetic energy from a ball to reduce NP size through grinding actions. This technique entails loading a cylindrical container with steel grinding balls and bulk powder. Subsequently, elevated pressure and temperature rotation transform bulk substances into nano-scale particles [45]. This transformation arises from the interplay between grinding balls, bulk materials, and container walls, generating heightened pressure and temperature. Multiple ball types and precursors promote friction within the container, driving NP phase transformation [46]. Predominantly employed by industries, this approach enables large-scale NP production, yielding amorphous NP variants. Its versatility encompasses a broad spectrum of NP types and metal alloys. For instance, zinc alloys with oxide and carbide reinforcement, graphene oxide-based nanocomposites, and nanoalloys involving aluminium, nickel, magnesium, and copper, alongside various other nanocomposites, can all be synthesized using mechanical ball milling. The method serves multiple purposes, including powder mixing, generating phase-change NPs, NP agglomeration, alloy compound synthesis, altering surface structures, promoting particle growth, and modifying mechanical properties such as density and work hardening [47]. Numerous factors influence nanoparticle properties (Table **4** offers a selection of key differentiating process parameters), including temperature elevation leading to mechanical deformation. Kinetic energy from the ball transforms into heat energy, supplemented by additional heat generated during NPs burn, inducing phase changes. Aside from temperature, NPs properties are also influenced by ball size, ball weight, rotational positioning, milling solution, and milling duration [48].

Table 4. Hybrid NPs synthesis by mechanical ball milling process.

NPs/Year	Chemical/ Precursor	Ball Milling Conditions	Synthesis Conditions	Challenges
ZnO/CNT (2019) [45]	$ZnCl_2$, KOH, CNT	150°C 15h - Autoclave conditions, 2h ball milled	NPC ball milled F-15Hz-2h then transferred to an autoclave	2 steps process, cost of production high
ZnO-V_2O5 (2020) [46]	x mol% of V_2O_5, (100-x) mol % of ZnO	18h at 300°C-Ball milling, 700°C -12h- Calcination	Different % of ZnO-V_2O_5 composite sintered at 900 °C	3 steps process, TC
Co-ZnO (2022) [47]	ZnO, cobalt oxide	250 rmp 10h	3-10 mol% ratio mixed with ethanol-1h, 200-300MPa pressure, annealed at 500°C for 2h.	TC, clean room required
ZnO-rGO (2022) [48]	Graphite, $NaNO_3$, $KMNO_4$, H_2SO_4, H_2O_2	15min-10h, 2100rpm	Modified hammer - GO, milling-ZnO, rGO-ZnO prepared by chemical process	4-steps process, contamination of impurity

Advantage

• For industrial-level production of NPs, this method is used.

• It can modify the surface properties of NPs.

• Alloying of nanomaterials is possible through this method [49].

• It is used to reduce the thickness of stacked materials [50].

Disadvantage

• The non-crystalline NPs formed which destructive, destroying or disordering [51].

• Times consuming process.

• Due to friction, material heating, wear on the balls and wall armour, and other factors, grinding consumes a lot of energy.

• Ball milling time produces lots of contamination and noise [52].

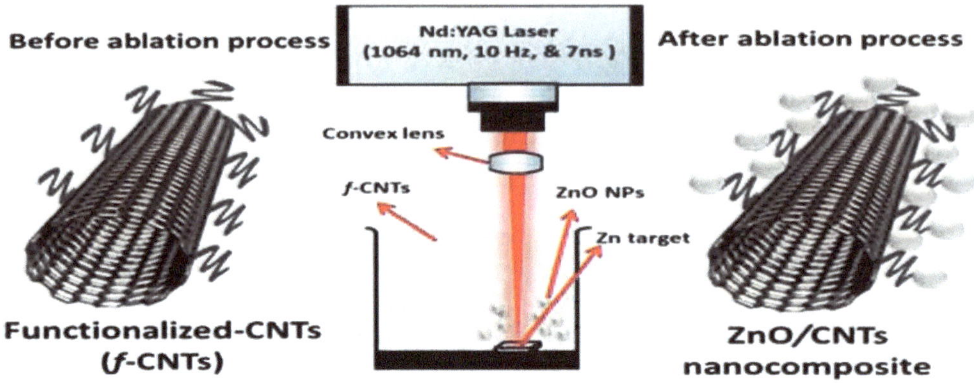

Fig. (6). Laser ablation of CNT and ZnO-based NPs [58].

Laser Ablation Method

The term laser refers to a high-intensity beam of electromagnetic radiation that is amplified by stimulated emission of radiation. Einstein was the first to propose the laser hypothesis. In 1960, Maiman *et al.* [52] first proposed the practical application of lasers. After that, they were widely used in medical industries. Laser ablation is converting solid materials into micro and NPs using the highly monochromatic light source into heat energy. Materials irradiation under laser ablation (due to high heat energy) is an eco-friendly, trouble-free method for generating highly pure graded NPs. This method converts solid materials in a specific environment, either gas or liquid, to obtain a definite composition or structure of nanomaterials. The laser ablation comprises the lens, power source of the laser, positioner and insulated environment target [53]. The gap control between the target material, solution liquid and laser beam plays a crucial role in achieving the uniform shape of NPs [54]. The laser beam ablated the metal NPs in the form of nano and micro-size metal droplets, which are further called nanoparticles. The nucleation and growth formed different shapes and sizes of NPs, which depend on the medium in which the laser is ablated [55, 56]. The ZnO Cu and carbon-based synthesis approach *via* laser ablation is discussed in Table **5**. The fluence, wavelength, and laser salt addition can be adjusted to modify nanoparticle parameters such as average size, shape and distribution shown in Fig. (**6**). It shows the formation of CNT and ZnO-based nanocomposite through Nd

Yag laser ablation. Some of the influencing factors that play an essential role in the synthesis of NPs and their composite are shown in Table **5**. The process parameter that influences the properties of NPs are (a) the wavelength, (b) the control of pulse duration, (c) the reflection and absorption of light from target materials, (d) the medium in which laser is ablated, e) control over lead time during pulse laser [57, 58].

Table 5. Synthesis of NPs *via* laser ablation method.

NPs/year	Laser Type	Laser Parameter	Time	Target Materials	Size, Shape	Challenges
ZnO@graphene (2019) [58]	Nd: YAG laser	WL-1064 nm, F-10Hz, LP-110mJ/pulse	1h	Zinc & Graphite	Spherical,	Go-ZnO bonding as core shell
copper-silica (2020) [59]	Fibre Pulse laser	WL-800nm F-1kHz LP-200μ J	1h	Cupper & silicon wafer	Cu-silica-5.4-10.4nm	Some number of core-shell NPs Form, amorphous Cu
Ni–CuO (2022) [60]	Nd: YAG nanosecond	WL-1064nm, 80 mJ/pulse	15-45min	Cu, Ni, Zn metal-based target, surfactant-based	Spherical, rod homogenous	Reverse ablation
Cu(2020) [61]	Nd: YAG nanosecond	WL-1064nm, 80 mJ/pulse	10min	Cu, Solvent based	Rod-400-500nm, Cuboids-5*5*5μm, Spherical	Convex lens focused control, Solvent selection

Advantages

- Capabilities to synthesize pure NPs for medicine industries.
- Low loss of energy.
- Synthesis from waste materials, irrespective of their conductivity.

Disadvantages

- High power source laser is required.
- The initial set-up cost is high.
- Maintenance cost is required.

• Some alloy compounds are difficult to ablate.

Electrochemical Discharge Methods

Over the last 15 years, a new class of solutions-based discharges and plasma-cathode interaction in ambient-gas arc discharges has been discovered. The various discharge methods explored are basically gas and liquid-based. The different name of discharge methods is (a) atmospheric pressure glow discharge generated (APGD), flowing liquid cathode/anode (FLC/FLA-APGD) [63], capillary liquid electrode discharge (CLED) [64], capillary liquid electrode microplasma (CLEP) [65], Solution anode glow discharge (SAGD), the electrolyte jet cathode glow discharge (ECGD), liquid drop anode-atmospheric pressure glow discharge (LDA-APGD) [66], liquid cathode glow discharge (LCGD) [67]. The discharge procedure, although the same (The electrolyte or gas used in a few micron gaps between cathode and anode for discharge), differs from their naming technique. The discharge is fundamentally formed in a gap between a conducting liquid and another electrode (cathode or anode) by transporting a high voltage (DC or AC) at atmospheric pressure. The synthesis of NPs through ECDP is different because it is low low-energy process that is performed in atmospheric conditions. In 1968, Kurafuji and Suda first explored the ECD process for machine-conducting and non-conducting materials [62]. The mechanism of spark generation in ECDP mainly depends on two principles, *i.e.*, the Electrochemical discharge process (EDP) and the Electrochemical process (ECP) process [68]. In -ECDP technology, both processes (EDP & ECP) exhibit their distinct properties. For example, -EDP induces thermal melting and vaporization of the workpiece, whereas -ECP promotes chemical etching. It differs from the EDP principle because in EDP, the dielectric is used as a working fluid, but in the case of ECDP, the electrolyte is used. Here temporary dielectric is formed due to hydrogen gas generation at the tip of the tool due to electrolysis of the ECM process [69]. Basically, in ECDP mechanism of spark formation is as follows [70].

(a) ECM- Electrolysis occurred between the workpiece and tool electrode.

(b) The electrolysis produces hydrogen bubbles at the cathode tip and oxygen at the anode tip. The oxygen bubbles evaporated, and hydrogen bubbles collected near the cathode electrode.

The following reaction occurred in the case of an alkaline solution.

The following reaction occurred when the DC pulse supply between the two electrodes, a potential difference develops between them, which leads to the initiation of the electrolysis process:

$$M \rightarrow M^{z+} + Ze^{-} \tag{1}$$

Where, M represents the anodic material, which is split into metallic ions (M+), (e-) represents electron and Z is the number of ions. Electrolyte fluid also dissociates, and these hydroxide ions combine with metallic ions and form the metal hydroxide, which is insoluble in the electrolyte.

$$M^{z+} + Z(OH)^{-} = M(OH)_{z} \tag{2}$$

$$2(OH)^{-} \rightarrow 2H_2O + O_2 \uparrow + 4e^{-} \text{ (in alkaline solution)} \tag{3}$$

Similarly, hydrogen gas is formed around the cathode surface as expressed through the following reaction:

$$2H_2O + 2e^{-} \rightarrow 2(OH)^{-} + H_2 \uparrow \text{(in alkaline solution)} \tag{4}$$

The small hydrogen gas bubbles, thus formed, accumulate at the tool surface and create a hindrance to the current flow at the interface of the cathode and electrolyte.

(c) Further increasing the voltage, smaller size of bubbles coalesce to form larger sizes, which temporarily insulates the cathode electrode.

(d) At a critical voltage, sudden discharge occurred between the cathode and anode, which generated a high temperature.

(e) By this melting and evaporation of cathode and anode forms NPs.

The spark between the cathode and anode is much less approx. Few microseconds, which leads to rapid heating at a small point of a workpiece. Such 1000s of spark is generated between the cathode and anode which is instantaneously heated to boiling point of materials. Further, it creates the vapour plume and sudden quenching in the electrolyte medium due to adiabatic expansion forms NPs. The materials removal mechanism due to spark or plasma formation is very limited. The schematic diagram of nanoparticle synthesis is shown in Fig. (7) and various materials synthesis parameters. The temperature generated at the sparking zone is 1297-10000K [71]. The gap between the workpiece, electrolyte and tool (in general 25-60mm), The electrolyte (acidic, alkaline and solvent) also affects the sparking zone, which further modifies the properties of synthesized NPs [72].

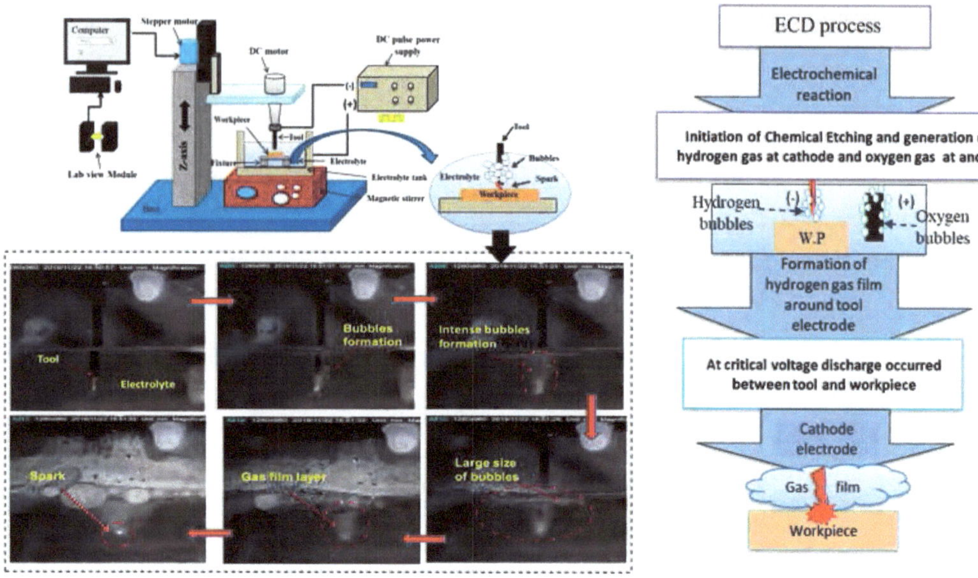

Fig. (7). Diagram of ECDM working principle and working set-up [73, 74].

Influencing Process Parameter of ECDP for NPs Synthesis

The electrochemical discharge process is basically the ablation of electrode materials using spark energy which is much more common than the laser ablation methods. The production of NPs on a large scale or output at a scalable level is also possible. Controlling the various process parameters is much better than laser ablation and other related processes. As the laser is not used for nanoparticle production, the cost of synthesis of NPs *via* ECDP is comparatively low. The process of synthesis is environmentally friendly and can be produced in an open environment without any chemical precursor [72]. The following process parameters used to tune the NPs synthesized by the ECDP process are also various pure and composite materials as shown in Table **6**.

Table 6. Synthesis of NPs *via* ECD process

NPs/year	Working Environment	Spark Parameter	Time	Anode & Cathode Materials	Size, Shape	Challenges
Si (2009) [73]	Ar Gas	30-40V,30-60A	10min	Si,Ni Mixed- anode W- cathode	Nano wire	Agglomeration of NPs, sealed Chamber, $5*10^4$ Pa
ZnO (20102) [74]	K_2CO_3	50-190V	1h	Pt wire-anode, Zn wire-Cathode	Flower	At low Voltage agglomeration of NPs
CuO-ZnO (2019) [75]	2M NaCl	150V 5A	-	Pt-anode, Brass-cathode	Flower	Agglomeration at a lower voltage

(Table 6) cont.....

NPs/year	Working Environment	Spark Parameter	Time	Anode & Cathode Materials	Size, Shape	Challenges
ZnO - GO(2022) [76]	2M NaOH	30-45V	1h	Graphite-anode, Zn-cathode	Flower, GO sheet	Temperature controller during spark

(i) Voltage (ii) Electrolyte type (iii) Duty cycle (iv) Hybridization (ultrasonic assisted, Magnetic assisted) (v) Temperature (vi) Powder mixing (vii) Annealing, freezing

Advantage

- Any materials can be synthesized irrespective of their conductivity.
- The costly waste materials converted into NPs.
- Environmentally friendly.
- Cost-effective.
- No Vacuumed environment required.

Disadvantage

- Agglomeration of NPs.
- Selection of electrolyte.
- Continuous spark at point deposition of NPs started.

CONCLUSION

Synthesis of NPs that are at the nanoscale in size (at least one dimension of materials in the range 1-100nm) is challenging in the nanoscience and nanotechnology field. Metal oxide NPs are increasingly being used in several sectors, such as material science, physics, chemistry, and biomedicine. Nanostructure synthesis is a dynamic and complex process. Tuning of NPs with proper morphology, several approaches have been used, which are main categories in top-down (Ball milling, laser ablation, ECDM) and bottom-up (Hydrothermal, chemical vapor deposition, sol-gel). Each method has its own set of process benefits and demerits. Green nanoparticle synthesis (ECDP & laser ablation) is environmentally friendly, affordable, and generally devoid of chemical pollutants, which is discussed in detail. This review includes a collection of commonly used ways to synthesize NPs, which aids in the selection of materials and the availability of facilities for good tools of process parameters analysis. Before choosing the suitable synthesis method, we need to understand how each one works, but we are still a long way from that. Nanotechnology can help solve problems that modern society will face in the future and current scenario. If the synthesis parameter can be controlled, it may solve various unknown facts.

ACKNOWLEDGMENTS

Thanks to IIT(ISM) for providing the platform to write this chapter. The chapter authors would like to acknowledge Dr. Nitesh Kumar (IIT ISM Dhanbad 826004), Dr. Santosh Kumar Mishra (Department of Production Engineering, NIT, Tiruchirappalli Tanjore Road, 620015, Tamil Nadu, India) and Dr. Purushottam Kumar Singh (Department of Mechanical Engineering, NIT Silchar. 788010, India) for their efforts and contribution to the manuscript.

REFERENCES

[1] Binnig, G.; Rohrer, H. Raster tunnel mikro skopie. *Helv. Phys. Acta,* **1982**, *55*, 726.

[2] Feynman, R.P. *There's plenty of room at the bottom: An invitation to enter a new field of physics. Miniaturisation*; Reinhold, **1961**.

[3] Heiligtag, F.J.; Niederberger, M. The fascinating world of nanoparticle research. *Mater. Today,* **2013**, *16*(7-8), 262-271. [http://dx.doi.org/10.1016/j.mattod.2013.07.004]

[4] National Research Council. A matter of size: Triennial review of the national nanotechnology initiative. **2006**.

[5] Mansoori, G.A.; Soelaiman, T.F. *Nanotechnology an introduction for the standards community*; ASTM International, **2005**.

[6] Delgado-Ramos, G.C. Ethical, social, environmental, and legal aspects of nanotechnologies: A reading from Mexico. *Int. J. Manag. Innov. Technol.,* **2013**, *10*(2), 1340001. [http://dx.doi.org/10.1142/S0219877013400014]

[7] Feynman, R.P. Plenty of Room at the Bottom. *APS annual meeting,* **1959**.

[8] Vollath, D. An introduction to synthesis, properties and application. *Management,* **2008**, *7*(6), 865-870.

[9] Thakur, P.; Thakur, A. Introduction to Nanotechnology. In: *Synthesis and Applications of NPs*; Springer: Singapore, **2022**; pp. 1-17.

[10] Chen, W.; Cai, W.; Zhang, L.; Wang, G.; Zhang, L. Sonochemical processes and formation of gold NPs within pores of mesoporous silica. *J. Colloid Interface Sci.,* **2001**, *238*(2), 291-295. [http://dx.doi.org/10.1006/jcis.2001.7525] [PMID: 11374924]

[11] Geethalakshmi, R.; Sarada, D.V.L. Gold and silver nanoparticles from Trianthema decandra: synthesis, characterization, and antimicrobial properties. *Int. J. Nanomedicine,* **2012**, *7*, 5375-5384. [http://dx.doi.org/10.2147/IJN.S36516] [PMID: 23091381]

[12] Vijayakumar, M.; Priya, K.; Nancy, F.T.; Noorlidah, A.; Ahmed, A.B.A. Biosynthesis, characterisation and anti-bacterial effect of plant-mediated silver nanoparticles using Artemisia nilagirica. *Ind. Crops Prod.,* **2013**, *41*, 235-240. [http://dx.doi.org/10.1016/j.indcrop.2012.04.017]

[13] Pacioni, N.L.; Borsarelli, C.D.; Rey, V.; Veglia, A.V. Synthetic routes for the preparation of silver NPs. In: *Silver nanoparticle applications*; Springer: Cham, **2015**; pp. 13-46. [http://dx.doi.org/10.1007/978-3-319-11262-6_2]

[14] Rajput, N. Methods of preparation of NPs-a review. *Int. J. Adv. Eng. Technol.,* **2015**, *7*(6), 1806.

[15] swathy, B. A review on metallic silver NPs. *IOSR J. Pharm.,* **2014**, *4*(7), 38-44. [http://dx.doi.org/10.9790/3013-0407038044]

[16] Horikoshi, S.; Serpone, N., Eds. *Microwaves in nanoparticle synthesis: fundamentals and applications*; John Wiley & Sons, **2013**. [http://dx.doi.org/10.1002/9783527648122]

[17] Ahmed, S.; Ahmad, M.; Swami, B.L.; Ikram, S. A review on plants extract mediated synthesis of silver nanoparticles for antimicrobial applications: A green expertise. *J. Adv. Res.,* **2016**, *7*(1), 17-28. [http://dx.doi.org/10.1016/j.jare.2015.02.007] [PMID: 26843966]

[18] Yadav, S.; Jaiswar, G. Review on undoped/doped TiO_2 nanomaterial; synthesis and photocatalytic and antimicrobial activity. *J. Chin. Chem. Soc. (Taipei),* **2017**, *64*(1), 103-116. [http://dx.doi.org/10.1002/jccs.201600735]

[19] Peerakiatkhajohn, P.; Onreabroy, W.; Chawengkijwanich, C.; Chiarakorn, S. Preparation of visible-light-responsive TiO_2 doped Ag thin film on PET plastic for BTEX treatment. *Journal of Sustainable Energy & Environment,* **2011**, *2*(3).

[20] Pan, H.; Wang, X.; Xiao, S.; Yu, L.; Zhang, Z. Preparation and characterization of TiO_2 NPs surface-modified by octadecyltrimethoxysilane. **2013**.

[21] Lakshmi, B.B.; Dorhout, P.K.; Martin, C.R. Sol− gel template synthesis of semiconductor nanostructures. *Chem. Mater.,* **1997**, *9*(3), 857-862. [http://dx.doi.org/10.1021/cm9605577]

[22] Shojaei, B.; Miri, R.; Bazyari, A.; Thompson, L.T. Asphaltene adsorption on MgO, CaO, SiO_2, and Al_2O_3 nanoparticles synthesized *via* the Pechini-type Sol–Gel method. *Fuel,* **2022**, *321*, 124136. [http://dx.doi.org/10.1016/j.fuel.2022.124136]

[23] Shojaei, B.; Miri, R.; Bazyari, A.; Thompson, L.T. Asphaltene adsorption on MgO, CaO, SiO_2, and Al_2O_3 nanoparticles synthesized *via* the Pechini-type Sol–Gel method. *Fuel,* **2022**, *321*, 124136. [http://dx.doi.org/10.1016/j.fuel.2022.124136]

[24] Chen, X.; Zhang, Z.; Zhao, C.; Fang, J.; Xia, D.; Zhang, Y.; Xiao, C.; Wu, Y.; You, S.; Li, W. Functional ligand-decorated ZnO NPs as cathode interlayers for efficient organic solar cells. *ACS Appl. Energy Mater.,* **2022**, *5*(1), 1291-1297. [http://dx.doi.org/10.1021/acsaem.1c03717]

[25] Rashid Al-Mamun, M.; Shofikul Islam, M.; Rasel Hossain, M.; Kader, S.; Shahinoor Islam, M.; Zaved Hossain Khan, M. A novel and highly efficient Ag and GO co-synthesized ZnO nano photocatalyst for methylene blue dye degradation under UV irradiation. *Environ. Nanotechnol. Monit. Manag.,* **2021**, *16*, 100495. [http://dx.doi.org/10.1016/j.enmm.2021.100495]

[26] Khatir, N.M.; Sabbagh, F. Green facile synthesis of silver-doped zinc oxide nanoparticles and evaluation of their effect on drug release. *Materials,* **2022**, *15*(16), 5536. [http://dx.doi.org/10.3390/ma15165536] [PMID: 36013672]

[27] Gan, Y.X.; Jayatissa, A.H.; Yu, Z.; Chen, X.; Li, M. Hydrothermal synthesis of nanomaterials. *J. Nanomater.,* **2020**, 2020.

[28] Yang, C.; Wen, J.; Chen, X.; Da, X.; Qiu, M.; Verweij, H.; Fan, Y. Modified hydrothermal treatment route for high-yield preparation of nanosized ZrO_2. *Ceram. Int.,* **2020**, *46*(12), 19807-19814. [http://dx.doi.org/10.1016/j.ceramint.2020.05.022]

[29] Vinay, S.P.; Udayabhanu, ; Nagaraju, G.; Chandrappa, C.P.; Chandrasekhar, N. Hydrothermal

synthesis of gold nanoparticles using spider cobweb as novel biomaterial: Application to photocatalytic. *Chem. Phys. Lett.*, **2020**, *748*, 137402.
[http://dx.doi.org/10.1016/j.cplett.2020.137402]

[30] Lertthanaphol, N.; Pienutsa, N.; Chusri, K.; Sornsuchat, T.; Chanthara, P.; Seeharaj, P.; Kim-Lohsoontorn, P.; Srinives, S. One-step hydrothermal synthesis of precious metal-doped titanium dioxide–graphene oxide composites for photocatalytic conversion of CO_2 to Ethanol. *ACS Omega,* **2021**, *6*(51), 35769-35779.
[http://dx.doi.org/10.1021/acsomega.1c05799] [PMID: 34984307]

[31] Boukhoubza, I.; Khenfouch, M.; Achehboune, M.; Leontie, L.; Carlescu, A.; Doroftei, C.; Mothudi, B.M.; Zorkani, I.; Jorio, A. Graphene oxide coated flower-shaped ZnO nanorods: Optoelectronic properties. *J. Alloys Compd.,* **2020**, *831*, 154874.
[http://dx.doi.org/10.1016/j.jallcom.2020.154874]

[32] Boukhoubza, I.; Khenfouch, M.; Achehboune, M.; Mothudi, B.M.; Zorkani, I.; Jorio, A. Graphene oxide/ZnO nanorods/graphene oxide sandwich structure: The origins and mechanisms of photoluminescence. *J. Alloys Compd.,* **2019**, *797*, 1320-1326.
[http://dx.doi.org/10.1016/j.jallcom.2019.04.266]

[33] Tantubay, K.; Das, P.; Baskey Sen, M. Ternary reduced graphene oxide–CuO/ZnO nanocomposite as a recyclable catalyst with enhanced reducing capability. *J. Environ. Chem. Eng.,* **2020**, *8*(4), 103818.
[http://dx.doi.org/10.1016/j.jece.2020.103818]

[34] Lv, S.; Wu, Q.; Xu, Z.; Yang, T.; Jiang, K.; He, M. Chirality distribution of single-walled carbon nanotubes grown from gold nanoparticles. *Carbon,* **2022**, *192*, 259-264.
[http://dx.doi.org/10.1016/j.carbon.2022.02.051]

[35] Yarbrough, W.A.; Messier, R. Current issues and problems in the chemical vapor deposition of diamond. *Science,* **1990**, *247*(4943), 688-696.
[http://dx.doi.org/10.1126/science.247.4943.688] [PMID: 17771885]

[36] Colli, A.; Fasoli, A.; Hofmann, S.; Ducati, C.; Robertson, J.; Ferrari, A.C. Deterministic shape-selective synthesis of nanowires, nanoribbons and nanosaws by steady-state vapour-transport. *Nanotechnology,* **2006**, *17*(4), 1046-1051.
[http://dx.doi.org/10.1088/0957-4484/17/4/034] [PMID: 21727380]

[37] Yan, K.; Fu, L.; Peng, H.; Liu, Z. Designed CVD growth of graphene *via* process engineering. *Acc. Chem. Res.,* **2013**, *46*(10), 2263-2274.
[http://dx.doi.org/10.1021/ar400057n] [PMID: 23869401]

[38] Ji, D.; Wen, X.; Foller, T.; You, Y.; Wang, F.; Joshi, R. Chemical vapour deposition of graphene for durable anticorrosive coating on copper. *Nanomaterials,* **2020**, *10*(12), 2511.
[http://dx.doi.org/10.3390/nano10122511] [PMID: 33327582]

[39] Ding, J.; Chen, S.; Han, N.; Shi, Y.; Hu, P.; Li, H.; Wang, J. Aerosol assisted chemical vapour deposition of nanostructured ZnO thin films for NO_2 and ethanol monitoring. *Ceram. Int.,* **2020**, *46*(10), 15152-15158.
[http://dx.doi.org/10.1016/j.ceramint.2020.03.051]

[40] Powell, M.J.; Potter, D.B.; Wilson, R.L.; Darr, J.A.; Parkin, I.P.; Carmalt, C.J. Scaling aerosol assisted chemical vapour deposition: Exploring the relationship between growth rate and film properties. *Mater. Des.,* **2017**, *129*, 116-124.
[http://dx.doi.org/10.1016/j.matdes.2017.05.017]

[41] Xu, D.H.; Shen, W.Z. Cu-doped ZnO hemispherical shell structures: synthesis and room-temperature ferromagnetism properties. *J. Phys. Chem. C,* **2012**, *116*(24), 13368-13373.
[http://dx.doi.org/10.1021/jp3003849]

[42] Singh, J.; Dutta, T.; Kim, K.H.; Rawat, M.; Samddar, P.; Kumar, P. 'Green' synthesis of metals and their oxide nanoparticles: applications for environmental remediation. *J. Nanobiotechnology,* **2018**, *16*(1), 84.

[http://dx.doi.org/10.1186/s12951-018-0408-4] [PMID: 29321058]

[43] Ghatule, M.P.; Devare, U.A. Study and analysis on synthesis of graphene and its applications in solar cells. *Int J Innov Res Adv Eng,* **2016**, *1*, 184.

[44] Zhang, Z.; Fraser, A.; Ye, S.; Merle, G.; Barralet, J. Top-down bottom-up graphene synthesis. *Nano Futures,* **2019**, *3*(4), 042003.
[http://dx.doi.org/10.1088/2399-1984/ab4eff]

[45] Arsalani, N.; Bazazi, S.; Abuali, M.; Jodeyri, S. A new method for preparing ZnO/CNT nanocomposites with enhanced photocatalytic degradation of malachite green under visible light. *J. Photochem. Photobiol. Chem.,* **2020**, *389*, 112207.
[http://dx.doi.org/10.1016/j.jphotochem.2019.112207]

[46] Singh, J.; Singh, R.C. Structural, optical, dielectric and transport properties of ball mill synthesized ZnO–V_2O_5 nano-composites. *J. Mol. Struct.,* **2020**, *1215*, 128261.
[http://dx.doi.org/10.1016/j.molstruc.2020.128261]

[47] Akgul, G.; Akgul, F.A. Impact of cobalt doping on structural and magnetic properties of zinc oxide nanocomposites synthesized by mechanical ball-milling method. *Colloid Interface Sci. Commun.,* **2022**, *48*, 100611.
[http://dx.doi.org/10.1016/j.colcom.2022.100611]

[48] Buldu-Akturk, M.; Toufani, M.; Tufani, A.; Erdem, E. ZnO and reduced graphene oxide electrodes for all-in-one supercapacitor devices. *Nanoscale,* **2022**, *14*(8), 3269-3278.
[http://dx.doi.org/10.1039/D2NR00018K] [PMID: 35166280]

[49] Li, L.H.; Chen, Y.; Behan, G.; Zhang, H.; Petravic, M.; Glushenkov, A.M. Large-scale mechanical peeling of boron nitride nanosheets by low-energy ball milling. *J. Mater. Chem.,* **2011**, *21*(32), 11862-11866.
[http://dx.doi.org/10.1039/c1jm11192b]

[50] Janot, R.; Guérard, D. Ball-milling: the behavior of graphite as a function of the dispersal media. *Carbon,* **2002**, *40*(15), 2887-2896.
[http://dx.doi.org/10.1016/S0008-6223(02)00223-3]

[51] Tavakoli, A.; Sohrabi, M.; Kargari, A. A review of methods for synthesis of nanostructured metals with emphasis on iron compounds. chemical papers, 61(3), 151-170.

[52] Ghorbani, H. A review of methods for synthesis of Al NPs. *Orient. J. Chem.,* **2014**, *30*(4), 1941-1949.
[http://dx.doi.org/10.13005/ojc/300456]

[53] Zhang, J.; Chaker, M.; Ma, D. Pulsed laser ablation based synthesis of colloidal metal nanoparticles for catalytic applications. *J. Colloid Interface Sci.,* **2017**, *489*, 138-149.
[http://dx.doi.org/10.1016/j.jcis.2016.07.050] [PMID: 27554172]

[54] Amendola, V.; Amans, D.; Ishikawa, Y.; Koshizaki, N.; Scirè, S.; Compagnini, G.; Reichenberger, S.; Barcikowski, S. Room-temperature laser synthesis in liquid of oxide, metal-oxide core-shells, and doped oxide NPs. *Chemistry,* **2020**, *26*(42), 9206-9242.
[http://dx.doi.org/10.1002/chem.202000686] [PMID: 32311172]

[55] Yang, G.W. Laser ablation in liquids: Applications in the synthesis of nanocrystals. *Prog. Mater. Sci.,* **2007**, *52*(4), 648-698.
[http://dx.doi.org/10.1016/j.pmatsci.2006.10.016]

[56] Sakka, T.; Iwanaga, S.; Ogata, Y.H.; Matsunawa, A.; Takemoto, T. Laser ablation at solid–liquid interfaces: An approach from optical emission spectra. *J. Chem. Phys.,* **2000**, *112*(19), 8645-8653.
[http://dx.doi.org/10.1063/1.481465]

[57] Naser, H.; Alghoul, M.A.; Hossain, M.K.; Asim, N.; Abdullah, M.F.; Ali, M.S.; Alzubi, F.G.; Amin, N. The role of laser ablation technique parameters in synthesis of nanoparticles from different target types. *J. Nanopart. Res.,* **2019**, *21*(11), 249.
[http://dx.doi.org/10.1007/s11051-019-4690-3]

[58] Shankar, P.; Ishak, M.Q.H.; Padarti, J.K.; Mintcheva, N.; Iwamori, S.; Gurbatov, S.O.; Lee, J.H.; Kulinich, S.A. ZnO@graphene oxide core@shell nanoparticles prepared via one-pot approach based on laser ablation in water. *Appl. Surf. Sci.,* **2020**, *531*, 147365. [http://dx.doi.org/10.1016/j.apsusc.2020.147365]

[59] John, M.G.; Tibbetts, K.M. Controlling the morphology of copper-silica nanocomposites from laser ablation in liquid. *Appl. Surf. Sci.,* **2020**, *510*, 145037. [http://dx.doi.org/10.1016/j.apsusc.2019.145037]

[60] Begildayeva, T.; Chinnadurai, D.; Lee, S.J.; Yu, Y.; Song, J.K.; Choi, M.Y. Implementation of novel pulsed laser ablation strategy to control the morphological growth and enrich the electrochemically active sites of multifunctional Ni–CuO electrocatalyst. *J. Alloys Compd.,* **2022**, *901*, 163446. [http://dx.doi.org/10.1016/j.jallcom.2021.163446]

[61] Begildayeva, T.; Lee, S.J.; Yu, Y.; Park, J.; Kim, T.H.; Theerthagiri, J.; Ahn, A.; Jung, H.J.; Choi, M.Y. Production of copper nanoparticles exhibiting various morphologies via pulsed laser ablation in different solvents and their catalytic activity for reduction of toxic nitroaromatic compounds. *J. Hazard. Mater.,* **2021**, *409*, 124412. [http://dx.doi.org/10.1016/j.jhazmat.2020.124412] [PMID: 33187798]

[62] Kurafuji, H. Electrical discharge drilling of glass-I. *Annals of the CIRP,* **1968**, *16*, 415.

[63] Greda, K.; Swiderski, K.; Jamroz, P.; Pohl, P. Flowing liquid anode atmospheric pressure glow discharge as an excitation source for optical emission spectrometry with the improved detectability of Ag, Cd, Hg, Pb, Tl, and Zn. *Anal. Chem.,* **2016**, *88*(17), 8812-8820. [http://dx.doi.org/10.1021/acs.analchem.6b02250] [PMID: 27476678]

[64] Xia, S.; Leng, A.; Lin, Y.; Wu, L.; Tian, Y.; Hou, X.; Zheng, C. Integration of flow injection capillary liquid electrode discharge optical emission spectrometry and microplasma-induced vapor generation: a system for detection of ultratrace Hg and Cd in a single drop of human whole blood. *Anal. Chem.,* **2019**, *91*(4), 2701-2709. [http://dx.doi.org/10.1021/acs.analchem.8b04222] [PMID: 30615435]

[65] Leng, A.; Lin, Y.; Tian, Y.; Wu, L.; Jiang, X.; Hou, X.; Zheng, C. Pump-and valve-free flow injection capillary liquid electrode discharge optical emission spectrometry coupled to a droplet array platform. *Anal. Chem.,* **2017**, *89*(1), 703-710. [http://dx.doi.org/10.1021/acs.analchem.6b03185] [PMID: 27982564]

[66] Jamroz, P.; Greda, K.; Dzimitrowicz, A.; Swiderski, K.; Pohl, P. Sensitive determination of Cd in small-volume samples by miniaturized liquid drop anode atmospheric pressure glow discharge optical emission spectrometry. *Anal. Chem.,* **2017**, *89*(11), 5729-5733. [http://dx.doi.org/10.1021/acs.analchem.7b01198] [PMID: 28478681]

[67] Yu, J.; Yang, S.; Lu, Q.; Sun, D.; Zheng, J.; Zhang, X.; Wang, X.; Yang, W. Evaluation of liquid cathode glow discharge-atomic emission spectrometry for determination of copper and lead in ores samples. *Talanta,* **2017**, *164*, 216-221. [http://dx.doi.org/10.1016/j.talanta.2016.11.015] [PMID: 28107920]

[68] Bishwakarma, H.; Kumar, N.; Anand, M.; Das, A.K.; Singh, P.K.; Mishra, S.K. Magnetic assisted synthesis of ZnO nanoparticle by electrochemical discharge method. *Mater. Today Proc.,* **2022**, *56*, 857-861. [http://dx.doi.org/10.1016/j.matpr.2022.02.511]

[69] Harugade, M.L.; Waigaonkar, S.D. Effect of different electrolytes on material removal rate, diameter of hole, and spark in electrochemical discharge machining. In: *Advances in Manufacturing*; Springer: Cham, **2018**; pp. 427-437. [http://dx.doi.org/10.1007/978-3-319-68619-6_41]

[70] Vickers, N.J. Animal communication: when i'm calling you, will you answer too? *Curr. Biol.,* **2017**, *27*(14), R713-R715. [http://dx.doi.org/10.1016/j.cub.2017.05.064] [PMID: 28743020]

[71] Abou Ziki, J.D.; Fatanat Didar, T.; Wüthrich, R. Micro-texturing channel surfaces on glass with spark assisted chemical engraving. *Int. J. Mach. Tools Manuf.,* **2012**, *57*, 66-72. [http://dx.doi.org/10.1016/j.ijmachtools.2012.01.012]

[72] Vons, V.A.; Anastasopol, A.; Legerstee, W.J.; Mulder, F.M.; Eijt, S.W.H.; Schmidt-Ott, A. Low-temperature hydrogen desorption and the structural properties of spark discharge generated Mg nanoparticles. *Acta Mater.,* **2011**, *59*(8), 3070-3080. [http://dx.doi.org/10.1016/j.actamat.2011.01.047]

[73] Feng, J.J.; Yan, P.X.; Zhuo, R.F.; Chen, J.T.; Yan, D.; Feng, H.T.; Li, H.J. Ultrafast synthesis of Si nanowires by DC arc discharge method and morphology control. *J. Alloys Compd.,* **2009**, *475*(1-2), 551-554. [http://dx.doi.org/10.1016/j.jallcom.2008.07.085]

[74] Saito, G.; Hosokai, S.; Akiyama, T. Synthesis of ZnO nanoflowers by solution plasma. *Mater. Chem. Phys.,* **2011**, *130*(1-2), 79-83. [http://dx.doi.org/10.1016/j.matchemphys.2011.05.084]

[75] Li, C.J.; Cao, X.; Li, W.H.; Zhang, B.W.; Xiao, L.Q. Co-synthesis of CuO-ZnO nanoflowers by low voltage liquid plasma discharge with brass electrode. *J. Alloys Compd.,* **2019**, *773*, 762-769. [http://dx.doi.org/10.1016/j.jallcom.2018.09.250]

[76] Bishwakarma, H.; Tyagi, R.; Kumar, N.; Das, A.K. Green synthesis of flower shape ZnO-GO nanocomposite through optimized discharge parameter and its efficiency in energy storage device. *Environ. Res.,* **2023**, *218*, 115021. [http://dx.doi.org/10.1016/j.envres.2022.115021] [PMID: 36495961]

CHAPTER 11

Heterogeneous Semiconductor Photocatalysis for Water Purification: Basic Mechanism and Advanced Strategies

Naorem Premjit Singh[1,*] and **Yengkhom Rangeela Devi**[2]

[1] *A. R. S. D. College, University of Delhi, Dhaula Kuan, New Delhi, India*

[2] *Department of Physics, Pachhunga University College (PUC), Aizawl, Mizoram, India*

Abstract: Water is essential for all living things, whether it is human beings, animals or plants. Around 70% of the total earth's surface is covered by water, however only a small fraction of it (2.5%) is found as fresh water. On the other hand, due to anthropogenic activities like industrialization, a huge increase in population, utilization of toxic chemicals in agricultural activities *etc*., the available freshwater bodies have been contaminated by various kinds of pollutants, including toxic chemicals released mainly from industries like textile, which causes hazardous to both human being and aquatic life. Therefore removal of these toxic chemicals before entering into fresh water bodies is of great importance. Heterogeneous semiconductor photocatalysis is the most effective green method in this regard because it enables to degrade the pollutants into non hazardous products like CO_2 and H_2O without releasing any harmful residue. Therefore, understanding the knowledge of photocatalysis mechanism is very significant to enable further improvement. Hence, this chapter presents the basic mechanism of photocatalysis, its drawbacks and the advanced strategies to improve the catalytic efficiency. Finally some of the important factors that provide strong influences on the catalytic activity also have been discussed.

Keywords: Basic mechanism, Heterostructure formation, Metal and non-metal doping, Photocatalysis, Surface Plasmon resonance.

INTRODUCTION

Water is one of the basic necessities for all forms of life. Around 70% of our total earth's surface is covered with water. However, only about 2.5% is found as fresh water in the form of river, glaciers, groundwater and water vapours in the atmosphere [1]. Furthermore, due to anthropogenic activities like industrialization, huge increase in population, use of different hazardous

* **Corresponding author Naorem Premjit Singh:** A. R. S. D. College, University of Delhi, Dhaula Kuan, New Delhi, India; Tel: +919366139124; E-mail: premjit.naorem@yahoo.in

Dibya Prakash Rai, Kingsley O. Obodo, & Jitendra Pal Singh (Eds.)

chemicals in agriculture and domestic waste *etc*., the available freshwater bodies are contaminated by different kinds of pollutants, including various toxic chemicals that lead hazardous to both aquatic and surface lives of our planet. Hence, water pollution is one of the major concerns all over the world at the present time. According to a literature report, approximately 3.2 million people die per year in the world due to the use of unsafe water, poor sanitation and inadequate hygiene [2]. The very important toxic chemicals include various coloring organic dyes released mainly from industries like textile, chlorinated and non-chlorinated organic aliphatic and aromatic compounds, inorganic compounds like heavy metals, insecticides, herbicides and byproducts of fertilizers from agricultural activities and antibiotics from agriculture and hospitals [3 - 5]. These toxic chemicals are non-biodegradable and stable in the natural cycle and cause both acute and chronic pollution to our environment [5, 6]. Hence, removal or degradation of these toxic chemicals before entering into freshwater bodies is highly essential. During the past few decades, several traditional treatment methods like adsorption, chlorination, chemical co-precipitation, reverse osmosis and electrochemical methods have been applied in order to solve the above problem. However, all these efforts resulted in some sorts of limitations like high cost of operation, generation of toxic byproducts and transfer of the pollutants from one dumping site to another site *etc*. [7 - 9]. Therefore, they are restricted from practical application. Nevertheless, fortunately, semiconductor heterogeneous photocatalysis is found to degrade these toxic chemicals into non-hazardous products like CO_2 and H_2O effectively without releasing any harmful residues [5]. So, it is considered as the promising green method for solving environmental problems, especially for the degradation of toxic chemicals in wastewater. In addition, applications of semiconductor photocatalysis were also found in other fields, such as the generation of hydrogen and oxygen from water, storage of sunlight energy in chemical bonds and anti-fogging and self-cleaning of windows and lenses [10]. Therefore, understanding the knowledge of photocatalysis mechanism is highly significant in order to enable further improvement. Hence, the present chapter presents the basic mechanism of heterogeneous semiconductor photocatalysis, its limitations and the advanced strategies that enable to improvement of the catalytic activity for the purification of waste water. Some of the important factors that have strong influences on catalytic activity have also been discussed.

PHOTOCATALYSIS AND ITS BASIC MECHANISM

Catalysis can be defined as a branch of science that deals with the acceleration of the rate of reaction by using a substance called a catalyst, and photocatalysis is a type of catalysis involving the utilization of light (photon) energy as a driving force for catalyzing the reaction and the catalyst substance is called photocatalyst.

Since then, the first semiconductor photocatalysis was reported by Fujishima and Honda in 1972 for water splitting into H_2 and O_2 by using semiconductor TiO_2 as anode and Pt as cathode [11, 12], extensive research on semiconductor photocatalysis has been conducted, and so far many semiconductor materials that enable light absorption in UV and visible regions have been developed for various applications including degradation of toxic organic pollutants in wastewater.

In the basic mechanism of semiconductor photocatalysis, the photocatalyst material absorbs light (UV, Visible or IR) energy greater than the width of the band gap energy of the semiconductor, which induces the excitation of electrons from the valence band to the conduction band and leaves holes behind in the valence band of semiconductor. These photo-induced electrons (Conduction band) and holes (Valence band) move to the surface of the catalyst and then promote redox reactions by reacting directly with the adsorbed toxic organic contaminants present in wastewater or by producing some other highly oxidizing and reducing species such as $OH^{\cdot}$ and $O_2^{\cdot-}$ *via* reacting with water as well as hydroxyl ion and adsorbed oxygen molecules [4, 9, 12, 13]. However, the direct participation of the photo-induced electrons and holes rarely occurs as the amount of water molecules present is very large compared to the contaminants [4, 13]. When the photo-excited electron reacts with the adsorbed dissolved oxygen on the surface of the catalyst, it reduces the dissolved oxygen and gives rise to a superoxide radical anion ($O_2^{\cdot-}$). Furthermore, this superoxide radical can get protonation and form hydroperoxyl radical ($HOO^{\cdot}$). On the other hand, the hole reacts with water as well as hydroxyl anion (HO^-) and forms hydroxyl radical ($HO^{\cdot}$) [14, 15]. These radical species can mineralize the organic pollutants. However, the generation of these radical species is allowed only if the potential of the corresponding reaction falls within the potential gap of the valence band and conduction band of the photocatalyst. For instance, TiO_2 has three crystalline phases: anatase, brookite and rutile. The anatase phase has a conduction band potential of E_{CB} = - 0.51V at pH = 7 and the valence band potential of E_{VB} = + 2.69V at pH = 7 [15]. So, both the reduction of adsorbed oxygen ($E^0_{O_2/O_2^{\cdot-}} = -0.33\,V,\ E^0_{O_2/HOO^{\cdot}} = -0.45V$) and oxidation of water and hydroxyl ion($E^0_{H_2O/HO^{\cdot}} = +0.82\,V,\ E^0_{HO^-/HO^{\cdot}} = +2.29V$)[15] can take place since their potential falls within the gap of conduction and valence band potentials of the catalyst. However, in the case of the rutile phase, it has the conduction band potential of E_{CB} = - 0.31V [15], which is less negative than the potential for reduction of dissolved oxygen, so the reduction reaction cannot take place. In addition, a scavenging reaction such as the recombination of the photo-induced electron-hole pair can also take place, which occurs at a very short time and retards the rate of redox reaction. So, for an efficient photocatalyst material, the photo-induced electron-hole pair would have a low recombination rate, high separation and high diffusion rate to the surface of the catalyst. Fig. (**1**) shows the schematic diagram of semiconductor photocatalysis showing the mechanism of photocatalysis.

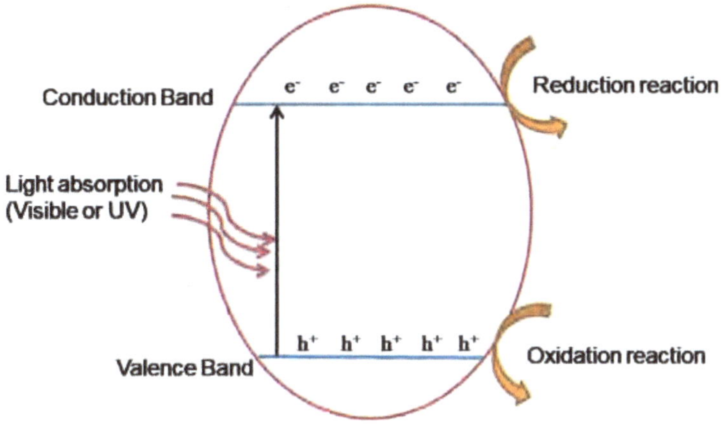

Fig. (1). Schematic mechanistic diagram of semiconductor photocatalysis.

$S + h\upsilon$ (Photon) $\rightarrow S^{\cdot} + h^{+}$ (VB) $+ e^{-}$ (CB)

e^{-} (CB) + Organic Compound $\xrightarrow{\text{Reduction}} CO_2 + H_2O$

e^{-} (CB) + Dissolved oxygen $\rightarrow O_2^{-\cdot}$ (Superoxide)

$O_2^{-\cdot} + H^{+} \rightarrow HOO^{\cdot}$ (Hydroperoxyl)

$HOO^{\cdot} + HOO^{\cdot} \rightarrow H_2O_2 + O_2$

$O_2^{-\cdot}$ + Organic Compound $\rightarrow$ Reaction intermediates $\rightarrow CO_2 + H_2O$

$HOO^{\cdot}$ + Organic Compound $\rightarrow$ Reaction intermediates $\rightarrow CO_2 + H_2O$

h^{+} (VB) + Organic Compound $\xrightarrow{\text{Oxidation}} CO_2 + H_2O$

h^{+} (VB) $+ H_2O \rightarrow OH^{\cdot} + H^{+}$

h^{+} (VB) $+ OH^{-} \rightarrow OH^{\cdot}$

$OH^{\cdot}$ + Organic Compound $\rightarrow$ Reaction intermediate $\rightarrow CO_2 + H_2O$

TiO_2, having a band gap (3.0-3.2 eV), is the most widely investigated semiconductor owing to its low cost, abundantly available, non-toxic and high thermal and chemical stability [12 - 18]. Due to the possession of a large band gap, it can absorb light only in the UV region, which brings low utilization of sunlight since sunlight reaching our planet Earth consists of approximately 5% UV light and 45% visible light [12 - 18]. Likewise, all other semiconductor materials having large band gap ($\geq$ 3.0 eV) including ZnO (3.37 eV) [19, 20], $ZnWO_4$ (3.4 eV) [21, 22], $CdWO_4$ (3.8 eV) [23, 24], SnO_2 (3.6 eV) [25, 26] and so on, also have the same limitation of visible light absorption; therefore they are hindered from practical application. Nevertheless, utilizing the semiconductor having low band gap energy (less than 3.0 eV) can bring light absorption to the visible region. Some of the commonly used visible light active semiconductors can include CdS (2.4 eV) [27, 28], MoS_2 (1.9 eV) [29], SnS_2 (2.31-2.46) [30], CdSe (2.17 eV) [31], CuO (2.2 eV) [32], Fe_2O_3 (2.3 eV) [33], WO_3 (2.4) [33], $FeWO_6$ (2.4) [33], $CoWO_6$ (2.5 eV) [16], $BiMoO_4$ (2.71) [34] *etc*. However, the utilization of pristine semiconductors in photocatalysis resulted in low quantum yield, a high recombination rate of photo-induced electron-hole pair and a lowering of carriers transfer rate to the surface of the catalyst, consequently leading to a reduction of redox reaction and retardation of photocatalytic efficiency. Furthermore, the reductive ability of photo-induced electron in the conduction band increases with the increase of the negative potential of the conduction band edge and the oxidative ability of the photo-induced hole in the valence band with the increase of the positive potential of the valence band edge [35]. So, although the light absorption can be extended to visible light by using the pristine visible light active semiconductor, the catalytic efficiency is lowered due to the fast recombination rate of the electron-hole pair and the low reductive and oxidative ability of the photo-induced electron and hole. Hence, high separation of photo-induced electron-hole pair and, hence, enhanced catalytic efficiency with light absorption in the visible region is of great challenge. Some of the work done on pristine photocatalyst is shown in Table **1**.

Table 1. Some of the works of pristine photocatalyst.

Catalyst	Analyzed Dyes	Irradiation Light	Synthetic Method	Refs.
TiO_2	Rhodamine B	UV	Hydrothermal + Calcination	[36]
	Acid red 27	UV	Sol-gel	[37]
	Benzene	VUV	Hydrolysis + Calcination	[38]
ZnO	Methylene blue	UV	Hydrothermal	[39]
	Methyl Orange	UV	Microwave	[40]
	Rhodamine B	UV	Microwave	[40]
	Methylene blue	UV	Combustion	[41]
SnO_2	Methylene blue	UV	Green method using plant extract	[26]
	-	-	Hydrothermal	-
$ZnWO_4$	Rhodamine B	UV	Hydrothermal	[21]
	Formaldehyde	UV	Hydrothermal	[21]
	Methylene Blue	UV	Microwave assisted	[22]
$CdWO_4$	Methyl Orange	Solvothermal	Solvothermal	[23]
CdS	Methylene Blue	Visible	Chemical bath deposition	[42]
SnS_2	Rhodamine B	Visible	Hydrothermal	[30]

(Table 1) cont.....

Catalyst	Analyzed Dyes	Irradiation Light	Synthetic Method	Refs.
CdSe	Rhodamine B	Visible	Hydrothermal	[31]
WO_3	Rhodamine B	Visible	Hydrothermal	[43]

STRATEGIES TO IMPROVE THE CATALYTIC EFFICIENCY AND LIGHT ABSORPTION

In the last few years, various strategies to improve the light absorption and catalytic activity have been adopted. Some of the important advanced strategies are:

1. Doping of metal and non-metal ions into the lattice of the pristine semiconductor.

2. Doping of noble metal (Ag., Au., Pt., Pd.) into the pristine semiconductor (Surface Plasmon Resonance).

3. Construction of heterostructure composites between the semiconductor materials.

DOPING OF METAL OR NON-METAL

The doping of a semiconductor with metal or non-metal ions leads to a change in the electronic structure of the semiconductor material by forming energy states between the conduction and valence bands of the semiconductor [3, 44]. Due to the insertion of this new energy state, the band gap of the semiconductor is reduced, which lowers the band gap energy and thus enables light absorption to the visible region. In this process, the doping ion may act either as an electron or hole trapping center and sometimes it also acts as both electron and hole trapping centers [44]. Furthermore, the doping ion may also act as a recombination center, depending on the mode of doping. When the dopant is doped near the surface of the semiconductor, it likely acts as an electron or hole trapping center, which increases the lifetime of the photoinduced electrons and holes, consequently the recombination rare of the photo-inducedelectron–hole pair is suppressed and promotes their mobility to the surface of the catalyst [4, 16]. Hence, the catalytic efficiency can be improved. On the other hand, if the doping material is doped inside at a deep level, it likely acts as a recombination center due to the transfer of carriers to the surface of the catalyst is difficult, therefore undermining the catalytic activity [4, 16].

Metal Ion Doping

Transition metal ions and rare earth ions are commonly used as dopant ions. If the transition metal ion is doped, the induced energy state usually originates from the d-orbital of the metal ion, while in the case of doping with rare earth ions, the induced energy state originates from the f-orbital [4, 16, 45]. Furthermore, the narrowing of the energy gap and the enhancement of photocatalytic activity can also be observed due to the creation of oxygen vacancy defects [46 - 48]. For instance, the doping of Al on TiO_2 reduces the band gap distance of TiO_2 due to the creation of oxygen vacancy [47]. Usually, the doping metal ion acts as either an electron trapping center or hole trapping center or both as electron and hole trapping centers.

When the doping ion captures the photo-excited electron from the conduction band, it reduces to a lower oxidation state, and if it captures a hole from the valence band, it oxidizes to a higher oxidation state. Because of this, the life span of the photo-induced electron and hole is increased, which leads to a reduction in the recombination rate and enhances the mobility of charges to the surface of the catalyst. As a consequence the redox reactions at the surface of the catalyst can be improved and hence the photocatalytic activity. So far, several metal ions have been reported to dope into different semiconductor photocatalyst materials to improve the catalytic efficiency. Some of the reported metal ions that improve the catalytic activity by suppressing the recombination of photo-induced electron-holes pair can include Cu [49], Fe [50, 51], Mo [52, 53], W [54], Pd [55, 56], Pt [57], Ag [58, 59], Co [60, 61], Eu [45], Er [62] *etc*. For instance, Kura Narsimha *et al*. [49] reported an enhancement in the photocatalytic activity of $CdWO_4$ on doping with Cu^{2+} for degradation of methyl orange under sunlight irradiation due to the enhanced separation of photo-induced charge carriers induced by the electron capture. Vaishali Soni *et al*. [58] reported enhanced photocatalytic activity of TiO_2 for degradation of rhodamine B under sunlight irradiation by doping with Ag^{+} ion due to the suppression of electron-hole recombination. D. Ye *et al*. [45] also reported enhanced photocatalytic activity of $CdWO_4$ for degradation of methyl orange by doping with Eu^{3+} ion. Sometimes, the metal ion captures both the electrons and holes. This normally occurs when the doping metal ion exists in two or more oxidation states in the lattice. In such cases, the species having a higher oxidation state generally captures the electron, and the species having a lower state captures the hole. For instance, Soonhyun Kim *et al*. [57] reported an enhancement in the catalytic activity of TiO_2 for the degradation of dichloro acetate and 4-chlorophenol under visible light irradiation by doping with Pt due to the suppression ofelectron–hole pair recombination induced by the capture of both electron and hole. Xiaofen Wang *et al*. [50] also reported an enhancement in the photocatalytic activity of $CdWO_4$ on the degradation of methyl orange dye by doping with Fe ion due to the capture of both electrons and holes. Fig. (**2**) shows the schematic diagram showing the reduction of band gap energy due to the doping of metal ions.

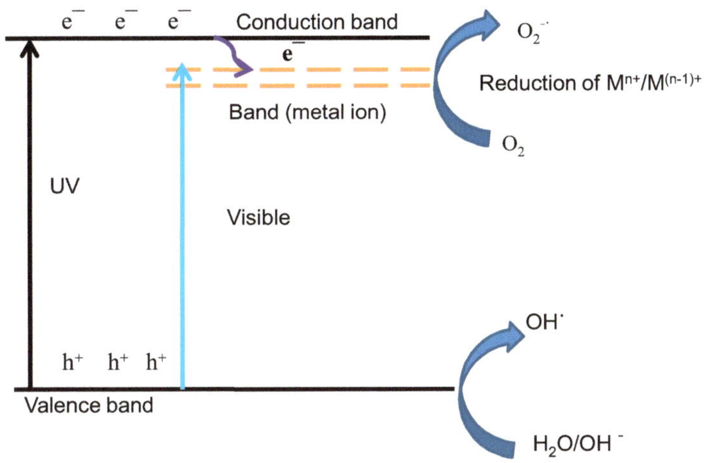

Fig. (2). Schematic band gap diagram showing the tuning of band gap energy due to metal ion doping.

Non-metal Doping

In this technique, non-metals such as C, N, S F *etc.*, are generally used as dopant material. The induced sub-band in non-metal doping usually originates from the 2p orbital of the doping non-metal [4, 63]. As non-metals generally have high electron density, they act as hole-trapping centres, and therefore, the induced energy level exists near the valence band of the substrate semiconductor. Fig. (**3**) shows the schematic diagram showing the tuning of band gap energy due to the doping of non-metal.

When the dopant non-metal captures the photo-induced hole from the valence band, the dopant non-metal oxidizes to a higher positive oxidation state, which increases the life time of the photo-induced hole. As a consequence, the recombination rate of the photo-induced electron-hole pair can be suppressed and improves the catalytic efficiency. Non-metals such as C [64, 65], N [66, 67], S [68, 69], B [70, 71], F [72, 73] *etc.*, have been reported to enhance the photocatalytic activity for degradation of toxic chemicals in wastewater by suppressing the electron-hole recombination. For instance, Asahi *et al.* [66] reported an enhancement in the catalytic activity of TiO_2 for degradation of methylene blue under visible light irradiation by doping with N. Ming Shen *et al.* [64] reported enhanced in photocatalytic activity of TiO_2 on degradation of trichloro acetic acid under visible light irradiation by doping with carbon. Apart from the sole doping of non-metal, co-doping such as metal and non-metal N-Cu [74], Fe-N, Ni-N [75] and non-metal-non metal C-N [76], N-P [77], N-S [78], C-N-S [79, 80] also have been reported to enhance the catalytic activity.

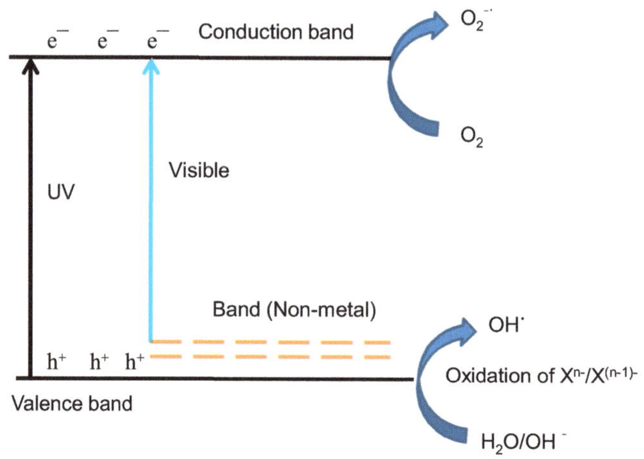

Fig. (3). Schematic band gap diagram showing tuning of band gap energy by non-metal doping.

SURFACE PLASMON RESONANCE

This process is one of the effective methods to enhance the semiconductor photocatalytic activity and, for extension, the light absorption. Surface Plasmon is a collective oscillation confined to the surface of conductive materials. Noble metals such as Ag, Au, Pt and Pd have localized surface Plasmon originating from the valence electron. Therefore, when the incident visible light falls on them, localized surface Plasmon resonance can be observed [81]. Hence, nanomaterials of noble metals are generally loaded onto the UV light active semiconductor, say TiO_2, to enhance the light absorption in the visible light and to improve the photocatalytic activity. Other metals, such as Al and Cu, can also exhibit localized surface Plasmon resonance, but due to their high instability under ambient conditions, their practical application is hindered [82]. The Plasmon resonance of the incident light at the plasmonic substance occurs when the oscillating frequency of the incident light is matched with the oscillating frequency of the surface Plasmon [83]. If the frequency of the incident light is lower than the Plasmon frequency, the light is reflected, and if the frequency is greater, then the light is transmitted [83]. Further, the frequency of localized surface Plasmon resonance of metal conductors is strongly affected by the nature of the metal, size, morphology and dielectric constant of the surrounding media [82]. Therefore, plasmonic semiconductor photocatalysts of light harvesting in various wavelengths can be constructed by varying these properties. When the incident light falls on the plasmonic substance, the following can be occurred.

1. Generation of hot/energetic electron.
2. Generation of enhanced electric field in the surrounding.
3. Scattering of light.

The enhancement in catalytic activity under visible light absorption by localized surface Plasmon resonance can be explained through the following three mechanisms [84].

1) Charge transfer mechanism: In this mechanism, the generated hot electron is transferred to the conduction band of the nearby semiconductor across the interfacial Schottky barrier, which lowers the recombination rate of the photo-induced electron-hole pair in metal [82]. As a result, the photocatalytic activity at the surface of the semiconductor is increased [84]. As the hot electrons play an important role in the activity, effective transfer of generated hot electrons to the conduction band of nearby semiconductors is of great significant to obtain efficient catalytic activity [82]. Further, in order to have effective electron transfer, there has to be good interfacial contact between the metal and semiconductor, and it is dependent on various parameters like morphology and growth site.

2) Near field electromagnetic mechanism: In this mechanism, the intense electromagnetic field generated in the vicinity of the plasmonic nanostructure on falling the incident visible light interacts with the nearby semiconductor, which induces electron and hole in conduction and valence of the semiconductor, respectively [84]. Thus, the redox reaction on the surface of the semiconductor can take place.

3) Enhanced light absorption: In this process, the incident light is scattered by the plasmonic substance to the nearby semiconductor, which provides more optical path lengths for incident photons to pass through the semiconductor and therefore yields higher excited charge carriers and hence, catalytic activity [83, 84].

So far, several photocatalysts that enable to enhance the degradation efficiency of toxic organic compounds *via* surface Plasmon resonance have been developed. Some of the recently developed examples can include SiO_2@Ag and SiO_2@Ag@$AgPO_4$ for phenol degradation [85], Ag-Bi_2WO_6 for Rhodamine B degradation [86], Au/g-C_3N_4 nanosheets for Rhodamine B degradation [81], Pt/TiO_2 for methyl orange degradation *etc.* [87]. However, unfortunately, noble metals are very expensive, which causes a high cost of catalyst production.

CONSTRUCTION OF SEMICONDUCTOR HETEROSTRUCTURE

This is one of the most effective strategies for the efficient absorption of sunlight and the separation of photo-induced electron-hole pairs. In this fabrication, two or more semiconductors having different band gap energy are combined together, and formed an interface at the junction. The heterostructure formation can be of different types such as n-p, n-n and p-p types and be classified into three types based on the band energies of the coupling semiconductors, namely Type I (Straddling gap), Type II (Staggered gap) and Type III (Broken gap) [12, 88]. Of these types, the type II heterostructure formation is the most effective in the separation of photo-induced electron-hole pair.

In the type II heterostructure composite, the band arrangement between the coupling semiconductors is done in such a way that the conduction and valence band of one of the semiconductors, say 1 is higher than the conduction and valence band of another semiconductor, say 2. Therefore, the conduction band of semiconductor 1 is more negative than the conduction band of semiconductor 2, whereas the valence band of semiconductor 2 is more positive than the valence band of semiconductor 1 [12]. As a result, the migration of photo-induced electrons can take place from the conduction band of semiconductor 1 to semiconductor 2 and holes from the valence band of semiconductor 2 to the valence band of semiconductor 1. Because of this migration, the photoinduced electrons and holes are accumulated at different semiconductor phases, which leads to effective separation of the photo-induced electron-hole pair and enhances catalytic efficiency. Fig. (**4**) shows the schematic diagram of type II heterostructure formation.

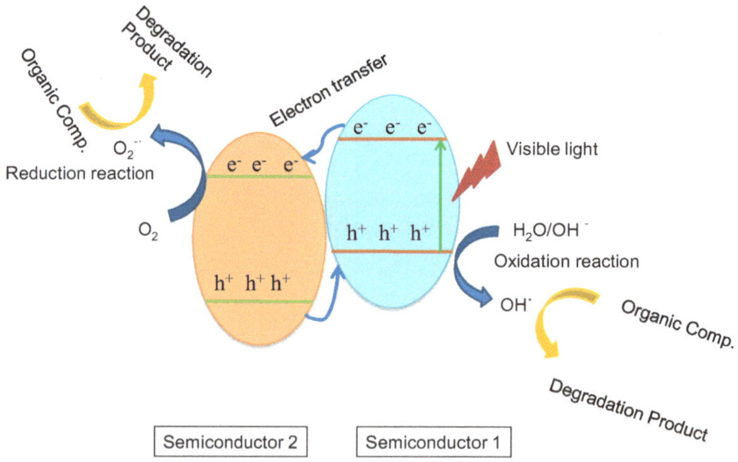

Fig. (4). Schematic diagram of type II heterostructure composite.

In order for the heterostructure composite to be visible light active, at least one of the coupling semiconductors should be enabled to absorb light in the visible region, and that semiconductor has to be present at the outer shell of the composite. Also, the band alignment at the interface of the coupling semiconductors needs to satisfy the conditions of type II heterostructure [12]. Further, if there is a difference in lattice spacing between the coupling semiconductors, lattice mismatch at the interface may be observed, which may induce defects that may trap the carriers. Hence, the diffusion of the carriers to the surface may be reduced [12]. So, choosing semiconductors for coupling is very important in the construction of a heterostructure composite. So far, several heterostructure composite photocatalysts of type II have been developed. Some of the recently developed examples include the ZnO-Cu_xO heterostructure composite synthesized by Prami Nandi and Debajyoti Das through the hydrothermal method for degradation of Rhodamine B dye under visible light irradiation [89]. Hydrothermally synthesized 2D layered structure TiO_2@MoS_2 heterostructure composite reported by Asmaa M A Omar *et al.* for degradation of methyl orange under visible light irradiation [90]. The authors reported remarkable degradation efficiency due to the formation of a unique interface between the MoS_2 and TiO_2. Ag_3PO_4/$CdWO_4$ heterostructure composite reported by Caiting Zhang *et al.* for degradation of rhodamine B and phenol under visible light irradiation [91]. They reported enhanced catalytic efficiency as compared to the pristine semiconductor and high stability and high recyclability. α-Fe_2O_3/g-C_3N_4 heterostructure reported by Fuxiang Ge *et al.* for degradation of phenol [92]. The authors reported a significant increase in catalytic efficiency after heterostructure formation due to the increase of visible light absorption and spatial separation of the photo-induced electron hole-pair. However, the shortcomings of this type II heterostructure formation is that the redox ability is weakened after charge transfer due to the electrons and holes being left at less negative conduction band potential and less positive valence band potential, respectively [35], which reduces the capability of the photoinduced electron and hole for redox reaction.

On the other hand, the heterostructure composite of type 1 and type 3 band alignments has a strong possibility of recombination of photo-induced electron-hole pair [93]. Therefore they are not considered as significant heterostructure in the photocatalytic process.

Z-SCHEME HETEROSTRUCTURE FORMATION

Nowadays, Z-scheme heterostructure formation is highly attractive due to its high effectiveness in the spatial separation of photo-induced electrons and holes. The band alignment is similar to the band alignment of type II heterostructure composite, however, the way of electron migration is different. The concept of z-scheme formation was taken from the photo-system 1 and 2 reactions of naturally occurring photosynthesis in plants, and it was first proposed by Bard in 1979 [94]. In the artificial z-scheme formation, the electrons on the valence bands of both the coupling semiconductors are excited to their respective conduction bands by absorption of incident light. Then, the excited electrons present in the lower conduction band potential are combined with the holes present in the higher valence band potential, which leaves strongly reducing electrons in the higher conduction potential and strongly oxidizing holes in the lower valence potential [28, 94, 95]. Consequently, the redox reaction can be improved and hence the photocatalytic activity. The electron-hole or charge carrier's recombination center was initially achieved by using an ionic redox mediator in solution such as Fe^{3+}/Fe^{2+} or IO_3^-/I^-, NO_3^-/NO_2^-. However, this type of z-scheme formation limited their application to the liquid phase only; therefore, they are prevented from wide practical applications [94]. In order to solve the above problem, an all-solid-state z-scheme heterostructure composite was developed by Tada *et al.* in 2006 [96, 97], and it is enable to operate in both liquid and gas phases.

The artificial all-solid-state Z-scheme heterostructure composite can be constructed in the following two ways:

1. By inserting an electron mediator between the coupling semiconductors.

2. By direct contact of the coupling semiconductors without insertion of any electron mediator.

An electron mediator is a substance that facilitates the recombination rate of electrons and holes present in different semiconductors of the heterostructure when inserted between the semiconductors. It acts as a charge transfer bridge or charge combination center. For an effective mediator, there should be less resistance to the flow of the electrons and holes; therefore, nanoparticles of highly conducting noble metals such as Ag, Au, Pt, and Pd *etc*., are generally used as electron mediators [94, 95]. Apart from acting as an electron mediator, the inserted noble metals may satisfy the Surface Plasmon Resonance (SPR) too, which will extend the light absorption to higher wavelengths and improve the spatial separation of photo-induced electron-hole pairs [95]. Fig. (**5**) shows the typical schematic diagram of the Z-scheme heterostructure formed by the insertion of an electron mediator.

Some of the developed composites for the degradation of toxic chemicals in wastewater can include AgBr-Ag-Bi_2WO_6 for degradation of Procion red MX- 5B and pentachlorophenol under visible light irradiation [98], SnO_2-Pt-In_2O_3 for degradation of 2, 4-dichloro phenol under visible light irradiation [99], $NaNbO_3$-Au-Sn_3O_4 for degradation of carbofuran [100] *etc*. Apart from metal, non-metal such as Bismuth, carbon-based materials like reduced graphene oxide (RGO), graphene, carbon dots and carbon nanotubes *etc*. can also act as electron mediators [96, 101, 102]. For instance, D. Ma *et al*. [101] developed RGO g-C_3N_4/Bi_2WO_6 composite with RGO electron mediator for degradation of 2, 4, 6 trichlorophenol. W-K Jo *et al.* [102] also developed CdS/ g-C_3N_4 composite with an RGO electron mediator for the degradation of atrazine.

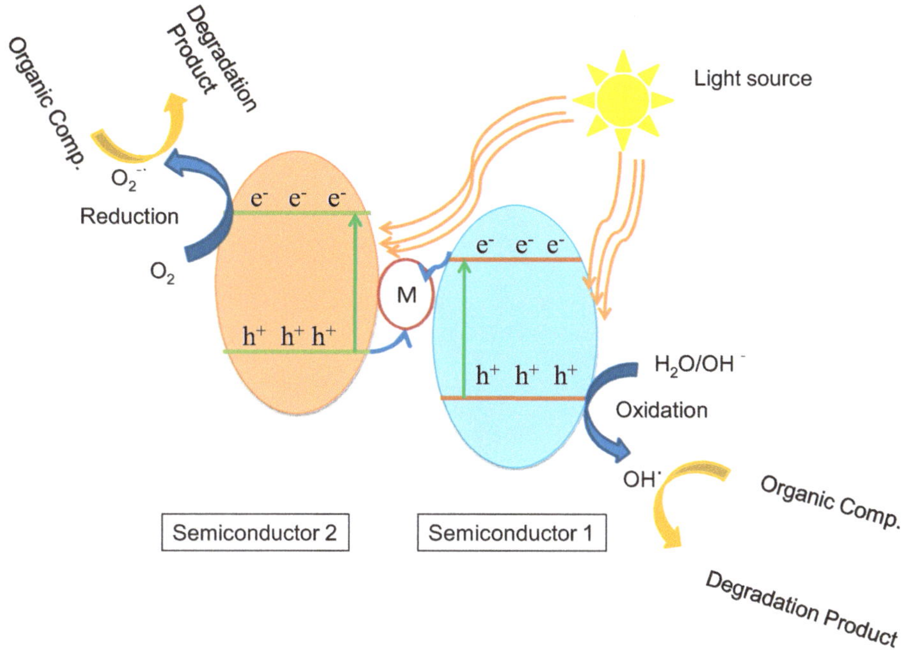

Fig. (5). Schematic diagram of Z-scheme heterostructure composite in the presence of an electron mediator.

The Z scheme heterostructure composite can also be constructed without the insertion of an electron mediator called the direct Z-scheme. In this type, the electron transfer takes place directly at the interface between the coupling semiconductors and the recombination is driven by the electrostatic interaction between the generated charge carriers. At the interface of the coupling semiconductors, a large number of defects are created at the interface, which is called an ohmic contact and it facilitates the transfer of electrons and holes. This ohmic contact acts as a mediator since there is a similarity of properties, such as low electrical resistance, between the ohmic contact and conductor [94]. The formation of defects at the interface relies on different parameters, including synthetic process and synthetic conditions, morphology, surface area *etc.* [103]. The schematic diagram of direct Z-scheme heterostructure formation is shown in Fig. (**6**).

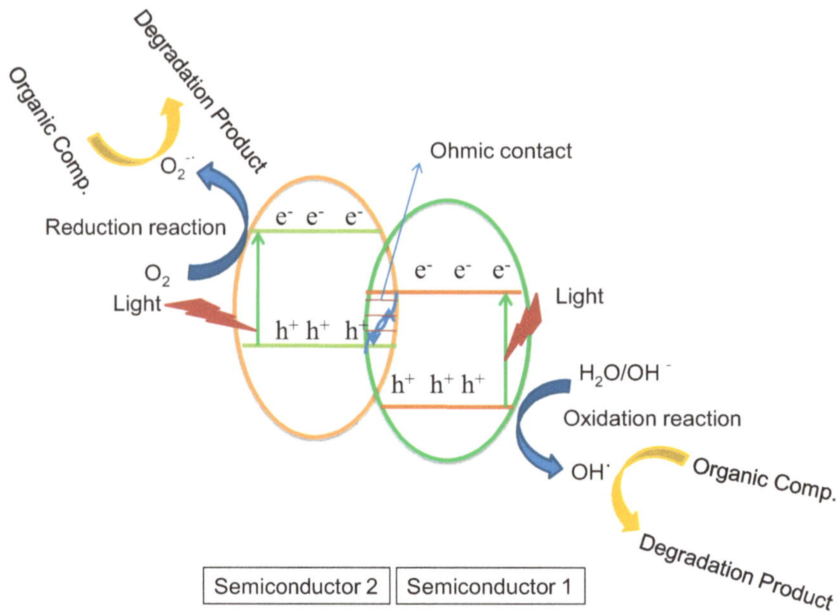

Fig. (6). Schematic diagram of direct Z-scheme heterostructure composite.

Some of the recently developed composites for the degradation of toxic organic chemicals include $Ag_3PO_4/CuBi_2O_4$ photocatalysts developed by Weilong Shi *et al.* using combined hydrothermal and precipitation methods for the degradation of tetracycline under visible light irradiation [104]. The authors reported significance enhanced in catalytic activity due to the strong separation of photo-induced electron-hole pair on the formation of composite, and the percentage degradation was reported to be 75% of the tetracycline within 60 minutes while only 32% and 45% by pure $CuBi_2O_4$ and Ag_3PO_4 respectively. The heterostructure composite of Cu_2O/WO_3 developed by Hassan Ali *et al.* through solid state method for degradation of methylene blue dye under visible light irradiation [103]. The authors reported 3.3 times increases in degradation efficiency after a composite of Cu_2O (10 wt %)/WO_3 is formed. The heterostructure composite of $CoWO_4/CdS$ synthesized by Haojie Cui *et al.* using in situ hydrothermal method for degradation of dyes like methylene green under visible light irradiation [105]. The authors reported outstanding degradation efficiency by forming the composite $CoWO_4$ (1.2%)/CdS with a degradation efficiency of 97% within 80 minutes. It also reported strong stability and recyclability of the composite, and the degradation efficiency was strong dependent on the amount of $CoWO_4$ contents.

Generally, electron mediator-based Z-scheme heterostructure composite shows greater ability of charge transfer as compared to the direct z-scheme type heterostructure composite, which is due to the lattice mismatch at the interface of the direct z-scheme [35, 103]. However, the demerit of electron mediator-based Z-scheme composite is that the noble metals used as electron mediators are highly expensive, which increases the cost of construction.

FACTORS AFFECTING THE CATALYTIC EFFICIENCY

Photocatalysis is a surface phenomenon, and the photocatalytic reactions take place at the surface of the semiconductor. So, apart from the rate of mobility or transport of the photoinduced electrons and holes from inside of the bulk to the surface of the catalyst and the rate of their recombination, the photocatalytic activity of semiconductor photocatalysis also depends on various factors, including the physical and chemical properties of the surface of the catalyst, amount of active sites present on the surface. Further, these properties are affected by variations in parameters such as morphology, particle size, crystallinity, interface, defects, synthetic methods *etc.* [106]. In addition, the catalytic efficiency depends on the amount of catalyst loading, pH of the solution, light intensity and the amounts of pollutants, too. Therefore, understanding the effect of these parameters on the photocatalytic efficiency is of great significance.

Effects of Particle Size, Morphology and Crystal Structure

Nanomaterials have particle sizes in the range of 1-100 nm in any dimension, and they possess unique properties that are different from their counterpart bulk particle. The catalytic efficiency usually increases with the reduction of catalyst particle size because of the increase of surface area to volume ratio, which provides significant amounts of active sites on the surface of the catalyst and also due to the shortening of charge transportation length [107]. On the other hand, the efficiency may also be increased with the increase of crystallinity due to the increases in lattice perfection, reduction in defects that sometimes behave as recombination centres and change in electronic structure [24]. So, there exists a contradiction between the particle size and crystallinity as particle size grows with the increase of crystallinity. Hence, for highly efficient activity, the catalyst has to be smaller in size with high crystallinity. Further, the catalytic efficiency can also be affected by the morphology of the particle. Generally, particle morphology that can provide more light absorption, more surface area and active sites, higher mobility and lower recombination rate exhibited higher catalytic activity. For instance, particles with one-dimensional morphology, such as nanotubes, nanowires, nanorods *etc.*, result in higher catalytic activity because of their higher mobility and lower recombination rate [108]. The porous structure has a large surface area and shortened diffusion pathway. Therefore generally showed enhanced catalytic activity [81]. Dong *et al.* reported the enhanced photocatalytic activity of branching morphology of Ag_3PO_4 as compared to other morphologies such as tetrapod, nanorod and triangular prism on the degradation of methylene blue and rhodamine B dyes due to the possession of large specific surface area [109].

Effect of pH

pH represents a negative log of hydrogen ion concentration of a solution, which indicates the acidity and alkalinity of the solution and its value is extended in the range of 0-14. It has a strong influence on the catalytic efficiency, especially in those compounds that carry charge in solution. For instance, dyes can be classified into two types such as cationic and anionic dyes depending on their dissociation charges in solution. Rhodamine B and Methylene blue are examples of cationic dyes, whereas methyl orange, Reactive black B, and Congo red are examples of cationic dyes. The effect of pH depends on the types of dyes to degrade whether cationic or anionic [110]. At the lower pH of the solution, the positive charge density on the surface of the catalyst will be increased due to the high concentration of H^+ ion, which will make an increase in the degree of adsorption and interaction of the anionic dyes, say methyl orange to the catalyst surface. Consequently, the degradation of the dye will lead to an increase. Similarly, at a higher pH of the solution, there will be an increase in adsorption and interaction of the cationic dye, say methylene blue, to the catalyst surface, therefore leading to an increase in degradation. For instance, K. Buvaneswari *et al.* [110] studied the effect of pH on the degradation of Methylene blue by $FeWO_4$/ZnO photocatalyst under visible light irradiation and reported that the activity was highest at pH = 10. This was attributed to the presence of excess hydroxyl ions around the surface of the semiconductor photocatalyst, which increases the interaction between photocatalyst and methylene blue since methylene blue is a cationic dye. Sushil Kumar Kansal *et al.* [111] reported an increase in degradation efficiency of anionic dye reactive black 5 (RB5) by the ZnO catalyst with the decreases of pH due to the increase of electrostatic interaction between the positive catalyst surface and the negative charge of anionic dye.

Effect of Catalyst Concentration

The efficiency of photocatalytic degradation is also strongly influenced by the amount of the catalyst suspended. Generally, the degradation efficiency increases with the increase of suspended catalyst amount due to the increase in absorption of light by semiconductor and adsorption of pollutant molecules on the surface of the catalyst, which results in an increase of the redox reaction and hence the catalytic activity [112]. However, this increase continues up to certain concentrations called the optimum concentrations. After that, further increases lead to retardation of degradation due to the increase of light opacity in the solution, and consequently, the light scattering is diminished [112]. For instance, Huayue Zhu *et al.* [113] reported an increase in methyl orange (15 mg/L) degradation efficiency with the increase of TiO_2/ZnO/chitosan nanocomposite thin-films catalyst amount under solar light irradiation and found a maximum at 0.5 g/l, after that further increases resulted to less effective. K. Buvaneswari *et al.* [110] reported an increase in methylene blue degradation efficiency with the increase of catalyst $FeWO_4$/ZnO amount up to 0.04 g/l due to the increase of active sites on the semiconductor after that further increase of catalyst amount resulted to a reduction in degradation efficiency due to the blocking of light penetration.

Effect of Pollutant Concentration

The rate of degradation and de-colorization of pollutant molecules is also strongly dependent on the initial concentration of the pollutants. Usually, the degradation efficiency increases first with the increase of pollutant concentration up to a certain concentration due to the greater amounts of pollutants being adsorbed on the surface of the catalyst. After a particular concentration, a further increase in pollutant concentration leads to decreases in activity. This is because in excess of pollutant amounts, the surface of the catalyst is heavily covered by the pollutant molecules, which results in decrease of light penetration into the semiconductor, consequently resulting in a decrease in the generation of oxidative and reductive species, the rate of degradation and de-colorization [110 - 112]. Xiaoqing Chen *et al.* [114] reported a significant decrease in methyl orange and congo red degradation efficiency by ZnO catalyst when the dye concentration changes from 10- 50 mg/l. Suhila Alkaykh *et al.* [115] showed an increase of methylene blue degradation first by 0.05 g of $MnTiO_3$ nanoparticle up to dye concentration of 1×10^{-5} M. After that, further increases of dye concentration resulted in to decrease in efficiency due to the blockage of light penetration into the catalyst.

Effect of Light Intensity

The intensity of light absorbed by the photocatalyst material for the excitation of valence electrons is also a very important parameter in photocatalysis because the initiation of photochemical reactions and the formation of electron-hole pair are strongly dependent on it [15, 116]. With the increase of light intensity, the photon for inducing excitation of the valence electron is increased, which eventually leads to the increase of electron-hole pair formation. Herrmann and Pichat [117] have reported that there is a direct relationship between incident flux and conversion rate. However, after all the active sites are activated, there will be no increase in degradation rate even though the light intensity is increased. Collazzo *et al.* [118] have reported an increase in the degradation rate of dye direct black 38 by the N-doped TiO_2 photocatalyst under visible light irradiation as compared to sunlight irradiation due to an increase in light intensity. Hung and Yuan [119] have reported an increase in the photo-degradation rate of Orange G dye with the increase in light intensity. Liu *et al.* [120] have reported increases in decolorisation of dye acid yellow 17 with the increase of light intensity under UV light irradiation.

Effect of Temperature

Reaction temperature also has a strong influence on the photocatalytic degradation rate. In general, with the increase in reaction temperature, the rate of recombination of the photo-induced electron-hole pair and desorption of the reactant molecules increases, which results from a decrease in the photocatalytic activity [121]. However, at a very low temperature, too, there occurs a decrease in desorption of the final reaction product molecules, which creates limitations on the adsorption of reactants on the catalyst surface and redox reaction, resulting in decreases in degradation rate [122]. Therefore, the optimum photocatalytic degradation rate generally occurs at temperatures not so high and not so low. Soares *et al.* [123] have reported the optimum temperature for photodegradation of dye rhodamine B by P-25 TiO_2 in the range of 40-50°C. Barakata N. A. M. *et al.* [124] have reported an increase in photo-degradation of dye rhodamine B by Ag-doped TiO_2 nanoparticle with the increase of solution temperature. They also reported that apart from the solution temperature, the activity was dependent on the morphology and amount of Ag contents.

Effect of O_2 Concentration

The concentration of dissolved oxygen in solution is also one of the important parameters affecting the photocatalytic activity since it can trap the excited conduction band electrons and form reduced species such as $O_2^{-\cdot}$ and $O_2^{2-\cdot}$, which are suggested to play an important role in some photocatalytic reactions [125]. However, it does not affect the adsorption of the pollutant molecules on the surface of the TiO_2 catalyst and its effect on catalytic activity (improve/hinder) relies on the mechanism of the reaction involved [15]. Wang Y. and Hong Ch-S [125]. have reported an increase in the mineralization of 2-chlorobiphenyl with the increase in oxygen concentrations. They reported 16%, 53%, 84% and 94% mineralization rate with the oxygen concentrations of 0.5, 4, 21 and 186 kPa under UV light irradiation for 5h. Shirayama H. *et al.* [126] studied the photodegradation rate of 7 chlorinated hydrocarbons C_2Cl_4, C_2HCl_3, $C_2H_2Cl_2$, 1,1,1 – $C_2H_3Cl_3$, 1, 1, 2 - $C_2H_3Cl_3$, $CHCl_3$, CCl_4 under UV light irradiation in presence and absence of dissolved oxygen. They reported that the degradation rate is faster in the absence of dissolved oxygen and the dissolved oxygen acts as scavenger which reduces the light intensity in the photoreactor.

CONCLUSION

This chapter summarizes the basic mechanism of the semiconductor photocatalysis which is considered as effective method for the degradation of toxic organic compounds present especially in industrial waste water. However the utilization of pure semiconductor leads to lower catalytic activity because of high recombination rate of photo-induced electron hole pair which makes limitation in practical application. So, in order to improve the catalytic activity and to extend the light absorption to longer wavelength such as visible light which accounts 45% of natural sunlight, several advanced strategies including metal or non-metal ion doping, surface Plasmon resonance and construction of heterostructure between the different semiconductors can be applied and the mechanism of how these strategies enhancing the light absorption and the catalytic efficiency can be understood from this study. Further, the catalytic efficiency is found to depend on various parameters such as size, morphology and crystallinity of the catalyst, intensity of light absorbed, amount of catalyst and pollutants, the concentration of dissolved oxygen, and temperature of solution on the catalytic activity. It is believed that this chapter will be very helpful to researchers in designing novel semiconductor photocatalyst to improve further activity.

REFERENCES

[1] Black, M. *The atlas of water: Mapping the world's most critical resource*; University of California Press: Berkeley, California, United States, **2016**.
[http://dx.doi.org/10.1525/9780520965607]

[2] Liu, L.; Johnson, H.L.; Cousens, S.; Perin, J.; Scott, S.; Lawn, J.E.; Rudan, I.; Campbell, H.; Cibulskis, R.; Li, M.; Mathers, C.; Black, R.E. Global, regional, and national causes of child mortality: An updated systematic analysis for 2010 with time trends since 2000. *Lancet,* **2012**, *379*(9832), 2151-2161.
[http://dx.doi.org/10.1016/S0140-6736(12)60560-1] [PMID: 22579125]

[3] Li, D.; Shi, W. Recent developments in visible-light photocatalytic degradation of antibiotics. *Chin. J. Catal.,* **2016**, *37*(6), 792-799.
[http://dx.doi.org/10.1016/S1872-2067(15)61054-3]

[4] Elsalamony, R.A. Advances in photo-catalytic materials for environmental applications. *Res. Rev.: J. Mat. Sci.,* **2016**, *4*, 26-50.

[5] Cheng, J.; Shen, Y.; Chen, K.; Wang, X.; Guo, Y.; Zhou, X.; Bai, R. Flower-like Bi_2WO_6/ZnO composite with excellent photocatalytic capability under visible light irradiation. *Chin. J. Catal.,* **2018**, *39*(4), 810-820.
[http://dx.doi.org/10.1016/S1872-2067(17)63004-3]

[6] Alvi, M.A.; Al-Ghamdi, A.A.; ShaheerAkhtar, M. Synthesis of ZnO nanostructures *via* low temperature solution process for photocatalytic degradation of rhodamine B dye. *Mater. Lett.*, **2017**, *204*, 12-15. [http://dx.doi.org/10.1016/j.matlet.2017.06.005]

[7] Zhang, D.; Zeng, F. Visible light-activated cadmium-doped ZnO nanostructured photocatalyst for the treatment of methylene blue dye. *J. Mater. Sci.*, **2012**, *47*(5), 2155-2161. [http://dx.doi.org/10.1007/s10853-011-6016-4]

[8] Jothivenkatachalam, K.; Prabhu, S.; Nithya, A.; Chandra Mohan, S.; Jeganathan, K. Solar, visible and UV light photocatalytic activity of $CoWO_4$ for the decolourization of methyl orange. *Desalination Water Treat.*, **2015**, *54*(11), 3134-3145. [http://dx.doi.org/10.1080/19443994.2014.906324]

[9] Saravanan, R. Francisco Gracia and A. Stephen. In: *Nonocomposites for visible light-induced photocatalysis; Mohammad Mansoob Khan, Debabrata Pradhan, Youngku Sohn*; Springer, **2017**; Vol. 1, pp. 19-40. [http://dx.doi.org/10.1007/978-3-319-62446-4_2]

[10] Hou, W.; Cronin, S.B. A review of surface plasmon resonance-enhanced photocatalysis. *Adv. Funct. Mater.*, **2013**, *23*(13), 1612-1619. [http://dx.doi.org/10.1002/adfm.201202148]

[11] Fujishima, A.; Honda, K. Electrochemical photolysis of water at a semiconductor electrode. *Nature,* **1972**, *238*(5358), 37-38. [http://dx.doi.org/10.1038/238037a0] [PMID: 12635268]

[12] Wang, Y.; Wang, Q.; Zhan, X.; Wang, F.; Safdar, M.; He, J. Visible light driven type II heterostructures and their enhanced photocatalysis properties: A review. *Nanoscale,* **2013**, *5*(18), 8326-8339. [http://dx.doi.org/10.1039/c3nr01577g] [PMID: 23873075]

[13] Chong, M.N.; Jin, B.; Chow, C.W.K.; Saint, C. Recent developments in photocatalytic water treatment technology: A review. *Water Res.*, **2010**, *44*(10), 2997-3027. [http://dx.doi.org/10.1016/j.watres.2010.02.039] [PMID: 20378145]

[14] Wetchakun, K.; Wetchakun, N.; Sakulsermsuk, S. An overview of solar/visible light-driven heterogeneous photocatalysis for water purification: TiO_2- and ZnO-based photocatalysts used in suspension photoreactors. *J. Ind. Eng. Chem.*, **2019**, *71*, 19-49. [http://dx.doi.org/10.1016/j.jiec.2018.11.025]

[15] Carolina, B.; Jorge, B.; Almudena, G.; Manuel, P.; Juan, J.R. *Nanoscale materials in water purification*; Sabu Thomas, Daniel Pasquini, Shao-Yuan Leu, Deepu A. Gopakumar, Elsevier., **2019**, pp. 581-651.

[16] Ke, J.; Adnan Younis, M.; Kong, Y.; Zhou, H.; Liu, J.; Lei, L.; Hou, Y. Nanostructured ternary metal tungstate-based photocatalysts for environmental purification and solar water splitting: A review. *Nano-Micro Lett.,* **2018**, *10*(4), 69. [http://dx.doi.org/10.1007/s40820-018-0222-4] [PMID: 30393717]

[17] Ibhadon, A.; Fitzpatrick, P. Heterogeneous photocatalysis: Recent advances and applications. *Catalysts,* **2013**, *3*(1), 189-218. [http://dx.doi.org/10.3390/catal3010189]

[18] Xu, H.; Ouyang, S.; Liu, L.; Reunchan, P.; Umezawa, N.; Ye, J. Recent advances in TiO_2 -based photocatalysis. *J. Mater. Chem. A Mater. Energy Sustain.*, **2014**, *2*(32), 12642-12661. [http://dx.doi.org/10.1039/C4TA00941J]

[19] Qi, K.; Cheng, B.; Yu, J.; Ho, W. Review on the improvement of the photocatalytic and antibacterial activities of ZnO. *J. Alloys Compd.*, **2017**, *727*, 792-820. [http://dx.doi.org/10.1016/j.jallcom.2017.08.142]

[20] Abed, C.; Ali, M.B.; Addad, A.; Elhouichet, H. Growth, structural and optical properties of ZnO-ZnMgO-MgO nanocomposites and their photocatalytic activity under sunlight irradiation. *Mater. Res. Bull.,* **2019**, *110*, 230-238.
[http://dx.doi.org/10.1016/j.materresbull.2018.10.041]

[21] Lin, J.; Lin, J.; Zhu, Y. Controlled synthesis of the $ZnWO_4$ nanostructure and effects on the photocatalytic performance. *Inorg. Chem.,* **2007**, *46*(20), 8372-8378.
[http://dx.doi.org/10.1021/ic701036k] [PMID: 17722916]

[22] Garadkar, K.M.; Ghule, L.A.; Sapnar, K.B.; Dhole, S.D. A facile synthesis of $ZnWO_4$ nanoparticles by microwave assisted technique and its application in photocatalysis. *Mater. Res. Bull.,* **2013**, *48*(3), 1105-1109.
[http://dx.doi.org/10.1016/j.materresbull.2012.12.002]

[23] Yan, T.; Li, L.; Tong, W.; Zheng, J.; Wang, Y.; Li, G. $CdWO_4$ polymorphs: Selective preparation, electronic structures, and photocatalytic activities. *J. Solid State Chem.,* **2011**, *184*(2), 357-364.
[http://dx.doi.org/10.1016/j.jssc.2010.12.013]

[24] Tong, W.; Li, L.; Hu, W.; Yan, T.; Li, G. Systematic control of monoclinic $CdWO_4$ nanophase for optimum photocatalytic activity. *J. Phys. Chem. C,* **2010**, *114*(3), 1512-1519.
[http://dx.doi.org/10.1021/jp910284u]

[25] Al-Hamdi, A.M.; Sillanpää, M.; Dutta, J. Photocatalytic degradation of phenol by iodine doped tin oxide nanoparticles under UV and sunlight irradiation. *J. Alloys Compd.,* **2015**, *618*, 366-371.
[http://dx.doi.org/10.1016/j.jallcom.2014.08.120]

[26] Elango, G.; Roopan, S.M. Efficacy of SnO_2 nanoparticles toward photocatalytic degradation of methylene blue dye. *J. Photochem. Photobiol. B,* **2016**, *155*, 34-38.
[http://dx.doi.org/10.1016/j.jphotobiol.2015.12.010] [PMID: 26724726]

[27] Chen, J.; Wu, X.J.; Yin, L.; Li, B.; Hong, X.; Fan, Z.; Chen, B.; Xue, C.; Zhang, H. One-pot synthesis of CdS nanocrystals hybridized with single-layer transition-metal dichalcogenide nanosheets for efficient photocatalytic hydrogen evolution. *Angew. Chem. Int. Ed.,* **2015**, *54*(4), 1210-1214.
[http://dx.doi.org/10.1002/anie.201410172] [PMID: 25470356]

[28] Cui, H.; Li, B.; Li, Z.; Li, X.; Xu, S. Z-scheme based $CdS/CdWO_4$ heterojunction visible light photocatalyst for dye degradation and hydrogen evolution. *Appl. Surf. Sci.,* **2018**, *455*, 831-840.
[http://dx.doi.org/10.1016/j.apsusc.2018.06.054]

[29] Li, Z.; Meng, X.; Zhang, Z. Recent development on MoS_2-based photocatalysis: A review. *J. Photochem. Photobiol. Photochem. Rev.,* **2018**, *35*, 39-55.
[http://dx.doi.org/10.1016/j.jphotochemrev.2017.12.002]

[30] Naorem, R. N.; Singh, N. P.; Singh, N. M. Synthesis of SnS_2 photocatalysts and investigation of their photocatalytic activities under sun light irradiation. *Adv. Mater. Process. Technol,* **2022**.
[http://dx.doi.org/10.1080/2374068X.2022.2079591]

[31] Ramalingam, G.; Magdalane, C.M.; Arjun Kumar, B.; Yuvakkumar, R.; Ravi, G.; Jothi, A.I.; Rotte, N.K.; Murugadoss, G.; Ananth, A. Enhanced visible light-driven photocatalytic performance of CdSe nanorods. *Environ. Res.,* **2022**, *203*, 111855.
[http://dx.doi.org/10.1016/j.envres.2021.111855] [PMID: 34384750]

[32] Tian, L.; Rui, Y.; Sun, K.; Cui, W.; An, W. Surface decoration of $ZnWO_4$ nanorods with Cu_2O nanoparticles to build heterostructure with enhanced photocatalysis. *Nanomaterials,* **2018**, *8*(1), 33.
[http://dx.doi.org/10.3390/nano8010033] [PMID: 29315264]

[33] Wang, H.; Wang, C.; Cui, X.; Qin, L.; Ding, R.; Wang, L.; Liu, Z.; Zheng, Z.; Lv, B. Design and facile one-step synthesis of $FeWO_4/Fe_2O_3$ di-modified WO_3 with super high photocatalytic activity toward degradation of quasi-phenothiazine dyes. *Appl. Catal. B,* **2018**, *221*, 169-178.
[http://dx.doi.org/10.1016/j.apcatb.2017.09.011]

[34] Li, N.; Zhu, L.; Zhang, W.D.; Yu, Y.X.; Zhang, W.H.; Hou, M.F. Modification of TiO_2 nanorods by

Bi_2MoO_6 nanoparticles for high performance visible-light photocatalysis. *J. Alloys Compd.*, **2011**, *509*(41), 9770-9775.
[http://dx.doi.org/10.1016/j.jallcom.2011.07.102]

[35] Zhou, P.; Yu, J.; Jaroniec, M. All-solid-state Z-scheme photocatalytic systems. *Adv. Mater.*, **2014**, *26*(29), 4920-4935.
[http://dx.doi.org/10.1002/adma.201400288] [PMID: 24888530]

[36] Tian, G.; Fu, H.; Jing, L.; Xin, B.; Pan, K. Preparation and characterization of stable biphase TiO_2 photocatalyst with high crystallinity, large surface area and enhanced photoactivity. *J. Phys. Chem. C*, **2008**, *112*(8), 3083-3089.
[http://dx.doi.org/10.1021/jp710283p]

[37] Gnanaprakasam, A.; Sivakumar, V.M.; Sivayogavalli, P.L.; Thirumarimurugan, M. Characterization of TiO_2 and ZnO nanoparticles and their applications in photocatalytic degradation of azodyes. *Ecotoxicol. Environ. Saf.*, **2015**, *121*, 121-125.
[http://dx.doi.org/10.1016/j.ecoenv.2015.04.043] [PMID: 25937630]

[38] Ji, J.; Xu, Y.; Huang, H.; He, M.; Liu, S.; Liu, G.; Xie, R.; Feng, Q.; Shu, Y.; Zhan, Y.; Fang, R.; Ye, X.; Leung, D.Y.C. Mesoporous TiO_2 under VUV irradiation: Enhanced photocatalytic oxidation for VOCs degradation at room temperature. *Chem. Eng. J.*, **2017**, *327*, 490-499.
[http://dx.doi.org/10.1016/j.cej.2017.06.130]

[39] Phuruangrat, A.; Kuntalue, B.; Thongtem, S.; Thongtem, T. Hydrothermal synthesis of hexagonal ZnO nanoplates used for photodegradation of methylene blue. *Optik*, **2021**, *226*, 165949.
[http://dx.doi.org/10.1016/j.ijleo.2020.165949]

[40] Wannapop, S.; Khawsaad, A.; Supanpong, A.; Janorat, Y.; Chuminjak, Y.; Tuantranont, A.; Phuruangrat, A.; Thongtem, T.; Thongtem, S.; Somdee, A. Photocatalytic study of metal oxide enhanced ZnO synthesized by a one-step cyclic-microwave method: The role of the p-n heterostructure. *Inorg. Chem. Commun.*, **2022**, *138*, 109210.
[http://dx.doi.org/10.1016/j.inoche.2022.109210]

[41] Surisa Sa-nguanprang, ; Phuruangrat, A.; Thongtem, T.; Thongtem, S. Synthesis of ZnO nanoparticles by tartaric acid solution combustion and their photocatalytic properties. *Russ. J. Inorg. Chem.*, **2020**, *65*(7), 1102-1110.
[http://dx.doi.org/10.1134/S0036023620070189]

[42] Ali, M.M.; Kudhier, M.A.; Sabry, R.S. Photocatalytic activity of CdS nanosheets prepared by chemical bath deposition method. *J. Phys. Conf. Ser.*, **2021**, *1963*(1), 012119.
[http://dx.doi.org/10.1088/1742-6596/1963/1/012119]

[43] Farhadian, M.; Sangpour, P.; Hosseinzadeh, G. Morphology dependent photocatalytic activity of WO_3 nanostructures. *Journal of Energy Chemistry*, **2015**, *24*(2), 171-177.
[http://dx.doi.org/10.1016/S2095-4956(15)60297-2]

[44] Huang, F.; Yan, A.; Zhao, H. *Semiconductor photocatalysis-Materials, Mechanisms and Applications*; Wenbin Cao, IntechOpen, **2016**, pp. 31-80.

[45] Ye, D.; Li, D.; Chen, W.; Shao, Y.; Xiao, G.; Sun, M.; Fu, X. Characterization and properties of Eu^{3+}-doped $CdWO_4$ prepared by a hydrothermal method. *Res. Chem. Intermed.*, **2009**, *35*(6-7), 675-683.
[http://dx.doi.org/10.1007/s11164-009-0104-y]

[46] Mohd Said, N.D.; Sahdan, M.Z.; Nayan, N.; Saim, H.; Adriyanto, F.; Bakri, A.S.; Morsin, M. Difference in structural and chemical properties of sol–gel spin coated Al doped TiO_2, Y doped TiO_2 and Gd doped TiO_2 based on trivalent dopants. *RSC Advances*, **2018**, *8*(52), 29686-29697.
[http://dx.doi.org/10.1039/C8RA03950J] [PMID: 35547299]

[47] Murashkina, A.A.; Murzin, P.D.; Rudakova, A.V.; Ryabchuk, V.K.; Emeline, A.V.; Bahnemann, D.W. Influence of the dopant concentration on photocatalytic activity: Al-doped TiO_2. *J. Phys. Chem. C*, **2015**, *119*(44), 24695-24703.
[http://dx.doi.org/10.1021/acs.jpcc.5b06252]

[48] Khlyustova, A.; Sirotkin, N.; Kusova, T.; Kraev, A.; Titov, V.; Agafonov, A. Doped TiO_2 : The effect of doping elements on photocatalytic activity. *Materials Advances,* **2020**, *1*(5), 1193-1201. [http://dx.doi.org/10.1039/D0MA00171F]

[49] Narsimha, K.; Babu, M.S.; Anuradha, N.; Guda, S.; Kumar, B.K.; Mallesh, D.; Upender, G.; Reddy, P.M.; Kumar, B.V. Preparation and characterization of $CdWO_4$:Cu nanorods with enhanced photocatalytic performance under sunlight irradiation. *New J. Chem.,* **2020**, *44*(6), 2380-2388. [http://dx.doi.org/10.1039/C9NJ05763C]

[50] Wang, X.; Su, W.; Hu, X.; Liu, H.; Sheng, M.; Zhou, Q. Synthesis and characterization of Fe-doped $CdWO_4$ nanoparticles with enhanced photocatalytic activity. *Mater. Res. Express,* **2018**, *6*(3), 035507. [http://dx.doi.org/10.1088/2053-1591/aaf432]

[51] Komaraiah, D.; Radha, E.; Kalarikkal, N.; Sivakumar, J.; Ramana Reddy, M.V.; Sayanna, R. Structural, optical and photoluminescence studies of sol-gel synthesized pure and iron doped TiO_2 photocatalysts. *Ceram. Int.,* **2019**, *45*(18), 25060-25068. [http://dx.doi.org/10.1016/j.ceramint.2019.03.170]

[52] Huang, J-g.; Guo, X-t.; Wang, B.; Li, L-y.; Zhao, M-x.; Dong, L-l.; Liu, X-j.; Huang, Y-t. Synthesis and photocatalytic activity of Mo-doped TiO_2 nanoparticles. *J. Spectrosc.,* **2015**, 681850.

[53] Wang, S.; Bai, L.N.; Sun, H.M.; Jiang, Q.; Lian, J.S. Structure and photocatalytic property of Mo-doped TiO_2 nanoparticles. *Powder Technol.,* **2013**, *244*, 9-15. [http://dx.doi.org/10.1016/j.powtec.2013.03.054]

[54] Sathasivam, S.; Bhachu, D.S.; Lu, Y.; Chadwick, N.; Althabaiti, S.A.; Alyoubi, A.O.; Basahel, S.N.; Carmalt, C.J.; Parkin, I.P. Tungsten doped TiO_2 with enhanced photocatalytic and optoelectrical properties *via* aerosol assisted chemical vapor deposition. *Sci. Rep.,* **2015**, *5*(1), 10952. [http://dx.doi.org/10.1038/srep10952] [PMID: 26042724]

[55] Jin, Y.; Xi, J.; Zhang, Z.; Xiao, J.; Xiao, F.; Qian, L.; Wang, S. An ultra-low Pd loading nanocatalyst with efficient catalytic activity. *Nanoscale,* **2015**, *7*(12), 5510-5515. [http://dx.doi.org/10.1039/C5NR00599J] [PMID: 25735825]

[56] Pan, X.; Xu, Y.J. Efficient thermal- and photocatalyst of Pd nanoparticles on TiO_2 achieved by an oxygen vacancies promoted synthesis strategy. *ACS Appl. Mater. Interfaces,* **2014**, *6*(3), 1879-1886. [http://dx.doi.org/10.1021/am404877e] [PMID: 24392776]

[57] Kim, S.; Hwang, S.J.; Choi, W. Visible light active platinum-ion-doped TiO_2 photocatalyst. *J. Phys. Chem. B,* **2005**, *109*(51), 24260-24267. [http://dx.doi.org/10.1021/jp055278y] [PMID: 16375422]

[58] Soni, V.; Singh, A.N.; Singh, P.; Gupta, A. Photocatalytic dye-degradation activity of nano-crystalline $Ti_{1-x} M_x O_{2-\delta}$ (M =Ag, Pd, Fe, Ni and $x = 0, 0.01$) for water pollution abatement. *RSC Advances,* **2022**, *12*(29), 18794-18805. [http://dx.doi.org/10.1039/D2RA02847F] [PMID: 35873333]

[59] Komaraiah, D.; Radha, E.; Sivakumar, J.; Ramana Reddy, M.V.; Sayanna, R. Photoluminescence and photocatalytic activity of spin coated Ag^+ doped anatase TiO_2 thin films. *Opt. Mater.,* **2020**, *108*, 110401. [http://dx.doi.org/10.1016/j.optmat.2020.110401]

[60] Safeen, A.; Safeen, K.; Ullah, R.; Zulfqar, ; Shah, W.H.; Zaman, Q.; Althubeiti, K.; Al Otaibi, S.; Rahman, N.; Iqbal, S.; Khan, A.; Khan, A.; Khan, R. Enhancing the physical properties and photocatalytic activity of TiO_2 nanoparticles *via* cobalt doping. *RSC Advances,* **2022**, *12*(25), 15767-15774. [http://dx.doi.org/10.1039/D2RA01948E] [PMID: 35685706]

[61] Xiao, Q.; Zhang, J.; Xiao, C.; Tan, X. Photocatalytic decolorization of methylene blue over $Zn_{1-x}Co_xO$ under visible light irradiation. *Mater. Sci. Eng. B,* **2007**, *142*(2-3), 121-125. [http://dx.doi.org/10.1016/j.mseb.2007.06.021]

[62] Obregón, S.; Colón, G. Heterostructured Er^{3+} doped $BiVO_4$ with exceptional photocatalytic performance by cooperative electronic and luminescence sensitization mechanism. *Appl. Catal. B,* **2014**, *158-159*, 242-249.
[http://dx.doi.org/10.1016/j.apcatb.2014.04.029]

[63] Guo, Y.; Qiu, X.; Dong, H.; Zhou, X. Trends in non-metal doping of the $SrTiO_3$ surface: A hybrid density functional study. *Phys. Chem. Chem. Phys.,* **2015**, *17*(33), 21611-21621.
[http://dx.doi.org/10.1039/C5CP03005F] [PMID: 26224623]

[64] Shen, M.; Wu, Z.; Huang, H.; Du, Y.; Zou, Z.; Yang, P. Carbon-doped anatase TiO_2 obtained from TiC for photocatalysis under visible light irradiation. *Mater. Lett.,* **2006**, *60*(5), 693-697.
[http://dx.doi.org/10.1016/j.matlet.2005.09.068]

[65] Irie, H.; Watanabe, Y.; Hashimoto, K. Carbon-doped anatase TiO_2 powders as a visible-light sensitive photocatalyst. *Chem. Lett.,* **2003**, *32*(8), 772-773.
[http://dx.doi.org/10.1246/cl.2003.772]

[66] Asahi, R.; Morikawa, T.; Ohwaki, T.; Aoki, K.; Taga, Y. Visible-light photocatalysis in nitrogen-doped titanium oxides. *Science,* **2001**, *293*(5528), 269-271.
[http://dx.doi.org/10.1126/science.1061051] [PMID: 11452117]

[67] Khore, S.K.; Tellabati, N.V.; Apte, S.K.; Naik, S.D.; Ojha, P.; Kale, B.B.; Sonawane, R.S. Green sol–gel route for selective growth of 1D rutile N–TiO_2 : A highly active photocatalyst for H_2 generation and environmental remediation under natural sunlight. *RSC Advances,* **2017**, *7*(52), 33029-33042.
[http://dx.doi.org/10.1039/C7RA01648D]

[68] Devi, L.G.; Kavitha, R. Enhanced photocatalytic activity of sulfur doped TiO_2 for the decomposition of phenol: A new insight into the bulk and surface modification. *Mater. Chem. Phys.,* **2014**, *143*(3), 1300-1308.
[http://dx.doi.org/10.1016/j.matchemphys.2013.11.038]

[69] Kim, C.; Doh, S.J.; Lee, S.G.; Lee, S.J.; Kim, H.Y. Visible-light absorptivity of a zincoxysulfide (ZnO_xS_{1-x}) composite semiconductor and its photocatalytic activities for degradation of organic pollutants under visible-light irradiation. *Appl. Catal. A Gen.,* **2007**, *330*, 127-133.
[http://dx.doi.org/10.1016/j.apcata.2007.07.016]

[70] Niu, P.; Wu, G.; Chen, P.; Zheng, H.; Cao, Q.; Jiang, H. Optimization of boron doped TiO_2 as an efficient visible light-driven photocatalyst for organic dye degradation with high reusability. *Front Chem.,* **2020**, *8*, 172.
[http://dx.doi.org/10.3389/fchem.2020.00172] [PMID: 32232026]

[71] Shah, A.A.; Chandio, A.D.; Sheikh, A.A. Boron doped ZnO nanostructures for photo degradation of methylene blue, methyl orange and rhodamine B. *J. Nanosci. Nanotechnol.,* **2021**, *21*(4), 2483-2494.
[http://dx.doi.org/10.1166/jnn.2021.19315] [PMID: 33500066]

[72] Chen, S.; Sun, S.; Sun, H.; Fan, W.; Zhao, X.; Sun, X. Experimental and theoretical studies on the enhanced photocatalytic activity of $ZnWO_4$ nanorods by fluorine doping. *J. Phys. Chem. C,* **2010**, *114*(17), 7680-7688.
[http://dx.doi.org/10.1021/jp911952v]

[73] Díez, A.M.; Núñez, I.; Pazos, M.; Sanromán, M.Á.; Kolen'ko, Y.V. Fluoride-doped TiO_2 photocatalyst with enhanced activity for stable pollutant degradation. *Catalysts,* **2022**, *12*(10), 1190.
[http://dx.doi.org/10.3390/catal12101190]

[74] Gupta, R.; KrishnaRao Eswar, N.; Modak, J.M.; Madras, G. Visible light driven efficient N and Cu co-doped ZnO for photoinactivation of Escherichia coli. *RSC Advances,* **2016**, *6*(89), 85675-85687.
[http://dx.doi.org/10.1039/C6RA16739J]

[75] Park, B-G. Photocatalytic activity of TiO_2-doped Fe, Ag, and Ni with N under visible light irradiation. *Gels,* **2022**, *8*(14), 1-10.

[76] Wang, M.; Han, J.; Hu, Y.; Guo, R.; Mesoporous, C. Mesoporous C, N-codoped TiO_2 hybrid shells with enhanced visible light photocatalytic performance. *RSC Advances,* **2017**, *7*(25), 15513-15520. [http://dx.doi.org/10.1039/C7RA00985B]

[77] Wang, F.; Ban, P.P.; Parry, J.P.; Xu, X.H.; Zeng, H. Enhanced photocatalytic properties of N–P co-doped TiO_2 nanosheets with 001 facets. *Rare Met.,* **2016**, *35*(12), 940-947. [http://dx.doi.org/10.1007/s12598-016-0807-3]

[78] Rawal, J.; Kamran, U.; Park, M.; Pant, B.; Park, S.J. Nitrogen and Sulfur co-doped graphene quantum dots anchored TiO_2 nanocomposites for enhanced photocatalytic activity. *Catalysts,* **2022**, *12*(5), 548. [http://dx.doi.org/10.3390/catal12050548]

[79] Sushma, C.; Kumar, S.G. C–N–S tridoping into TiO_2 matrix for photocatalytic applications: observations, speculations and contradictions in the codoping process. *Inorg. Chem. Front.,* **2017**, *4*(8), 1250-1267.
[http://dx.doi.org/10.1039/C7QI00189D]

[80] Kadam, S.R.; Mate, V.R.; Panmand, R.P.; Nikam, L.K.; Kulkarni, M.V.; Sonawane, R.S.; Kale, B.B. A green process for efficient lignin (biomass) degradation and hydrogen production *via* water splitting using nanostructured C, N, S-doped ZnO under solar light. *RSC Advances,* **2014**, *4*(105), 60626-60635.
[http://dx.doi.org/10.1039/C4RA10760H]

[81] Tonda, S.; Kumar, S.; Shanker, V. Surface plasmon resonance-induced photocatalysis by Au nanoparticles decorated mesoporous g-C_3 N_4 nanosheets under direct sunlight irradiation. *Mater. Res. Bull.,* **2016**, *75*, 51-58.
[http://dx.doi.org/10.1016/j.materresbull.2015.11.011]

[82] Li, J.; Lou, Z.; Li, B. Nanostructured materials with localized surface plasmon resonance for photocatalysis. *Chin. Chem. Lett.,* **2022**, *33*(3), 1154-1168.
[http://dx.doi.org/10.1016/j.cclet.2021.07.059]

[83] Liu, M.; Kang, Q.; Xie, Z.; Lu, L.; Dai, K.; Dawson, G. Heterostructure nanocomposite with local surface plasmon resonance effect enhanced photocatalytic activity—a critical review. *J. Phys. D Appl. Phys.,* **2022**, *55*(4), 043002.
[http://dx.doi.org/10.1088/1361-6463/ac2cac]

[84] Ingram, D.B.; Christopher, P.; Bauer, J.L.; Linic, S. Predictive model for the design of plasmonic metal/semiconductor composite photocatalysts. *ACS Catal.,* **2011**, *1*(10), 1441-1447.
[http://dx.doi.org/10.1021/cs200320h]

[85] Mohanty, S.; Babu, P.; Parida, K.; Naik, B. Surface-plasmon-resonance-induced photocatalysis by core - shell SiO_2@Ag NCs@Ag_3PO_4 toward water-splitting and phenol oxidation reactions. *Inorg. Chem.,* **2019**, *58*(15), 9643-9654.
[http://dx.doi.org/10.1021/acs.inorgchem.9b00218] [PMID: 31339037]

[86] Guo, Y.; Wei, J.; Yang, T.; Lv, Z.; Xu, Z. Manipulation of surface plasmon resonance for high photocatalytic activity of Ag-Bi_2WO_6 hetero-architecture. *Optik,* **2019**, *180*, 285-294.
[http://dx.doi.org/10.1016/j.ijleo.2018.11.094]

[87] Shuang, S.; Lv, R.; Xie, Z.; Zhang, Z. Surface plasmon enhanced photocatalysis of Au/Pt-decorated TiO_2 nanopillar arrays. *Sci. Rep.,* **2016**, *6*(1), 26670.
[http://dx.doi.org/10.1038/srep26670] [PMID: 27215703]

[88] Peleyeju, M.G.; Arotiba, O.A. Recent trend in visible-light photoelectrocatalytic systems for degradation of organic contaminants in water/wastewater. *Environ. Sci. Water Res. Technol.,* **2018**, *4*(10), 1389-1411.
[http://dx.doi.org/10.1039/C8EW00276B]

[89] Nandi, P.; Das, D. ZnO-Cu O heterostructure photocatalyst for efficient dye degradation. *J. Phys. Chem. Solids,* **2020**, *143*, 109463.

[http://dx.doi.org/10.1016/j.jpcs.2020.109463]

[90] Omar, A.M.A.; Hassen, A.; Metwalli, O.I.; Saber, M.R.; Mohamed, S.R.E.; Khalil, A.S.G. Construction of 2D layered TiO_2 @MoS_2 heterostructure for efficient adsorption and photodegradation of organic dyes. *Nanotechnology,* **2021**, *32*(33), 335605. [http://dx.doi.org/10.1088/1361-6528/abff8a] [PMID: 33971641]

[91] Zhang, C.; Wang, L.; Yuan, F.; Meng, R.; Chen, J.; Hou, W.; Zhu, H. Construction of p-n type $Ag_3PO_4/CdWO_4$ heterojunction photocatalyst for visible-light-induced dye degradation. *Appl. Surf. Sci.,* **2020**, *534*, 147544. [http://dx.doi.org/10.1016/j.apsusc.2020.147544]

[92] Ge, F.; Li, X.; Wu, M.; Ding, H.; Li, X. A type II heterojunction α-Fe_2 O_3 /g-C_3 N_4 for the heterogeneous photo-Fenton degradation of phenol. *RSC Advances,* **2022**, *12*(14), 8300-8309. [http://dx.doi.org/10.1039/D1RA09282K] [PMID: 35424783]

[93] Ahmed, S.N.; Haider, W. Heterogeneous photocatalysis and its potential applications in water and wastewater treatment: A review. *Nanotechnology,* **2018**, *29*(34), 342001. [http://dx.doi.org/10.1088/1361-6528/aac6ea] [PMID: 29786601]

[94] Huang, D.; Chen, S.; Zeng, G.; Gong, X.; Zhou, C.; Cheng, M.; Xue, W.; Yan, X.; Li, J. Artificial Z-scheme photocatalytic system: What have been done and where to go? *Coord. Chem. Rev.,* **2019**, *385*, 44-80. [http://dx.doi.org/10.1016/j.ccr.2018.12.013]

[95] Jia, X.; Tahir, M.; Pan, L.; Huang, Z.F.; Zhang, X.; Wang, L.; Zou, J.J. Direct Z-scheme composite of CdS and oxygen-defected $CdWO_4$: An efficient visible-light-driven photocatalyst for hydrogen evolution. *Appl. Catal. B,* **2016**, *198*, 154-161. [http://dx.doi.org/10.1016/j.apcatb.2016.05.046]

[96] Yuan, Y.; Guo, R.; Hong, L.; Ji, X.; Lin, Z.; Li, Z.; Pan, W. A review of metal oxide-based Z-scheme heterojunction photocatalysts: actualities and developments. *Mater. Today Energy,* **2021**, *21*, 100829. [http://dx.doi.org/10.1016/j.mtener.2021.100829]

[97] Tada, H.; Mitsui, T.; Kiyonaga, T.; Akita, T.; Tanaka, K. All-solid –state Z-scheme in CdS-Au-TiO_2 three component nanojunction system. *Nat Mater.,* **2006**, *5*(10), 782-786.

[98] Zhang, L.; Wong, K.H.; Chen, Z.; Yu, J.C.; Zhao, J.; Hu, C.; Chan, C.Y.; Wong, P.K. AgBr-Ag-Bi_2WO_6 nanojunction system: A novel and efficient photocatalyst with double visible-light active components. *Appl. Catal. A Gen.,* **2009**, *363*(1-2), 221-229. [http://dx.doi.org/10.1016/j.apcata.2009.05.028]

[99] Sun, Y.; Zhu, Q.; Bai, B.; Li, Y.; He, C. Novel all-solid-state Z-scheme $SnO_2/Pt/In_2O_3$ photocatalyst with boosted photocatalytic performance on water splitting and 2,4-dichlorophenol degradation under visible light. *Chem. Eng. J.,* **2020**, *390*, 124518. [http://dx.doi.org/10.1016/j.cej.2020.124518]

[100] Li, S.; Liu, Z.; Qu, Z.; Piao, C.; Liu, J.; Xu, D.; Li, X.; Wang, J.; Song, Y. An all-solid-state Z-scheme $NaNbO_3$-Au-Sn_3O_4 photocatalyst for effective degradation of carbofuran under sunlight irradiation. *J. Photochem. Photobiol. Chem.,* **2020**, *389*, 112246. [http://dx.doi.org/10.1016/j.jphotochem.2019.112246]

[101] Ma, D.; Wu, J.; Gao, M.; Xin, Y.; Ma, T.; Sun, Y. Fabrication of Z-scheme g -C_3 N_4 /RGO/Bi_2 WO_6 photocatalyst with enhanced visible-light photocatalytic activity. *Chem. Eng. J.,* **2016**, *290*, 136-146. [http://dx.doi.org/10.1016/j.cej.2016.01.031]

[102] Jo, W.K.; Selvam, N.C.S. Z-scheme CdS/g-C 3 N 4 composites with RGO as an electron mediator for efficient photocatalytic H 2 production and pollutant degradation. *Chem. Eng. J.,* **2017**, *317*, 913-924. [http://dx.doi.org/10.1016/j.cej.2017.02.129]

[103] Ali, H.; Guler, A.; Masar, M.; Urbanek, P.; Urbanek, M.; Skoda, D.; Suly, P.; Machovsky, M.; Galusek, D.; Kuritka, I. Solid-state synthesis of direct Z-scheme Cu_2O/WO_3 nanocomposites with

enhanced visible-light photocatalytic performance. *Catalysts,* **2021**, *11*(2), 293. [http://dx.doi.org/10.3390/catal11020293]

[104] Shi, W.; Guo, F.; Yuan, S. *In situ* synthesis of Z-scheme $Ag_3PO_4/CuBi_2O_4$ photocatalysts andenhanced photocatalytic performance for the degradation oftetracycline under visible light irradiation. *Appl. Surf. Sci.,* **2017**, *209*, 720-728.

[105] Cui, H.; Li, B.; Zhang, Y.; Zheng, X.; Li, X.; Li, Z.; Xu, S. Constructing Z-scheme based $CoWO_4$/CdS photocatalysts with enhanced dye degradation and H_2 generation performance. *Int. J. Hydrogen Energy,* **2018**, *43*(39), 18242-18252.
[http://dx.doi.org/10.1016/j.ijhydene.2018.08.050]

[106] Xie, X.; Liu, M.; Wang, C.; Chen, L.; Xu, J.; Cheng, Y.; Dong, H.; Lu, F.; Wang, W.H.; Liu, H.; Wang, W. Efficient photo-degradation of dyes using $CuWO_4$ nanoparticles with electron sacrificial agents: A combination of experimental and theoretical exploration. *RSC Advances,* **2016**, *6*(2), 953-959.
[http://dx.doi.org/10.1039/C5RA18788E]

[107] Lv, S.; Du, Y.; Wu, F.; Cai, Y.; Zhou, T. Review on LSPR assisted photocatalysis: Effects of physical fields and opportunities in multifield decoupling. *Nanoscale Adv.,* **2022**, *4*(12), 2608-2631.
[http://dx.doi.org/10.1039/D2NA00140C] [PMID: 36132289]

[108] Tian, J.; Zhao, Z.; Kumar, A.; Boughton, R.I.; Liu, H. Recent progress in design, synthesis, and applications of one-dimensional TiO_2 nanostructured surface heterostructures: A review. *Chem. Soc. Rev.,* **2014**, *43*(20), 6920-6937.
[http://dx.doi.org/10.1039/C4CS00180J] [PMID: 25014328]

[109] Dong, P.; Wang, Y.; Li, H.; Li, H.; Ma, X.; Han, L. Shape-controllable synthesis and morphology-dependent photocatalytic properties of Ag_3PO_4 crystals. *J. Mater. Chem. A Mater. Energy Sustain.,* **2013**, *1*(15), 4651-4656.
[http://dx.doi.org/10.1039/c3ta00130j]

[110] Buvaneswari, K.; Karthiga, R.; Kavitha, B.; Rajarajan, M.; Suganthi, A. Effect of $FeWO_4$ doping on the photocatalytic activity of ZnO under visible light irradiation. *Appl. Surf. Sci.,* **2015**, *356*, 333-340.
[http://dx.doi.org/10.1016/j.apsusc.2015.08.060]

[111] Kansal, S.K.; Kaur, N.; Singh, S. Photocatalytic degradation of two commercial reactive dyes in aqueous phase using nanophotocatalysts. *Nanoscale Res. Lett.,* **2009**, *4*(7), 709-716.
[http://dx.doi.org/10.1007/s11671-009-9300-3] [PMID: 20596421]

[112] Ajmal, A.; Majeed, I.; Malik, R.N.; Idriss, H.; Nadeem, M.A. Principles and mechanisms of photocatalytic dye degradation on TiO_2 based photocatalysts: A comparative overview. *RSC Advances,* **2014**, *4*(70), 37003-37026.
[http://dx.doi.org/10.1039/C4RA06658H]

[113] Zhu, H.; Jiang, R.; Fu, Y.; Guan, Y.; Yao, J.; Xiao, L.; Zeng, G. Effective photocatalytic decolorization of methyl orange utilizing TiO_2/ZnO/chitosan nanocomposite films under simulated solar irradiation. *Desalination,* **2012**, *286*, 41-48.
[http://dx.doi.org/10.1016/j.desal.2011.10.036]

[114] Chen, X.; Wu, Z.; Liu, D.; Gao, Z. Preparation of ZnO photocatalyst for the efficient and rapid photocatalytic degradation of azo dyes. *Nanoscale Res. Lett.,* **2017**, *12*(1), 143.
[http://dx.doi.org/10.1186/s11671-017-1904-4] [PMID: 28235375]

[115] Alkaykh, S.; Mbarek, A.; Ali-Shattle, E.E. Photocatalytic degradation of methylene blue dye in aqueous solution by $MnTiO_3$ nanoparticles under sunlight irradiation. *Heliyon,* **2020**, *6*(4), e03663.
[http://dx.doi.org/10.1016/j.heliyon.2020.e03663] [PMID: 32322705]

[116] Cassano, A.E.; Alfano, O.M. Reaction engineering of suspended solid heterogeneous photocatalytic reactors. *Catal. Today,* **2000**, *58*(2-3), 167-197.
[http://dx.doi.org/10.1016/S0920-5861(00)00251-0]

[117] Herrmann, J.M.; Pichat, P. Heterogeneous photocatalysis. Oxidation of halide ions by oxygen in ultraviolet irradiated aqueous suspension of titanium dioxide. *J. Chem. Soc., Faraday Trans. I,* **1980**, *76*(0), 1138-1146. [http://dx.doi.org/10.1039/f19807601138]

[118] Collazzo, G.C.; Foletto, E.L.; Jahn, S.L.; Villetti, M.A. Degradation of direct black 38 dye under visible light and sunlight irradiation by N-doped anatase TIO_2 as photocatalyst. *J. Environ. Manage.,* **2012**, *98*, 107-111. [http://dx.doi.org/10.1016/j.jenvman.2011.12.029] [PMID: 22257572]

[119] Hung, C.H.; Yuan, C. Reduction of Azo-dye *via* TiO_2– photocatalysis. *J. Chin. Inst. Environ. Eng.,* **2000**, *10*, 209-216.

[120] Liu, C.; Hsieh, Y.; Lai, P.; Li, C.; Kao, C. Photodegradation treatment of azo dye wastewater by UV/TiO process. *Dyes Pigments,* **2006**, *68*(2-3), 191-195. [http://dx.doi.org/10.1016/j.dyepig.2004.12.002]

[121] Gaya, U.I.; Abdullah, A.H. Heterogeneous photocatalytic degradation of organic contaminants over titanium dioxide: A review of fundamentals, progress and problems. *J. Photochem. Photobiol. Photochem. Rev.,* **2008**, *9*(1), 1-12. [http://dx.doi.org/10.1016/j.jphotochemrev.2007.12.003]

[122] Reza, K.M.; Kurny, A.S.W.; Gulshan, F. Parameters affecting the photocatalytic degradation of dyes using TiO_2: A review. *Appl. Water Sci.,* **2017**, *7*(4), 1569-1578. [http://dx.doi.org/10.1007/s13201-015-0367-y]

[123] Soares, E.T.; Lansarin, M.A.; Moro, C.C. A study of process variables for the photocatalytic degradation of rhodamine B. *Braz. J. Chem. Eng.,* **2007**, *24*(1), 29-36. [http://dx.doi.org/10.1590/S0104-66322007000100003]

[124] Barakat, N.A.M.; Kanjwal, M.A.; Chronakis, I.S.; Kim, H.Y. Influence of temperature on the photodegradation process using Ag-doped TiO_2 nanostructures: Negative impact with the nanofibers. *J. Mol. Catal. Chem.,* **2013**, *366*, 333-340. [http://dx.doi.org/10.1016/j.molcata.2012.10.012]

[125] Wang, Y.; Hong, C.S. TiO_2-mediated photomineralization of 2-chlorobiphenyl: The role of O_2. *Water Res.,* **2000**, *34*(10), 2791-2797. [http://dx.doi.org/10.1016/S0043-1354(00)00009-9]

[126] Shirayama, H.; Tohezo, Y.; Taguchi, S. Photodegradation of chlorinated hydrocarbons in the presence and absence of dissolved oxygen in water. *Water Res.,* **2001**, *35*(8), 1941-1950. [http://dx.doi.org/10.1016/S0043-1354(00)00480-2] [PMID: 11337840]

SUBJECT INDEX

A

B

C

Dibya Prakash Rai, Kingsley O. Obodo, & Jitendra Pal Singh (Eds.)

D

E

F

G

H

I

K

L

M

N

O

P

Q

R

S

T

U

V

W

www.ingramcontent.com/pod-product-compliance
Lightning Source LLC
LaVergne TN
LVHW070946210926
848482LV00002B/107